RELIABILITY ENGINEERING

Reliability Engineering

by

K. K. AGGARWAL
Centre for Excellence in Reliability Engineering,
Regional Engineering College,
Kurukshetra, India

Library of Congress Cataloging in Publication Data

Aggarwal, K. K.
Reliability engineering / K. K. Aggarwal.
p. cm. - (Topics in safety, reliability, and quality : v. 3)
Includes bibliographical references and index.
ISBN 0-7923-2524-9 (hb)
1. Reliability (Engineering) I. Title II. Series.
TA 168.A33 1993
620' .00452—dc20 93-33130

ISBN 0-7923-2524-9

Aggarwal: Reliability Engineering

First Indian Reprint 2007
Second Indiasn Reprint 2014

ISBN 978-81-8128-557-7

This edition is manufactured in India for sale in India, Pakistan, Bangladesh, Nepal and Sri Lanka and any other country as authorized by the publisher.

This edition is published by Springer (India) Private Limited,
A part of Springer Science+Business Media, Registered Office: 7th Floor,
Vijaya Building, 17, Barakhamba Road, New Delhi – 110 001, India.

Printed in India by Rajkamal Electric Press, Kundli, Haryana.

CONTENTS

PREFACE

The growth of present day societies in population, transportation, communication and technology points towards the use of larger and more complex systems. It took man 250,000 years to arrive at agricultural revolution; 25,000 years to arrive at industrial revolution; and only 150 years to attain the space age and we really do not know where will we go from here. We know for certain, however, that the coming age will involve the use of still larger and more complex systems.

The importance of reliability has assumed new dimensions in the recent years primarily because of the complexity of larger systems and the implications of their failure. Unreliability in the modern age of technology besides causing the operational inefficiency and uneconomical maintenance can also endanger human life. The transition towards thinking about Reliability is obviously difficult because it requires a change of mentality; but difficult does not mean impossible.

In its most wider sense, the word Reliability has a very important meaning: Re-Liability which simply means that it is liability, not once but again and again, from designers, manufacturers, inspectors, vendors to users and on all those who are involved with a system in any way to make it reliable. Much attention is being paid, more than ever before, to the quality and reliability of engineering systems.

It is of course not easy to decide how much reliability is required and how much one can afford to pay for it. However, Defence, Aeronautical, Space, and Nuclear Power generation systems are some prime examples where compromise with quality and reliability just cannot be made; for a compromise here could mean much not only in terms of money but also the loss of many precious lives.

Author's 20 years of experience as Reliability Educator, Researcher and Consultant made it abundantly clear that although every Industrial Organisation desired to equip its Scientists, Engineers and Managers with the knowledge of Reliability Concepts and Applications, yet this has not been achieved. A detailed study reveals that the major bottlenecks for this situation are the non-availability of qualified faculty for their continuing education programs, the non-availability of reliability related courses at most Universities or Schools, as well as the lack of availability of a good text. This made the author to put his knowledge and experience in the form of the present text book.

This text is intended to be useful for senior undergraduate and graduate students in engineering schools as also for professional engineers, reliability administrators and managers. It was difficult for the author to keep away from the temptation of including many of his research papers published in the reputed journals but a very concerted effort has been made to keep the book ideally suitable for a first course or even for a novice in the discipline of reliability engineering. This text has primarily emerged from the lecture notes which the author used for delivering courses to the students at his Institution and also several courses which were organized for the engineers in the Industrial Organizations. The text has therefore, been class room tested till the qualification of acceptance testing stage. A number of solved examples have been added to make the subject clear during studies. Many problems have also been given at the end so that the reader could voluntarily test himself/herself. The answers to odd numbered problems have been given as test verification.

Much of the subject matter for the text has been taken from the lecture notes of the courses which the author co-ordinated for the benefit of practising engineers. Some of the contributors to these lecture notes deserve my special acknowledgment. These are: Professor Krishna Gopal, Dr.V.K.Sharma, Ms.Shashwati and Ms.Namrata of Regional Engineering College, Kurukshetra; Professor N.Viswanadham, and Professor V.V.S.Sarma of Indian Institute of Science, Bangalore; Shri A.K.Sinha and Shri P.K.Rao of Centre for Reliability Engineering, Madras; Shri Siharan De and Shri Chandragupta from Indian Space Research Organization. In addition to these lecture notes, I have drawn very heavily from several books and papers already published in the field of reliability engineering. It is my pleasure to specially mention my obligation to Balagurusamy, Dhillon, Bazovsky, Ireson,

Musa and Shooman. I regret any omission which I assure is unintentional only. Also, I have given a list of references at the end of the book which I realize is in no way complete. As a matter of fact I decided to draw a line taking the acceptable level of risk not to make the list unwieldy. The present shape of the book is primarily due to my co-workers in the Centre for Excellence for man power development in Reliability Engineering namely Ch.Rajendra Prasada Rao, Shri Shakti Kumar, Dr.Brijendra Singh, and Shri Yogesh Singh.

The author has tried his level best to make the text complete and free of mistakes. Nonetheless, as a student of reliability engineering he does realize that failures can only be minimized and their effects mitigated but these can not be completely eliminated. I thank all those who helped me directly and indirectly to reduce the failures and own full responsibility for all those which still remain. I shall be grateful if any such shortcomings or mistakes are brought to my notice.

I thank the authorities of my institution, Regional Engineering College, Kurukshetra, India for providing me the necessary facilities to undertake this project. Finally, I acknowledge whole heartedly the support given by my wife, Dr.Saroj, son Ashutosh and daughter Vidhu, without whose understanding and patience it would not have been possible for me to complete this book.

K K AGGARWAL

1
RELIABILITY FUNDAMENTALS

1.1 INTRODUCTION

In the earlier times, the problems connected with the development and operation of the systems were serious but the consequences of failures were not as dramatic or as catastrophic. From the beginning of the industrial age reliability problems had to be considered rather seriously. At first, reliability was confined to mechanical equipment. However, with the advent of electrification considerable effort went into making the supply of electric power reliable. With the use of aircraft came the reliability problems connected with airborne equipment, which were more difficult to solve than reliability problems of stationary or land-transportation equipment. Reliability entered a new era with the onset of the electronic age, the age of jet aircraft flying at sonic and supersonic speeds and the age of missiles and space-vehicles. In the early days, the reliability problems had been approached by using:

1. Very high safety factors which tremendously added to the cost and weight of the equipment.

2. By extensive use of redundancy which again added to the overall cost and weight.

3. By learning from the failures and breakdowns of previous designs when designing new equipments and systems of a similar configuration

The above approaches suddenly became impractical for the new types of airborne and electronic equipment. The intuitive approach and the redesign approach had to make way for an entirely new approach to reliability - statistically defined, calculated and designed.

The overall scientific discipline that deals with general methods and procedures during the planning, design, acceptance, transportation and testing of manufactured products to ensure their maximum effectiveness during use and provides general methods for estimating reliability of complex systems from component reliabilities has received the name ***Reliability Engineering.*** Designing equipment with specified reliability figures, demonstration of reliability values, issues of maintenance, inspection, repair and replacement and the notion of maintainability as a design parameter come under the purview of Reliability Engineering. It is thus obvious that the reliability theory needed for achieving the above mentioned tasks is a precise mathematical theory based on probability and mathematical statistics. Also there exist conflicting requirements of cost, performance, safety and reliability needing system-theoretic techniques of optimization and simulation. The complexity of modern systems however demands computer aided approaches to reliability assessment.

1.2 NEED FOR RELIABILITY ENGINEERING

During the World War II reliability was considered to be one of the pressing needs in order to study the behaviour of various systems used by the military. Several studies carried out during this period revealed startling results.

(a) A study uncovered the fact that for every vacuum tube in use, there was one in spare and seven tubes in transit for which orders had already been placed.

(b) Approximately one electronics technician was required for every 250 tubes.

(c) An army study revealed that between two thirds and three fourths of equipments were out of commission or under repair.

(d) An air force study conducted over a five year period disclosed that repair and maintenance costs were about 10 times the original cost.

(e) A navy study made during manoeuvres which showed that the electronic equipment was operative only 30% of the time.

(f) A recent study showed the composition of skilled workers for mechanical, electrical and vehicle body repairs is in the ratio of 3: 5: 2, in a field workshop.

(g) Twenty- four maintenance man -hours per flight hour were required in Navy aircraft in 1949. It was estimated that this rose to 80 in 1965, primarily because of an increase in electronic equipment complexity from 120 parts in 1949 to 8,900 in 1960 to an estimated 33,000 in 1965.

(h) A study revealed that a pre- World War II civil aeroplane had about $4,000 worth of electronic control, navigation and communication apparatus. The post- war commercial DC-6 required in excess of $50,000 worth of electronic apparatus while a contemporary jet bomber has over $1,000,000 worth of electronic gear, a twenty fold increase over DC-6 and over 200 times that of pre- World War II aeroplanes.

These findings served as an impetus for further studies and investigations.

The size of the system, the intricacy of the specified functions, the length of the useful interval of the life variable, and the degree of hostility of the system's environment all influence the reliability.

It will be clear that the tendency towards larger systems, i.e. systems with larger numbers of components, would decrease the reliability if the development of more reliable system components and structures does not keep in step. There are many such systems with a large quantitative complexity, such as energy distribution networks, telecommunication systems, digital computer networks, and space probes.

In addition, there is a tendency towards the use of more complex system functions to be performed by a single system, the functions are more involved (which is expressed in more specified properties), and the allowed tolerances become smaller. This increase in qualitative complexity also causes the reliability to drop if no adequate counter measures are taken. We may think of: Multi- function measuring equipment with a higher, required accuracy, automatic landing systems for aeroplanes, process control equipment, and so on.

Further, the correct functioning of a system over a longer interval of the life variable is increasingly important as we become dependent on such systems (energy generation systems, pacemakers and the like). These so-called critical systems require a high reliability, often over long periods (e.g. 25 years for telecommunication systems). A source of concern in pacemakers, for instance, is the energy source, since circuit failures in pacemakers occur with a probability of less than 140x 10^{-9} per hour.

Besides this, our technical systems are more and more put to use in hostile environments; they have to be suitable for a wider variety of environments. Just think of applications in the process industry (heat, humidity, chemical substances), mobile applications in aircraft, ships, and vehicles (mechanical vibrations, shocks, badly defined power supply voltages, high electromagnetic interference level).

All in all, these are sufficient reasons for reliability engineering to be so much in the limelight these days. Add to that the emphasis on reliability in situations where no maintenance is possible, because of an isolated location (unmanned arctic weather stations, remote space probes, underwater amplification stations in transatlantic cables, etc). Even if maintenance were possible, it is often better (more cost -effective) to increase the initial reliability of a system because of the high costs associated with that system being down for repairs. Despite the higher initial costs, the life cycle cost may turn out to be lower. This is called the *invest now, save later* principle of reliability.

Also the socio-ethical aspects of products with a reliability that is too low cannot be underestimated. These low- reliability disposable products lead to a waste of labour, energy, and raw materials that are becoming more and more scarce.

1.3 DEFINITION

The concept of reliability has been interpreted in many ways in numerous works. Since many of these do not agree in content, it is expedient to examine the main ones.

The following definitions of reliability are most often mot with in the literature.

1. Reliability is the integral of the distribution of probabilities of failure - free operation from the instant of switch- on to the first failure.

2. The reliability of a component (or a system) is the probability that the component (or a system) will not fail for a time t.

3. Reliability is the probability that a device will operate without failure for a given period of time under given operating conditions.

4. Reliability is the mean operating time of a given specimen between two failures.

5. The reliability of a system is called its capacity for failure -free operation for a definite period of time under given operating conditions, and for minimum time lost for repair and preventive maintenance.

6. The reliability of equipment is arbitrarily assumed to be the equipment's capacity to maintain given properties under specified operating conditions and for a given period of time.

One of the definitions which has been accepted by most contemporary reliability authorities is given by the Electronics Industries Association, (EIA) USA (formerly known as RETMA) which states:

The *reliability* of an item (a component, a complex system, a computer program or a human being) is defined as the *probability* of performing its *purpose adequately* for the *period of time intended* under the *operating and environmental conditions encountered.*

This definition stresses four elements:

1. Probability
2. Adequate performance
3. Time
4. Operating and environmental conditions.

The true reliability is never exactly known, but numerical estimates quite close to this value can be obtained by the use of statistical methods and probability calculations. How close the statistically estimated reliability comes to the true reliability depends on the amount of testing, the completeness of field service reporting all successes and failures, and other essential data. For the statistical evaluation of an equipment, the equipment has to be operated and its performance observed for a specified time under actual operating conditions in the field or under well-simulated conditions in a Laboratory. Criteria of what is considered an *adequate performance* have to be exactly spelled out for each case, in advance.

Measurement of the adequate performance of a device requires measuring all important performance parameters. As long as these parameters remain within the specified limits, the equipment is judged as operating *satisfactorily.* When the performance parameters drift out of the specified tolerance limits, the equipment is judged as having malfunctioned or failed. For instance, if the gain of an electronic amplifier reduces to a value K_1 from the designed value K its performance may have to be considered unsuitable for a control system application but may still be quite acceptable for a consumer electronics equipment.

In the probability context, satisfactory performance is directly connected to the concepts of failure or malfunction. The relation between these two is that of mutually exclusive events-which means the equipment when in operation, is either operating satisfactorily or has failed or malfunctioned. Sometimes, it may be simpler to specify first what is regarded as *failure* and *satisfactory performance* is then every other operating condition which is not a failure. The frequency at which failures occur is called the *failure rate* (λ). It is usually measured in number of failures per unit operating hour. Its reciprocal value is called the *mean time between failures (m)* and this is measured in hours.

It is true that only in some simple cases, where devices of the *go-no-go* type are involved, the distinction between adequate performance and failure is a very simple matter. For instance, a switch either works or does not work - it is good or bad. But there are many more cases where such a clear-cut decision can not be made so easily and a number of performance parameters and their limits must first be specified.

Since reliability is a yardstick of capability to perform within required limits when in operation, it normally involves a parameter which measures time. This may be any time unit which is preferable in cases where continuous operation is involved; it may be number of cycles when the equipment operates only sporadically, in regular or irregular periods, or a combination of both. It is meaningful to speak of the operating hours of an engine, generator, aircraft, etc. But for a switch or relay it may be more meaningful to speak of the number of operations which such a device has to perform. The probability that no failure will occur in a number of operations (cycles) may in these cases tell much more than the probability of no failure in a number of hours. Thus, a switch measures its *time* in cycles of operation rather than in hours. Similarly, a vehicle may more meaningfully measure its *time* in miles or kilometers rather than in hours.

In addition to the conventional systems approach to reliability studies, we also frequently use Failure mode and effects analysis (FMEA), and Fault tree analysis (FTA) approaches. Failure mode and effects analysis is a preliminary design evaluation procedure used to identify design weakness that may result in safety hazards or reliability problems. The FMEA procedure may be termed a *what if* approach in that it starts at component level and asks *what if this component fails*. The effects are then traced on to system level. Any component failures that could have a critical effect on the system are identified and either eliminated or controlled, if possible. Fault tree analysis begins with the definition of an undesirable event and traces this event down through the system to identify basic causes. In systems parlance, the FMEA is a *bottom-up* procedure while the FTA is a *top-down* technique.

1.4 CAUSES OF FAILURES

The specific causes of failures of components and equipments in a system can be many. Some are known and others are unknown due to the complexity of the system and its environment. A few of them are listed below:

1. Poor Design, Production and Use

Poor design and incorrect manufacturing techniques are obvious reasons of the low reliability. Some manufacturers hesitate to invest more money on an improved design and modern techniques of manufacturing and testing. Improper selection of materials is another cause for poor design.

Components and equipments do not operate in the same manner in all conditions. A complete knowledge of their characteristics, applications, and limitations will avoid their misuse and minimize the occurrence of failures. All failures have a cause and the lack of understanding these causes is the primary cause of the unreliability of a given system.

2. System Complexity

In many cases a complex and sophisticated system is used to accomplish a task which could have been done by other simple schemes. The implications of complexity are costly. First it employs more components thereby decreasing overall reliability of the system. Second, a complex scheme presents problems in terms of users' understanding and maintenance. On the other hand, simplicity costs less, causes less problems, and has more reliability. A basic rule of reliability with respect to complexity is: *Keep the system as simple as is compatible with the performance requirements.*

3. Poor Maintenance

The important period in the life cycle of a product or a system is its operating period. Since no product is perfect, it is likely to fail. However its life time can be increased if it can be repaired and put into operation again. In many cases preventive-measures are possible and a judiciously designed preventive-maintenance policy can help eliminate failures to a large extent. The adage *Prevention is better than cure* applies to products and equipments as well.

4. Communication and Coordination

Reliability is a concern of almost all departments of an organization. It is essentially a birth-to-death problem involving such areas as raw material and parts, conceptual and detailed engineering design, production, test and quality control, product shipment and storage, installation, operation and maintenance. A well-organized management with an efficient system of communication is required to share the information and experiences about components. Sufficient opportunity should be available for the people concerned to discuss the causes of failures. In some organizations, rigidity of rules and procedures prohibits the creative-thinking and design.

5. Human Reliability

In spite of increased application of automation techniques in industries and other organisations, it is impossible to completely eliminate the human involvement in the operation and maintenance of systems. The contribution of human-errors to the unreliability may be at various stages of the product cycle. Failures due to the human- error can be due to:

- Lack of understanding of the equipment
- Lack of understanding of the process
- carelessness
- forgetfulness
- poor judgemental skills
- absence of correct operating procedures and instructions
- physical inability

Although, it is not possible to eliminate all human-errors, it is possible to minimize some of them by the proper selection and training of personnel, standardization of procedures, simplification of control schemes and other incentive measures. The designer should ensure that the operation of the equipment is as simple as possible with practically minimum probability for error. The operator should be comfortable in his work and should be free from unnecessary stresses. The following checklist should prove useful to the design engineer:

- Is the operator position comfortable for operating the controls?
- Do any of the operations require excessive physical effort?
- Is lighting of the workplace and surrounding area satisfactory?
- Does the room temperature cause any discomfort to the operator?
- Are noise and vibration within the tolerable limits?
- Does the layout ensure the required minimum movement of operator?
- Can the operator's judgement be further minimized?

With all this care, human operators are still likely to make errors. A human error may or may not cause a failure. Consequently, the quantitative measurement of the human reliability is required in order to present a correct picture of the total system reliability.

1.5 CATASTROPHIC FAILURES AND DEGRADATION FAILURES

When the ability of an item to perform its required function is terminated the item is said to have failed. As failure is an ill-defined term, we have tried to cross-reference some of the more important kinds of failures by way of a contingency Table 1.1. A failure may be *complete or partial* depending upon how complete the lack of the required function is. If we follow a particular item in time as it functions and finally fails we will see that it may fail in one of two ways, by a *catastrophic failure* or by a *degradation failure.*

Table 1.1: Failures

	Sudden failures: Failures that could not be anticipated by prior examination. (Sudden failures are similar to random failures. A random failure is any failure whose time of occurrence is unpredictable).	Gradual failures: Failures that could be anticipated by prior examination.
Complete failures: Failures resulting from deviations in characteristic (s) beyond specified limits.	Catastrophic failures: Failures that are both sudden and complete.	This state of affairs may be the end result when degradation failures are left unattended.
Partial failures: Failures resulting from deviations in characteristic (s) beyond specified limits but not such as to cause complete lack of required function.	We define marginal failures as failures which are observed at time t=0, when the item has just been finished. Sudden and partial failures are rarely seen later in life of an item.	Degradation failures: Failures that are both gradual and partial.

Catastrophic failures are characterized as being *both complete and sudden.* Complete in the sense that the change in output is so gross as to cause complete lack of the required function, and sudden in the sense that the failure could not be anticipated. For example, at the system level the event of the gain of an amplifier suddenly going to zero would be a catastrophic failure.

Degradation failures often called *drift failures*, require further categorization. We can distinguish between monotonic and non-monotonic drift. *Monotonic drift* is characterized by an output variable continuously varying in the same direction as illustrated in the Fig1.1. At some point in time the value of the output crosses one of the constraints, giving rise to failure. *Non- monotonic drift* is characterized by both positive and negative excursions of an output variable as shown in Fig1.2(a), the excursions being somewhat similar to Brownian movements. The definition of unsatisfactory performance (especially failure) in the case of non-monotonic drift is not quite so straightforward as for monotonic drift. Of course, violation of the constraints at any point must strictly speaking be classified as a failure.

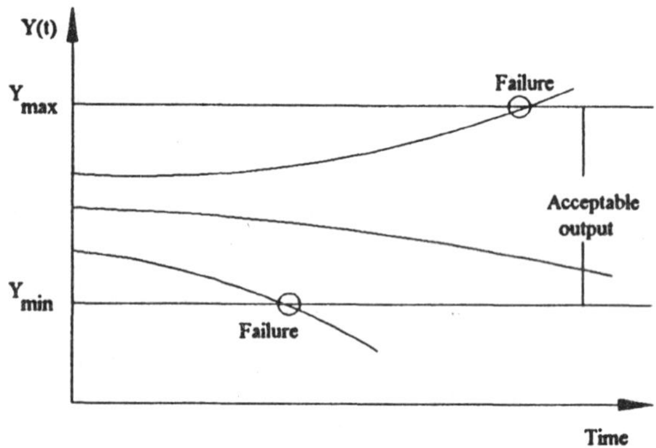

Fig.1.1 Three examples of monotonic drift two of which give rise to failures.

However, in the case of non-monotonic drift, it may happen that the output drifts back into the acceptable region shortly afterwards-if so the short-lasting excursion into the region of unsatisfactory performance may not have harmed the system performance appreciably. Depending on the system, this consequence of drift may more properly be defined in terms of the accumulated amount of resulting degradation. As an example, consider the definition of a possible function v(t) for measuring the accumulated degradation as shown in Fig 1.2(b). Only when the accumulated amount of degradation defined by this function exceeds a specified level, V_f, is the system deemed to have performed unsatisfactorily. Other indications of unsatisfactory performance are also possible in the case of non- monotonic drift. We might for example use the area of Y(t) above or below the limits for acceptable performance as an indicator. Unsatisfactory performance would then be evidenced when the area exceeds a specified amount. A third possibility would be to use the number of crossings of the limits as an indicator of unsatisfactory performance.

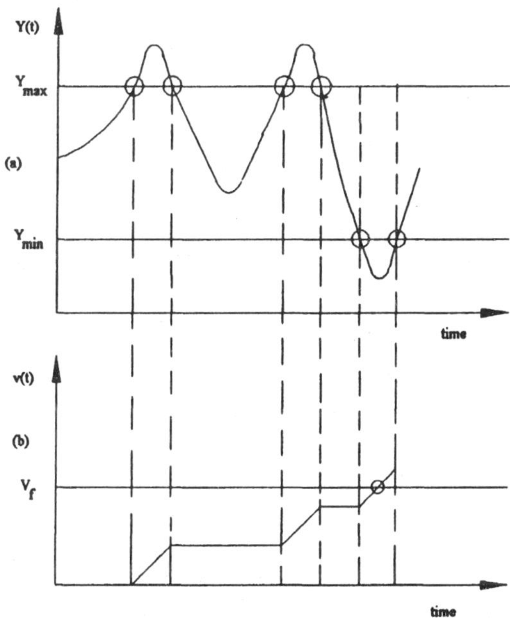

Fig. 1.2 (a) Non- monotonic drift of a variable.
(b) v(t) is the total time Y(t) has spent in the region of degradation.

1.6 CHARACTERISTIC TYPES OF FAILURES

Reliability Engineering distinguishes three characteristic types of failures (excluding damage caused by careless handling, storing, or improper operation by the users) which may be inherent in the equipment and occur without any fault on the part of the operator.

First, there are the failures which occur early in the life of a component. They are called *early failures.* Some examples of early failures are:

* Poor welds or seals
* Poor solder joints
* Poor connections
* Dirt or contamination on surfaces or in materials
* Chemical impurities in metal or insulation
* Voids, cracks, thin spots in insulation or protective coatings
* Incorrect positioning of parts

Many of these early failures can be prevented by improving the control over the manufacturing process. Sometimes, improvements in design or materials are required to increase the tolerance for these manufacturing deviations, but fundamentally these failures reflect the *manufacturability* of the component or product and the control of the manufacturing processes. Consequently, these early failures would show up during:

* In-process and final tests
* Process audits
* Life tests
* Environmental tests.

Early failures can be eliminated by the so-called *debugging or burn-in* process. The debugging process consists of operating an equipment for a number of hours under conditions simulating actual use. The weak or substandard components fail in these early hours of the equipment's operation and they are replaced by good components. Similarly poor solder connections or other assembly faults show up and they are corrected. Only then is the equipment released for service.

Secondly, there are failures which are caused by wearout of parts. These occur in an equipment only if it is not properly maintained-or not maintained at all. *Wearout failures* are due primarily to deterioration of the design strength of the device as a consequence of operation and exposure to environmental fluctuations. Deterioration results from a number of familiar chemical and physical phenomena:

* Corrosion or oxidation
* Insulation breakdown or leakage
* Ionic migration of metals in vacuum or on surfaces
* Frictional wear or fatigue
* Shrinkage and cracking in plastics

In most cases wearout failures can be prevented. For instance, in repeatedly operated equipment one method is to replace at regular intervals the accessible parts which are known to be subject to wearout, and to make the replacement intervals shorter than the mean wearout life of the parts. Or, when the parts are inaccessible, they are designed for a longer life than the intended life of the equipment. This second method is also applied to so-called *one-shot* equipment, such as missiles, which are used only once during their lifetime.

Third, there are so-called *chance failures* which neither good debugging techniques nor the best maintenance practices can eliminate. These failures are caused by sudden stress accumulations beyond the design strength of the component. Chance failures occur at random intervals, irregularly and unexpectedly. No one can predict when chance failures will occur. However, they obey certain rules of collective behaviour so that the frequency of their occurrence during sufficiently long periods is approximately constant. Chance failures are sometimes called *catastrophic* failures, which is inaccurate because early failures and wearout failures can be as catastrophic as chance failures. It is not normally easy to eliminate chance failures. However, reliability techniques have been developed which can reduce the chance of their occurrence and, therefore, reduce their number to a minimum within a given time interval.

Reliability engineering is concerned with eliminating early failures by observing their distribution and determining accordingly the length of the necessary debugging period and the debugging methods to be followed. Further, it is concerned with preventing wearout failures by observing the statistical distribution of wearout and determining the overhaul or preventive replacement periods for the various parts or their design life. Finally, its main attention is focused on chance failures and their prevention, reduction, or complete elimination because it is the chance failure phenomenon which most undesirably affects after the equipment has been debugged and before parts begin to wear out.

1.7 USEFUL LIFE OF COMPONENTS

If we take a large sample of components and operate them under constant conditions and replace the components as they fail, then approximately the same number of failures will occur in sufficiently long periods of equal length. The physical mechanism of such failures is a sudden accumulation of stresses acting on and in the component. These sudden stress accumulations occur at random and the randomness of the occurrence of chance failures is therefore an obvious consequence.

If we plot the curve of the failure rate against the lifetime T of a very large sample of a homogeneous component population, the resulting failure rate graph is shown in Fig 1.3. At the time T=0 we place in operation a very large number of new components of one kind. This population will initially exhibit a high failure rate if it contains some proportion of substandard, weak specimens. As these weak components fail one by one, the failure rate decreases comparatively rapidly during the so-called *burn-in* or *debugging* period, and stabilizes to an approximately constant value at the time T_b when the weak components have died out. The component population after having been burned in or debugged, reaches its lowest failure rate level which is approximately constant. This period of life is called the *useful life* period and it is in this period that the exponential law is a good approximation. When the components reach the life T_w wearout begins to make itself noticeable. From this time on, the failure rate increases rather rapidly. If upto the time T_w only a small percentage of the component population has failed of the many components which survived up to the time T_w, about one-half will fail in the time period from T_w to M. The time M is the mean wearout life of the population. We call it simply mean life, distinguished from the mean time between failures, $m = 1/\lambda$ in the useful life period.

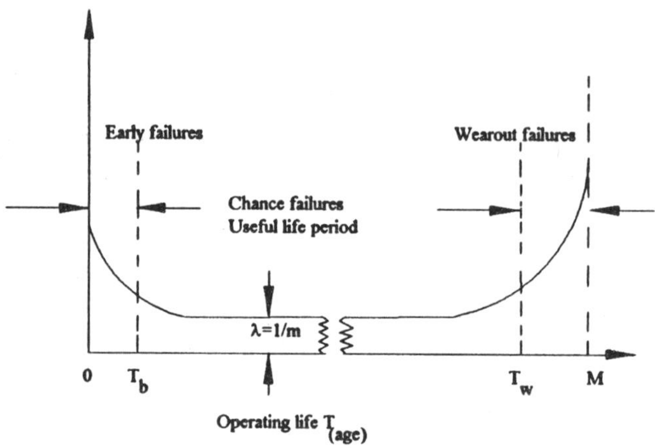

Fig. 1.3 Component failure rate as a function of age.

If the chance failure rate is very small in the useful life period, the mean time between failures can reach hundreds of thousands or even millions of hours. Naturally, if a component is known to have a mean time between failures of say 100,000 hours (or a failure rate of 0.00001) that certainly does not mean that it can be used in operation for 100,000 hours.

The mean time between failures tells us how reliable the component is in its useful life period, and such information is of utmost importance. A component with a mean time between failures of 100,000 hours will have a reliability of 0.9999 or 99.99 percent for any 10-hour operating period. Further if we operate 100,000 components of this quality for 1 hour, we would expect only one to fail. Equally, would we expect only one failure if we operate 10,000 components under the same conditions for 10 hours, or 1000 components for 100 hours, or 100 components for 1000 hours.

Chance failures cannot be prevented by any replacement policy because of the constant failure rate of the components within their useful life. If we try to replace good nonfailed components during useful life, we would improve absolutely nothing. We would more likely do harm, as some of the components used for replacement may not have been properly burned in, and the presence of such components could only increase the failure rate. Therefore, the very best policy in the useful life period of components is to replace them only as they fail. However, we must stress again that no component must be allowed to remain in service beyond its wearout replacement time T_W. Otherwise, the component probability of failure increases tremendously and the system probability of failure increases even more.

The *golden rule* of reliability is, therefore: Replace components as they fail within the useful life of the components, and replace each component preventively, even if it has not failed, not later than when it has reached the end of its useful life. The burn-in procedure is an absolute must for missiles, rockets, and space systems in which no component replacements are possible once the vehicle takes off and where the failure of any single component can cause the loss of the system. Component burn-in before assembly followed by a debugging procedure of the system is, therefore, *another golden rule* of reliability.

1.8 THE EXPONENTIAL CASE OF CHANCE FAILURES

In the simplest case, when a device is subject only to failures which occur at random intervals, and the expected number of failures is the same for equally long operating periods, its reliability is mathematically defined by the well-known exponential formula

$$R(t) = \exp(-\lambda t) \tag{1.1}$$

In this formula λ is a constant called the failure rate, and t is the operating time. The failure rate must be expressed in the same time units as time, t- usually in hours. However, it may be better to use *cycles or miles* in same cases. The reliability R is then the probability that the device, which has a constant failure rate λ will not fail in the given operating time t.

This reliability formula is correct for all properly debugged devices which are not subject to early failures, and which have not yet suffered any degree of wearout damage or performance degradation because of their age.

To illustrate the important fact of an equal chance of survival for periods of equal length throughout the useful life, let us assume that a device with a 1000-hour useful life has a constant failure rate $\lambda=0.0001$ per hour. Its reliability for any 10 hours' operation within these 1000 hours is

$$R = \exp(-0.0001 \times 10) = 0.9990 \text{ (or 99.9 percent)}$$

The probability that the device will not fail in its entire useful life period of 1000 hours is

$$R = \exp(-0.0001 \times 1000) = 0.9048 \text{ (or 90.48 percent)}$$

Thus, it has a chance of 90 percent to survive up to 1000 hours counted from the moment when first put into operation. But if it survives up to 990 hours, then its chance to survive the last 10 hours (from 990 to 1000 hours) of its useful life is again 99.9 percent.

We often use the reciprocal value of the failure rate, which is called the mean time between failures, m. The mean time between failures, abbreviated MTBF can be measured directly in hours. By definition, in the exponential case, the mean time between failures, or MTBF is

$$m = 1/\lambda \tag{1.2}$$

The reliability function can , therefore, also be written in the form

$$R(t) = \exp(-t/m) \tag{1.3}$$

When plotting this function, with Reliability values on the ordinate and the corresponding time values on the abscissa, we obtain a curve which is often referred to as the survival characteristic and is shown in Fig 1.4.

It is important to understand that the time t on the abscissa is not a measure of the calendar life. It counts only the hours of any arbitrarily chosen operating period with $t = 0$ designating the beginning of the considered operating period. Therefore, 't' in this formula is often called *mission time.* It is assumed that the device has survived previous missions, and it will not reach the end of its useful life in the mission now under consideration. The first assumption is written as $R = 1$ at $t = 0$, which means that the device has survived to the beginning of the mission. The second assumption is contained in the original assumption of λ = constant. Second, it is seen that the time t in the graph extends to infinity, which seems to make no sense. However, when only chance failures are considered, the certainty that a device will fail because of a chance failure exists only for an infinitely long operating period.

There are a few points on this curve which are easy to remember and which help greatly in rough predicting work. For an operating time $t = m$, the device has a probability of only 36.8 percent (or approximately 37 percent) to survive. For $t = m/10$, the curve shows a reliability of $R = 0.9$ and for $t = m/100$, the reliability is $R = 0.99$; for $t = m/1000$, it is 0.999.

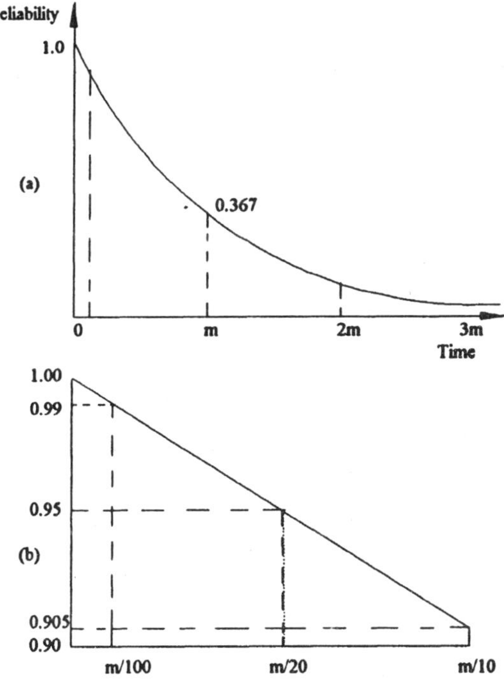

Fig. 1.4 The standardised Reliability curve

(a) The curve

(b) The upper portion of the reliability curve.

For fast reliability calculations, we can use a Nomogram as shown in Fig 1.5. If we know any two of the following three parameters, the third can be directly read on the straight line joining the first two.

(i) Failure rate (or MTBF)
(ii) Reliability
(iii) Operating Time

Example 1.1

Consider the failure rate of an instrument as 0.0001/hr. What will be its reliability for an operating period of 100 hours?

Solution

$\lambda = 0.0001/\text{hr}$

Therefore, $m = 1/\lambda = 10{,}000$ hr

t = 100 hours

Using relation (1.1) or (1.3),

R = 0.99 (or 99 percent)

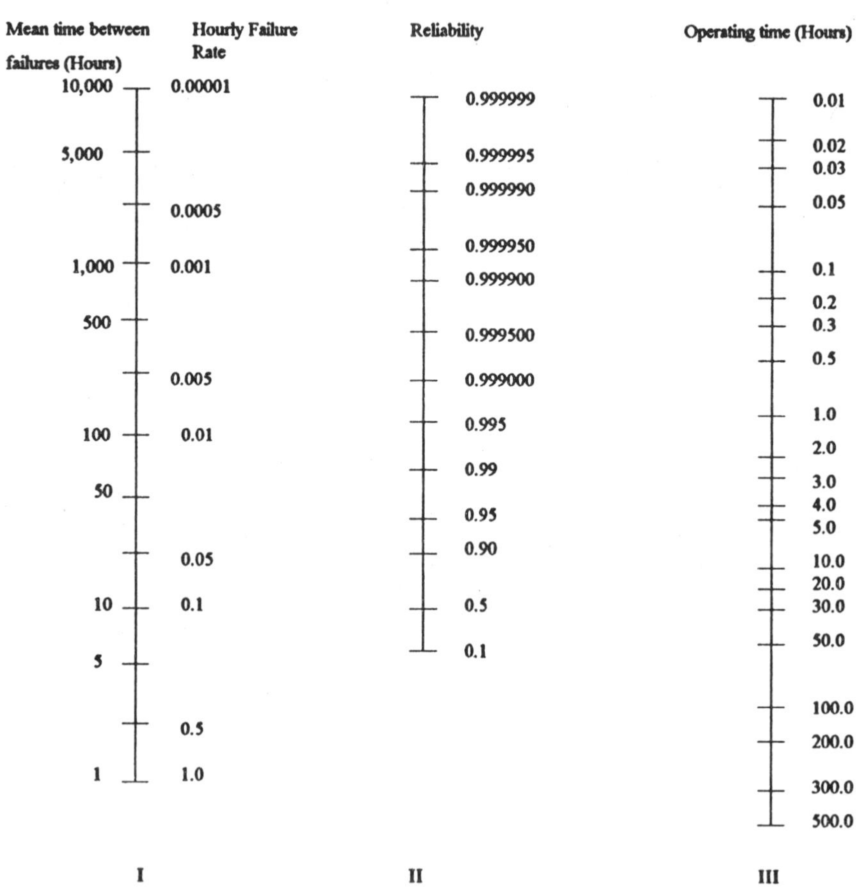

Fig. 1.5 Reliabillity Nomogram.

Alternatively, mark the points on scale I & III of Nomogram corresponding to the given values of λ & t. Join these two points and read the value of the reliability directly on scale II as 0.99.

* * *

1.9 RELIABILITY MEASURES

The reliability of a component can be interpreted as the fraction of the number of components surviving a test to the total number of components present at the beginning of the test.

If a fixed number N_o of components are tested, there will be, after a time t, $N_s(t)$ components which survive the test and $N_f(t)$ components which fail. Therefore, $N_o = N_s(t) + N_f(t)$ is a constant throughout the test. The reliability, expressed as a fraction by the probability definition at any time t during the test is:

$$R(t) = (N_s(t)/ N_o) = N_s(t)/ (N_s(t) + N_f (t)) \qquad (1.4)$$

In the same way, we can also define the probability of failure Q (called unreliability) as

$$Q(t) = (N_f (t)/ N_o) = N_f (t)/ (N_s (t) + N_f (t)) \qquad (1.5)$$

It is at once evident that at any time t,

$$R(t) + Q(t) = 1 \qquad (1.6)$$

The events of component survival and component failure are called complementary events because each component will either survive or fail. These are also called mutually exclusive events because if a component has failed, it has not survived, and vice versa.

The reliability can also be written as

$$R(t) = (N_o-N_f(t))/N_o = 1-(N_f(t)/N_o) \qquad (1.7)$$

By differentiation of this equation we obtain

$$dR(t)/dt = -(1/N_o)(dN_f(t)/dt) \qquad (1.8)$$

Rearranging,

$$dN_f(t)/dt = - N_o\, dR(t)/dt \qquad (1.9)$$

The term $dN_f(t)/dt$ can be interpreted as the number of components failing in the time interval dt between the times t and t+dt, which is equivalent to the rate at which the component population still in test at time t is failing.

At the time t, we still have $N_s(t)$ components in test; therefore, $dN_f(t)/dt$

components will fail out of these $N_s(t)$ components. When we now divide both sides of the equation (1.9) by $N_s(t)$, we obtain the rate of failure or the instantaneous probability of failure per one component, which we call the failure rate:

$$\lambda(t) = (1/N_s(t))(dN_f(t)/dt) = -(N_o/N_s(t))(dR(t)/dt) \tag{1.10}$$

Using (1.4) we get

$$\lambda(t) = -(1/R(t))(dR(t)/dt) \tag{1.11}$$

which is the most general expression for the failure rate because it applies to exponential as well as non-exponential distributions. In the general case, λ is a function of the operating time t, for both R and dR/dt are functions of t. Only in one case will the equation yield a constant, and that is when failures occur exponentially at random intervals in time. By rearrangement and integration of the above equation, we obtain the general formula for reliability,

$$\lambda(t)dt = -(dR(t)/R(t))$$

or, $$\ln (R(t)) = -\int_0^t \lambda(t)\, dt$$

Solving for R(t) and knowing that at t=0, R(t)=1, we obtain

$$R(t) = \exp[-\int_0^t \lambda(t)\, dt] \tag{1.12}$$

So far in this derivation, we have made no assumption regarding the nature of failure rate and therefore it can be any variable and integrable function of the time t. Consequently, in the equation (1.12), R(t) mathematically describes reliability in a most general way and applies to all possible kinds of failure distributions.

When we specify that failure rate is constant in the above equation, the exponent becomes

$$-\int_0^t \lambda(t)\, dt = -\lambda t$$

and the known reliability formula for constant failure rate results,

$$R(t) = \exp(-\lambda t) \tag{1.13}$$

In the above derivation, $dN_f(t)/dt$ is the frequency at which failures occur at any time during a non-replacement test. When $dN_f(t)/dt$ is plotted as a graph against t, we obtain the time distribution of the failures of all the original N_o components. And when we plot $(1/N_o)(dN_f(t)/dt)$ as a graph, we have the distribution of failures in time on a per component basis, or the failure frequency curve per component. Therefore, the graph $(1/N_o)(dN_f(t)/dt)$ is a unit frequency curve which is called the *failure density function f(t).*

$$f(t) = (1/N_o)(dN_f(t)/dt) = -dR(t)/dt \tag{1.14}$$

It may be observed that the total area under this curve equals unity because

$$A = -\int_0^\infty (dR/dt)\, dt = -\int_0^\infty dR = -[R(\infty) - R(0)] = 1$$

The failure rate can, also be written as

$$\lambda(t) = -[1/R(t)](dR(t)/dt) = f(t)/R(t) \tag{1.15}$$

which means the failure rate at any time t equals the f(t) value divided by the reliability, both taken at the time t. This equation again applies to all possible distributions and reliabilities, whether or not they are exponential. In the special case when λ is constant, the distribution is

$$f(t) = \lambda \exp(-\lambda t) \tag{1.16}$$

We also have

$$f(t) = (1/N_o)(dN_f(t)/dt) = dQ(t)/dt \tag{1.17}$$

By integration, we obtain,

$$Q(t) = \int_0^t f(t)\, dt \tag{1.18}$$

which means that the probability of failure Q(t) at time t is equivalent to the area under the density curve taken from t = o to t. Thus Q(t) is the cumulative probability of failure function. Also,

$$R(t) = 1 - \int_0^t f(t)\, dt \tag{1.19}$$

but because the area under the density curve is always unity, we can write

$$R(t) = \int_0^\infty f(t)dt - \int_0^t f(t)dt = \int_t^\infty f(t)dt \tag{1.20}$$

This is shown in Fig1.6, the graph of the density function for the exponential case.

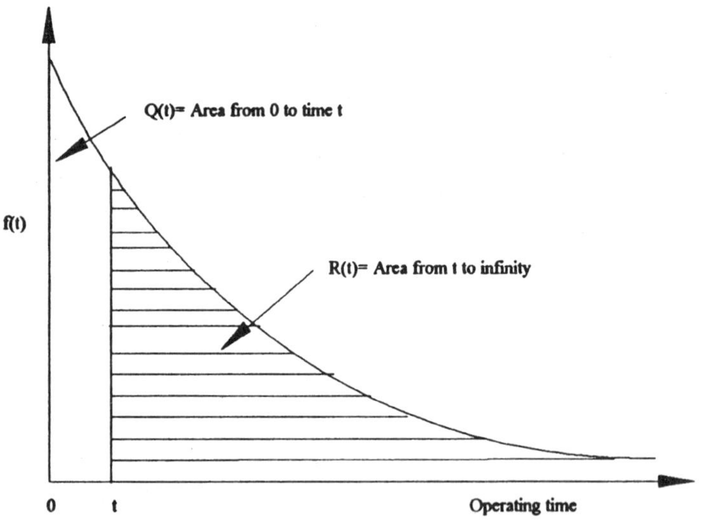

Fig. 1.6 The exponential density function.

The failure rate can be written also in terms of Q(t):

$$\lambda(t) = f(t)/R(t) = f(t)/[1-Q(t)]$$
$$= [1/(1-Q(t))]\ (dQ(t)/dt) = [1/R(t)]\ dQ(t)/dt \tag{1.21}$$

The important point we have made here is that the failure rate is always equal to the ratio of density to reliability. In the exponential case this ratio is constant. However, in the case of non- exponential distributions, the ratio changes with time and, therefore, the failure rate is then a function of time. We have thus specified relationships between four important reliability measures:

- Reliability function, R(t)
- Unreliability function, Q(t)
- Failure rate, $\lambda(t)$
- Failure density, f(t)

A summary of these relationships is given in the form of Table 1.2.

The mean time between failures can also be expressed in terms of reliability as

$$m = \int_0^\infty R(t)dt \tag{1.22}$$

Referring to the component testing experiment described earlier, let dN_f components fail during the interval t to t+dt. As all these dN_f components have already survived for t hours, MTBF can be expressed as:

Table 1.2: Relations between Reliability Measures

Given	R(t)	Q(t)	λ(t)	f(t)
R(t)	*	1-Q(t)	$\exp[-\int_0^t \lambda(t)dt]$	$\int_t^\infty f(t)dt$
Q(t)	1-R(t)	*	$1-\exp[-\int_0^t \lambda(t)dt]$	$\int_0^t f(t)dt$
λ(t)	[-1/R(t)]dR(t)/dt	[1/(1-Q(t))]dQ(t)/dt	*	$f(t)/[\int_0^t f(t)dt]$
f(t)	-dR(t)/dt	dQ(t)/dt	$\lambda(t)\exp[-\int_0^t (t)dt]$	*

$$m = (1/N_o) \int_0^{N_o} t\, dN_f$$

$$m = (1/N_o) \int_0^\infty t\, N_o\, f(t)\, dt = \int_0^\infty t\, f(t)dt \tag{1.23}$$

As f(t) = -dR/dt

$$m = \int_0^1 t\ dR \tag{1.24}$$

From the reliability curve Fig 1.7, this can be easily interpreted as

$$m = \int_0^\infty R(t)\, dt \tag{1.25}$$

Hence, MTBF can always be expressed as the total area under the reliability curve.

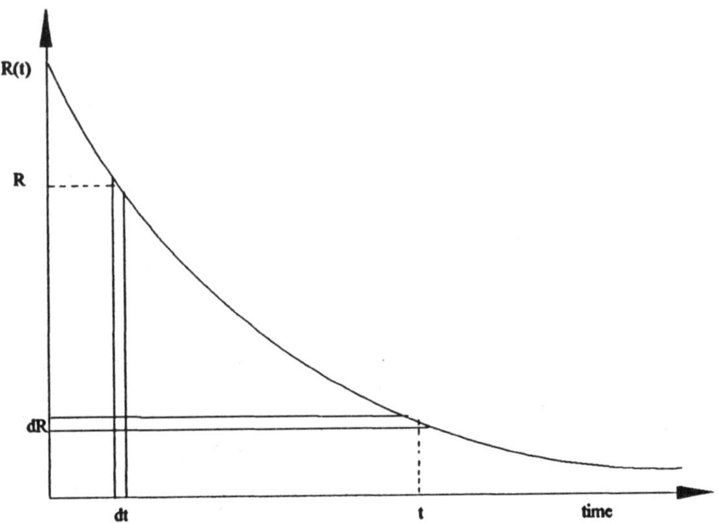

Fig. 1.7 Area under the reliability curve.

For the special case of an exponential distribution

$$\lambda(t) = \lambda$$

Hence, $R(t) = \exp(-\lambda t)$

$$Q(t) = 1-\exp(-\lambda t)$$

$$f(t) = \lambda \exp(-\lambda t)$$

$$m = \int_0^{\infty} \exp(-\lambda t)\, dt = 1/\lambda \qquad (1.26)$$

Similarly all these parameters can be evaluated for other distributions if any one of these parameters is known.

Example 1.2

Rayleigh distribution is characterized by constantly increasing failure rate. Determine the reliability parameters.

Solution

In this case,

$$\lambda(t) = kt$$

Hence, $$R(t) = \exp[-\int_0^t kt\,dt] = \exp(-kt^2/2)$$

$$Q(t) = 1 - R(t) = 1 - \exp(-kt^2/2)$$

$$f(t) = \lambda(t)\,R(t) = kt\exp(-kt^2/2)$$

$$m = \int_0^\infty R(t)\,dt = (\pi/2k)^{1/2}$$

* * *

1.10 FAILURE DATA ANALYSIS

The pattern of failures can be obtained from life test results, i.e. by testing a fairly large number of models until failure occurs and observing failure-rate characteristics as a function of time. The first step, therefore, is to link reliability with experimental or field-failure data. Suppose we make observations on the system at times t_1, t_2,....etc. Then we can define the failure density function as follows:

$$f(t) = \frac{N_s(t_i) - N_s(t_{i+1})}{N_o[t_{i+1} - t_i]}; \qquad t_i < t \le t_{i+1} \tag{1.27}$$

It is the ratio of number of failures occurring in the interval to the size of the original population divided by the length of the interval. Similarly, failure rate over the interval $t_i < t \le t_{i+1}$ is defined as the ratio of the number of failures occurring in the time interval to the number of survivors at the beginning of the time interval, divided by the length of the time interval.

$$\lambda(t) = \frac{N_s(t_i) - N_s(t_{i+1})}{N_s(t_i)\,[t_{i+1} - t_i]}; \qquad t_i < t < t_{i+1} \tag{1.28}$$

The failure density function f(t) is a measure of the overall speed at which failures are occurring whereas $\lambda(t)$ is a measure of the instantaneous speed of failure. The choice of t_i is unspecified and is best discussed by means of examples.

Example 1.3:

The failure data for ten electronic components is as given in Table1.3. Compute and plot failure density, failure rate, reliability and unreliability functions.

Table 1.3: Data for Example 1.3

Failure No	1	2	3	4	5	6	7	8	9	10
Operating time hrs.	8	20	34	46	63	86	111	141	186	266

Solution

The computation of failure density and failure rate is shown in Table 1.4. Similarly the computation of reliability and unreliability function is shown in Table 1.5. These results are also shown in Fig 1.8. As shown, we can compute R(t) for this example using the formula $R(t) = N_s(t_i)/N_o$ at each value of t_i and connecting these points by a set of straight lines. In the data analysis one usually finds it convenient to work with $\lambda(t)$ curve and deduce the reliability and density functions theoretically. For example, in this illustration, we can see that the hazard rate can be modeled as a constant.

* * *

Table 1.4 Computation of failure density and failure rate

Time Interval (Hours)	Failure density	Failure rate
0-8	1/(10 x 8) = 0.0125	1/(10 x 8) = 0.0125
8-20	1/(10 x 12) = 0.0084	1/(9 x 12) = 0.0093
20-34	1/(10 x 14) = 0.0072	1/(8 x 14) = 0.0096
34-46	1/(10 x 12) = 0.0084	1/(7 x 12) = 0.0119
46-63	1/(10 x 17) = 0.0059	1/(6 x 17) = 0.0098
63-86	1/(10 x 23) = 0.0044	1/(5 x 23) = 0.0087
86-111	1/(10 x 25) = 0.0040	1/(4 x 25) = 0.0100
111-141	1/(10 x 30) = 0.0033	1/(3 x 30) = 0.0111
141-186	1/(10 x 45) = 0.0022	1/(2 x 45) = 0.0111
186-266	1/(10 x 80) = 0.0013	1/(1 x 80) = 0.0125

We now show how can we measure the constant failure rate of a component population very conveniently. Referring to the previous experiment, if λ is constant, the product $(1/N_s(t))\,(dN_f(t)/dt)$ must also be constant throughout a test.

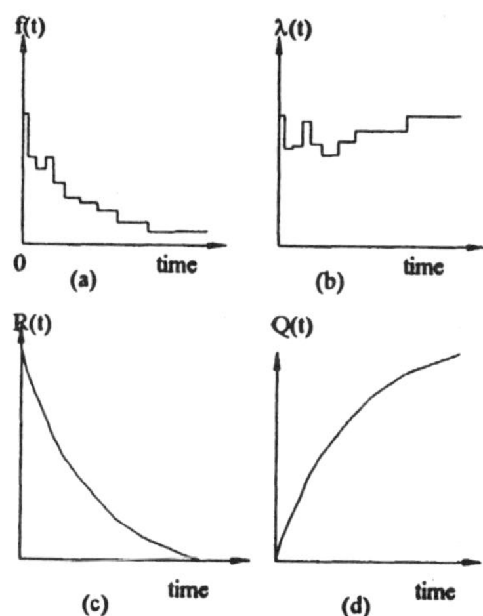

Fig. 1.8 Reliability Parameters for Example 1.3.

Table 1.5 Computation of Reliability and Unreliability

Time(hrs)	Reliability	Unreliability
0	1.0	0.0
8	0.9	0.1
20	0.8	0.2
34	0.7	0.3
46	0.6	0.4
63	0.5	0.5
86	0.4	0.6
111	0.3	0.7
141	0.2	0.8
186	0.1	0.9
266	0.0	1.0

That means that $1/N_s(t)$ and $dN_f(t)/dt$ must either decrease at the same rate or must be held constant through the entire test. A simple way to measure a constant failure rate is to keep the number of components in the test constant by immediately replacing the failed components with good ones. The number of alive components $N_s(t)$ is then equal to N_o throughout the test. Therefore, $1/N_s(t) = 1/N_o$ is constant, and $dN_f(t)/dt$ in this test must also be constant if the failure rate is to be constant. But $dN_f(t)/dt$ will be constant only if the total number of failed components $N_f(t)$ counted from the beginning of test increases linearly with time. If N_f components have failed in time t at a constant rate, the number of components failing per unit time becomes N_f/t and in this test we can substitute N_f/t for $dN_f(t)/dt$ and $1/N_o$ for $1/N_s(t)$. Therefore,

$$\lambda = (1/N_s(t))(dN_f(t)/dt) = (1/N_o)(N_f/t) \qquad (1.29)$$

Thus, we need to count only the number of failures N_f and the straight hours of operation t. The constant failure rate is then the number of failures divided by the product of test time t and the number of components in test which is kept continuously at N_o. This product $N_o.t$ is the number of unit-hours accumulated during the test. Of course, this procedure for determining the failure rate can be applied only if λ is constant.

If only one equipment ($N_o = 1$) is tested but is repairable so that the test can continue after each failure, the failure rate becomes $\lambda = N_f/t$ where the unit-hours t amount to the straight test time.

Example 1.4:

Consider another example wherein the time scale is now divided into equally spaced intervals called class intervals. The data is tabulated in the Table 1.6 in class intervals of 1000 hours. Compute the failure density and failure rate functions.

Table 1.6: Data for Example 1.4

Time interval hours	Failures in the interval
0000 - 1000	59
1001 - 2000	24
2001 - 3000	29
3001 - 4000	30
4001 - 5000	17
5001 - 6000	13

Solution:

The solution for this example is shown in Table 1.7.

Table 1.7 Computation of failure density and failure rate

Interval	Failure density	Failure rate
0000 - 1000	59/(172 x1000) = 0.000343	59/(172 x 1000) = 0.000343
1001 - 2000	24/(172 x1000) = 0.000140	24/(113 x 1000) = 0.000212
2001 - 3000	29/(172 x1000) = 0.000169	29/(89 x 1000) = 0.000326
3001 - 4000	30/(172 x1000) = 0.000174	30/(60 x 1000) = 0.000500
4001 - 5000	17/(172 x1000) = 0.000099	17/(30 x 1000) = 0.000569
5001 - 6000	13/(172 x1000) = 0.000076	13/(13 x 1000) = 0.001000

It can be seen that the failure rate in this case can be approximated by a linearly increasing time function.

Example 1.5 :

A sample of 100 electric bulbs was put on test for 1500 hrs. During this period 20 bulbs failed at 840, 861, 901, 939, 993, 1060, 1100, 1137, 1184, 1200, 1225, 1251, 1270, 1296, 1314, 1348, 1362, 1389, 1421, and 1473 hours. Assuming constant failure rate, determine the value of failure rate.

Solution:

In this case,

$$N_f = 20$$

$$N_o t = 840 + 861 + 901 + 939 + 993 + 1060 + 1100 + 1137 + 1184 + 1200 + 1225 + 1251 + 1270 + 1296 + 1314 + 1348 + 1362 + 1389 + 1421 + 1473 + 80(1500) = 143,564 \text{ hrs.}$$

Hence, $\lambda = N_f/N_o t = 20/143{,}564 = 0.139 \times 10^{-4}$ /hr.

* * *

2

RELIABILITY MATHEMATICS

2.1 FUNDAMENTALS OF SET THEORY

A *set* is a collection of objects viewed as a single entity. The individual objects of the set are called the *elements* of the set. Sets usually are denoted by capital letters: A,B,C,.......Y.Z, and elements are designated by lower-case letters; a,b,c,....y.z. If a is an element of the set A, we write $a \in A$, and we write $a \notin A$ for a is not an element of A. A set is called a *finite* set when it contains a finite number of elements and an infinite set otherwise. The *null set* ϕ is the set that contains no elements. The total or *universal set* $\cup$ is the set which contains all the elements under consideration.

We say a set A is a *subset* of set B if each element of A is also an element of B and write as $A \subseteq B$. The relation I is referred to as *set inclusion.*

2.11 The Algebra of Sets

(i) The *union* of the two sets A and B, denoted by $A \cup B$ is the set of all elements of either set, that is, $c \in (A \cup B)$ means $c \in A$, or $c \in B$, or both.

(ii) The *intersection* of the two sets A and B, denoted by $A \cap B$, is the set of all elements common to both A and B, that is, $c \in (A \cap B)$ means $c \in A$ and $c \in B$.

(iii) The *complement* of a set A, denoted by A' is the set of elements of the universal set that do not belong to A.

(iv)The two sets are said to be *disjoint or mutually exclusive* if they have no elements in common, i.e. $A \cap B = \phi$.

2.12 Venn Diagrams

When considering sets and operations on sets, Venn diagrams can be used to represent sets diagrammatically. Fig 2.1(a) shows a Venn diagram for $A \cap B$ and Fig 2.1(b) shows a Venn diagram for $A \cup B$. Fig 2.1(c) shows a Venn diagram with three sets A, B and C.

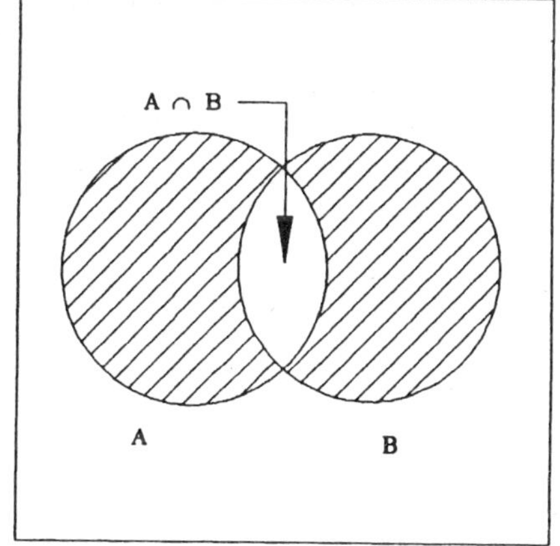

(a)

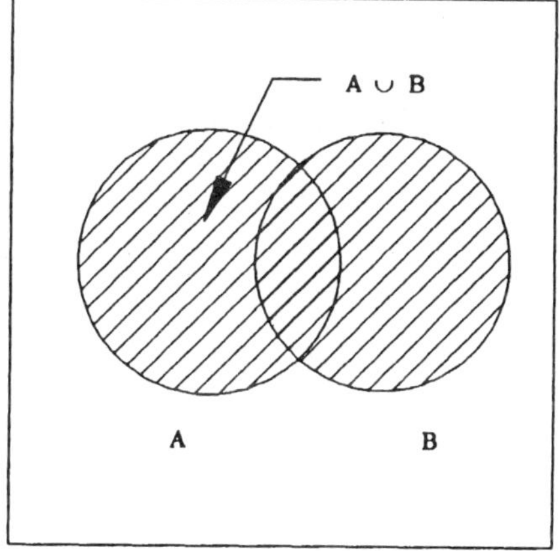

(b)

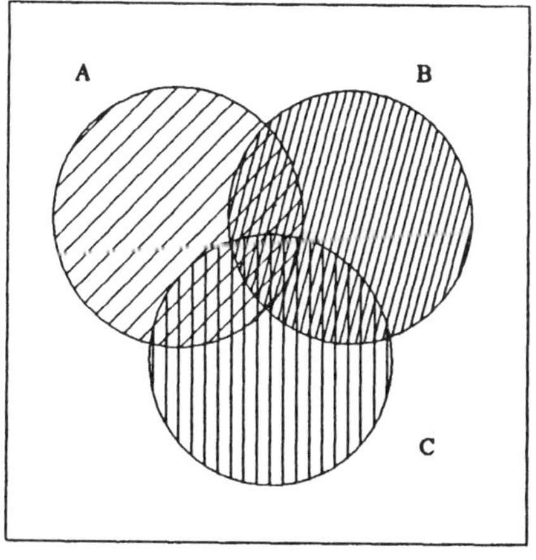

(c)

Fig. 2.1 Venn Diagrams

Example 2.1

A group of 10 men and 8 women are administered a test for high blood pressure. Among the men, 4 are found to have high blood pressure, whereas 3 of the women have high blood pressure. Use a Venn diagram to illustrate this idea.

Solution

The Venn diagram is shown in Fig 2.2. The circle labeled H represents the 7 people having high blood pressure, and the circle labeled W represents the 8 women. The numbers placed in the various regions indicate how many people there are in the category corresponding to the region. For example, there are 4 people who have high blood pressure and are not women. Similarly there are 5 women who do not have high blood pressure.

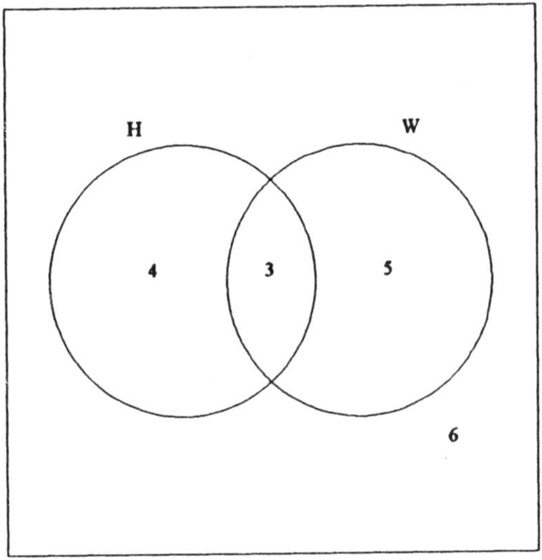

Fig. 2.2 Venn diagram for Example 2.1.

* * *

2.2 PROBABILITY THEORY

There is a natural relation between probability theory and set theory based on the concept of a random experiment for which it is impossible to state a particular outcome, but we can define the set of all possible outcomes. The *sample space* of an experiment, denoted by S, is the set of all possible outcomes of the experiment. An *event* is any collection of outcomes of the experiment or subset of the sample space S. An event is said to be simple if it consists of exactly one outcome, and compound if it consists of more than one outcome.

The objective of probability is to assign to each event A of the sample space S associated with an experiment a number Pr(A), called the probability of event A, which will give a precise measure of the chance that A will occur. The function Pr(.) has the following properties:

1. $0 \leq Pr(A) \leq 1$ for each event A of S.
2. $Pr(S) = 1$.
3. For any finite number k of mutually exclusive events defined on S.

$$Pr\left(\bigcup_{i=1}^{k} A_i\right) = \sum_{i=1}^{k} Pr(A_i)$$

4. If $A_1, A_2, A_3, \ldots\ldots$ is a denumerable or countably infinite sequence of mutually exclusive events defined on S, then

$$Pr(A_1 \cup A_2 \cup A_3 \cup \ldots) = Pr(A_1) + Pr(A_2) + Pr(A_3) + \ldots.$$

We can also use the concept *relative frequency* to develop the function Pr(.). If we repeat an experiment n times and event A occurs n_A times, $0 < n_A < n$, then the value of the relative frequency $f_A = n_A/n$ approaches Pr(A) as n increases to infinity.

2.21 Properties of Probability

1. If f is the empty or null set, then $Pr(f) = 0$
2. $Pr(A') = 1 - Pr(A)$
3. $Pr(A \cup B) = Pr(A) + Pr(B) - Pr(A \cap B)$
4. $$Pr(A_1 \cup A_2 \cup \ldots\ldots \cup A_n) = \sum_{i=1}^{k} Pr(A_i) - \sum_{i=1}^{n-1} \sum_{j=i+1}^{n} Pr(A_i \cap A_j)$$

$$+ \sum_{i=1}^{n-2} \sum_{j=i+1}^{n-1} \sum_{k=j+1}^{n} Pr(A_i \cap A_j \cap A_k)$$

$$+ \ldots + (-1)^{n+1} Pr(A_1 \cap A_2 \cap \ldots \cap A_n) \quad (2.1)$$

2.22 Conditional Probability

We will frequently be interested in evaluating the probability of events where the event is conditioned on some subset of the sample space. The conditional probability of event A given event B is defined as

$$Pr(A/B) = \frac{Pr(A \cap B)}{Pr(B)} \quad \text{if } Pr(B) > 0 \tag{2.2}$$

This statement can be restated to what is often called the multiplication rule, that is

$$Pr(A \cap B) = Pr(A/B)\, Pr(B), \quad Pr(B) > 0 \tag{2.3}$$

$$Pr(A \cap B) = Pr(B/A)\, Pr(B), \quad Pr(A) > 0 \tag{2.4}$$

Two events A and B are called ***independent events*** if and only if

$$Pr(A \cap B) = Pr(A)\, Pr(B) \tag{2.5}$$

This definition leads to the following statement. If A and B are independent events, then

$$Pr(A/B) = Pr(A) \quad \text{and} \quad Pr(B/A) = Pr(B) \tag{2.6}$$

2.23 Total Probability

If $A_1,..........A_n$ are disjoint subsets of S (Mutually exclusive events) and if $A_1 \cup A_2 \cup..... \cup A_n = S$, then these subsets are said to form a partition of S. The total probability of any other event B is given by

$$Pr(B) = \sum_{i=1}^{n} Pr(B/A_i)\, Pr(A_i) \tag{2.7}$$

Another important outcome of total probability is Bayes' theorem. If A_1, A_2, - - - A_k constitute a partition of the sample space S and B is an arbitrary event, then Bayes' theorem states that

$$Pr(A_i /B) = \frac{Pr(A_i \cap B)}{Pr(B)}; \quad i = 1,2,......., n \tag{2.8}$$

Example 2.2

Consider a lot of 10 relays 2 of which are defective. Find the probability that a sample of 2 relays will not contain any defective relay.

Solution

Think of the relays as being drawn one at a time. Let A be the event that the first is good, and B the event that the second is good. Then the probability that both are good is

$$\begin{aligned} Pr(A \cap B) &= Pr(A)\ Pr(B/A) \\ &= (8/10) \times (7/9) = 28/45 \end{aligned}$$

The reason that Pr(B/A) = 7/9 is that knowing that the first one is good means that there are now 7 good ones left among the 9 possible ones that might be chosen second.

* * *

Example 2.3

A manufacturing company uses two machines for production of devices. Machine 1 produces 40% of the total output, and machine 2 produces the other 60%. Of the devices produced by machine 1, 95% are good and 5% are defective. The output of machine 2 is 90% good and 10% defective. If a device is randomly selected by a customer, what is the probability that the device will be good?

Solution

Let B denote the event that the randomly selected device is good, and let A_1 and A_2 be the events that it comes from machines 1 and 2 respectively. Then using (2.7),

$$\begin{aligned} Pr(B) &= Pr(B \cap A_1) + Pr(B \cap A_2) \\ &= Pr(A_1)\ Pr(B/A_1) + Pr(A_2)\ Pr(B/A_2) \\ &= (0.4)\ (0.95) + (0.6)\ (0.9) \\ &= 0.92 \end{aligned}$$

* * *

Example 2.4

Three boxes contain two coins each. Box 1 contains two gold coins; box 2, one gold and one silver coin; and box 3, two silver coins. A box is selected at random, and then a coin is selected at random from the box. The coin turns out to be gold. What is the probability that the other coin in the box is gold?

Solution

Using the theorem of total probability of equation (2.7),

$$Pr(gold) = \sum_{i=1}^{3} Pr(box\text{-}i)\, Pr(gold/box\text{-}i)$$

$$= (1/3)(1) + (1/3)(1/2) + (1/3)(0) = 1/2$$

Now using Bayes' theorem i.e relation (2.8),

$$Pr(box\text{-}1/gold) = \frac{Pr(box\text{-}1)\, Pr(gold/box\text{-}1)}{Pr(gold)}$$

$$= \frac{(1/3)(1)}{(1/2)} = 2/3$$

* * *

2.3 RANDOM VARIABLES

As discussed already, the result of random experiment is not the same at each performance and depends upon "chance". The number of defective articles in a batch of 10 by a random selection from a lot depends on chance. Similarly, the duration of uninterrupted operation of a communication transmitter drawn at random from a lot that are manufactured apparently under the same conditions and out of the same materials as well as the time involved in the repair of a TV set selected at random from a batch of identical TV sets, both depend on chance.

A variable quantity which denotes the result of a given random experiment is called a *random variable.* It is also known as the stochastic variable or variate. A random variable will, in general, assume different values in different performances of the random experiment (depending upon chance).

Let X be a random variable associated with a random experiment. Suppose 'a' is a value that has been observed at an actual performance of the experiment. Then we say that in this trial the random variable X has assumed the value 'a'. Then the probability of the event that the variable X assumes the value 'a' is denoted by Pr(X = a).

Random variables may be either *discrete* or *continuous.* A random variable X is discrete if its range forms a discrete (countable) set of real numbers. A random variable X is continuous if its range forms a continuous (uncountable) set of real numbers and the probability of X equalling any single value in its range is zero.

2.31 Distribution Functions

If X is a random variable, then for any real number x, the probability that X will assume a value less than or equal to x is called the probability distribution function of the random variable X and is denoted by F(x), i.e.

$$F(x) = Pr(X \leq x), \quad -\infty < x < +\infty \tag{2.9}$$

Clearly, $Pr(X \leq x)$ depends on the choice of x and therefore it is function of x.

Every distribution function possesses the following properties:

1. Since every value of F(x) is a probability, its range is a set of numbers between 0 and 1.
2. The distribution function of a random variable X is a nondecreasing function of x. It implies that if x increases, then F(x) must also increase or, at least not decrease.
3. It has the limits:
   $F(+\infty) = 1$
   $F(-\infty) = 0$
4. For any arbitrary value c, we can show that if c is a point of discontinuity of F(x), then the probability that X=c is the size of the jump at that point.

2.4 DISCRETE DISTRIBUTIONS

A random variable and its probability distribution are said to be discrete if the random variable assumes only finite or atmost countably infinite set of values for which the random variable has a positive probability. Let x_1, x_2,.... be the values contained in the domain of X and p_1, p_2,....... be the corresponding probabilities. Then the probability that X will assume a given value x_i is

$$Pr(X = x_i) = p_i, \; i = 1,2....$$

We now define a new function

$$f(x) = p_i, \quad \text{for } x = x_i, \; i = 1,2.... \tag{2.10}$$

called the *probability density function* of the discrete random variable X.

The function f(x) has the following properties:

1. $f(x) = 0$ unless x is one of x_1, x_2 ....
2. $0 \leq f(x) \leq 1$ for each x_i in the range.
3. $\sum_i f(x_i) = \sum_i Pr(X=x_i) = 1.$ (2.11)

where the sums are to be taken over the entire range of x.

If we know the probability-density function of a discrete random variable X, then the probability that $X \leq x$ is

$$Pr(X \leq x) = F(x) = \sum_{x_i \leq x} f(x_i) \tag{2.12}$$

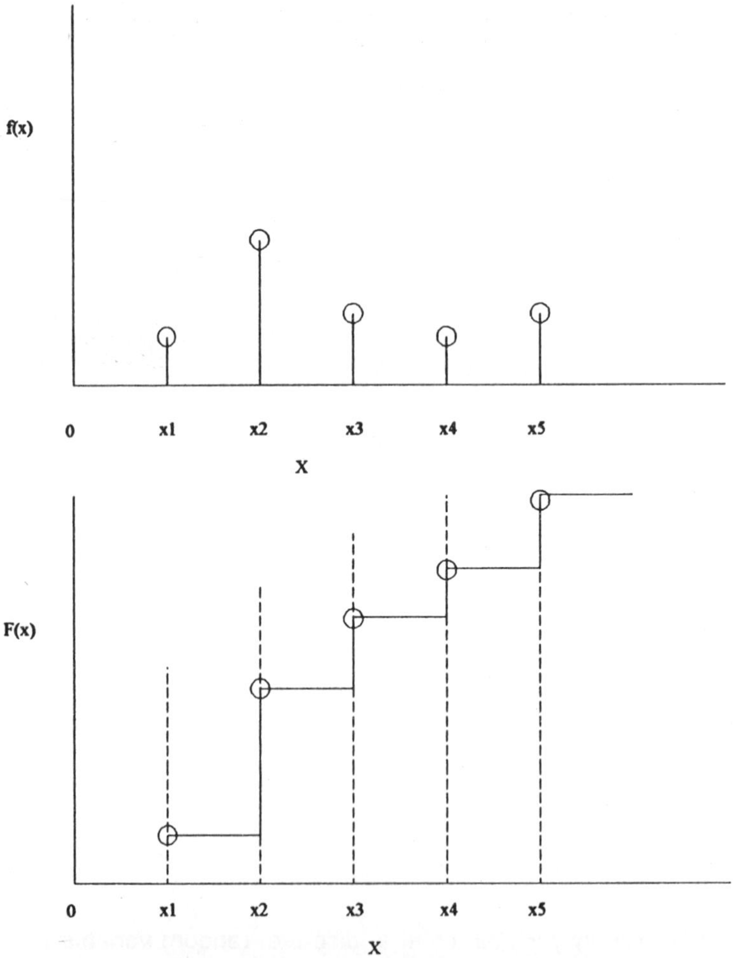

Fig. 2.3 Probability density function f(x) and distribution function F(x) for a discrete random variable.

where the summation is extended over all indices for which $x_i \le x$. It is clear that F(x) is the distribution function of the random variable X. Since the distribution function is a cumulative probability, it is often called the cumulative-distribution function. The distribution function and probability-density function for a discrete random variable are shown in Fig 2.3.

Example 2.5

Suppose that 100 people have been checked by a dentist, and the breakdown of the number of cavities found is as follows:

No. of cavities	0	1	2	3	4	5	6	7
No. of people with this many cavities	40	25	15	12	4	2	0	2

Sketch a graph of the distribution function for this random variable.

Solution

The values of probabilities are easily read from the data given as:

$$\Pr(x=0) = 0.40\,,$$
$$\Pr(x=1) = 0.25\,, \text{ and so on.}$$

Hence, the distribution function is shown in Fig 2.4.

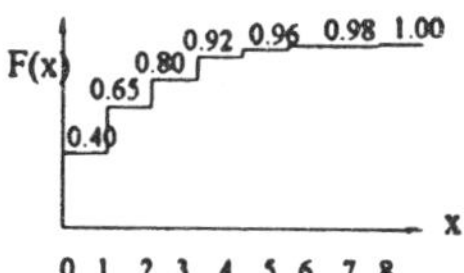

Fig. 2.4 Distribution function for example 2.5

* * *

2.41. Binomial Distribution

In many situations we are interested in the number of times a particular event occurs in a series of repetitions of a random experiment. For example, we may be interested to know the probability that at least five out of ten tubelights will last for 10,000 hours or the probability that at least two out of four engines of an aircraft are good after 1000 hours of operation. In all such cases we are interested in the random variable X, which denotes the number of times the given event occurs in a set of trials. Such problems can be solved by using the so-called binomial distribution provided they satisfy the following assumptions:

1. There are only two possible outcomes, success or failure, for each trial.
2. The probability of success is constant from trial to trial.
3. There are m trials, where m is a constant.
4. The m trials are independent, i.e. they do not influence each other.

If the probability of success on any one trial is p, then the probability of failure is q=1-p. Suppose that we are interested in the probability of getting k successes out of m trials.

$$Pr(m,k) = {}^{m}C_{k}\ p^{k}\ (1-p)^{m-k}, \quad k = 0,1,2....,m \tag{2.13}$$

The probability function is called the *binomial-density* function. This defines a family of probability densities, with each member of this family being characterized by the parameters m and p.

The binomial coefficients can recursively be evaluated by using Table 2.1.

Table 2.1: Computation of $^{n}C_{x} = {}^{n}C_{n-x}$

x

n	0	1	2	3	4	5	6	7	8	9	10
0	1										
1	1	1									
2	1	2	1								
3	1	3	3	1							
4	1	4	6	4	1						
5	1	5	10	10	5	1					
6	1	6	15	20	15	6	1				
7	1	7	21	35	35	21	7	1			
8	1	8	28	56	70	56	28	8	1		
9	1	9	36	84	126	126	84	36	9	1	
10	1	10	45	120	210	252	210	120	45	10	1

Example 2.6

An aircraft uses three active and identical engines in parallel. All engines fail independently. At least one engine must function normally for the aircraft to fly successfully. The probability of success of an engine is 0.8. Calculate the probability of the aircraft crashing. Assume that one engine can only be in two states, i.e., operating normally or failed.

Solution

The probability of success of an engine, $p = 0.8$.
Hence, the probability of failure of an engine, $q = 0.2$

Therefore, the probability of success of 0 engine out of a total of 3, using equation (2.13) is :

$$Pr(3,0) = {}^3C_0 \, (0.8)^0 \, (0.2)^3 = 0.008$$

Hence, the probability of aircraft crashing is 0.008.

* * *

Example 2.7

Among the products of a certain manufacturer, 30% are defective. If we pick 10 off the assembly line for testing, what is the probability that exactly 2 are defective?

Solution

In this case,

$$m = 10$$
$$k = 2$$
$$p = 0.30$$
$$\text{or, } q = 0.70$$

Hence , using (2.13),

$$Pr(10,2) = {}^{10}C_2 \, (0.30)^2 \, (0.70)^8 = 0.233474.$$

* * *

2.42 Poisson Distribution

In reliability engineering we often come across problems that consist of observing discrete events in a continuous interval. For example, we may be interested in determining the number of replacements of a transistor in a TV set for a certain length of time. Or, we may be required to determine the number of imperfections (or defects) in a length L of a cable. Other examples may be, the number of parts produced, number of equipments repaired, number of accidents occurring in a manufacturing plant in some fixed interval of time, and so on. In each of these, the occurrences are discrete and the interval (time, length, etc) is continuous. Such a process is known as the

Poisson process if it satisfies the following conditions:

1. For each small interval of time Δt, the probability of occurrence of one event is $\lambda\Delta t$; λ is a constant.
2. The probability of two or more occurrences in the interval Δt is negligible.
3. Each occurrence is independent of all other occurrences.

Let X be the number of occurrences. Then X, if it satisfies the above conditions, is called the ***Poisson random variable.*** We are interested in deriving the probability function for X.

Suppose there are n intervals of time Δt in a time interval of t. We can think of the n intervals as forming a sequence of n independent trials with a probability of success equal to $\lambda \Delta t$. Then X is approximately a binomial random variable and therefore the binomial probability is

$$Pr(n,x) = {}^{n}C_{x} (\lambda \Delta t)^{x} (1-\lambda \Delta t)^{n-x}, \quad x=0,1,2......,n \qquad (2.14)$$
$$= 0, \text{ otherwise}$$

The limit approached by the above probability when $t \rightarrow 0$ is the desired probability of getting x occurrences in time t. Substituting t/n for Δt in the above equation and letting $n \rightarrow \infty$ (thus $\Delta t \rightarrow 0$), we find that the binomial probability approaches

$$f(x) = \frac{(\lambda t)^{x} \exp(-\lambda t)}{(x)!}; \quad x=0,1,........,n \qquad (2.15)$$

The function f(x) is said to be the Poisson probability-density function. Substituting $\lambda t=\mu$, the equation for the Poisson-density function in general is

$$f(x) = \frac{\exp(-\mu)\ \mu^{x},}{(x)!}; \quad x= 0,1,2,... \qquad (2.16)$$

It can be seen that it is a limiting form of the binomial distribution for large n and small p, where $np=\mu$ is the most probable number of occurrences.

The distribution function of the Poisson distribution is

$$F(x) = \exp(-\mu) \sum_{s \leq x} \mu^{s}/s! \quad x \geq 0 \qquad (2.17)$$
$$= 0 \qquad x < 0$$

Example 2.8

A sample contains 1500 units of an engineering product. The failure probability of a unit is 0.0005. Calculate the probability of 5 units failing out of the entire sample.

Solution

We observe that in this case, the failure probability of a unit is very small and the sample is large. It is thus a fit case for Poisson distribution, Now,

$$\mu = [1500]\,[0.0005] = 0.75$$

Hence, using relation (2.16),

$$f(5) = \frac{\mu^5 e^{-\mu}}{5!} = \frac{(0.75)^5 e^{-0.75}}{5!} = 0.00093$$

* * *

Example 2.9

Suppose the number of cars entering a certain parking lot during a 30-second time period is known to be a random variable having a Poisson mass function with parameter $\mu = 5$. What is the probability that during a given 30 seconds period exactly 7 cars will enter the lot.

Solution

Using relation (2.16);

$$\Pr(x = 7) = f(7) = \frac{\mu^7 e^{-\mu}}{7!} = \frac{5^7 e^{-5}}{7!} = 0.104445.$$

* * *

Let us extend the last example a little more. Suppose it is known that during a time period of 100 minutes, exactly 1000 cars entered the parking lot. (This is again an average of 5 cars every 30 seconds.) A particular subinterval of 30 seconds duration constitutes 1/200 of the total time. So we might think of the 1000 cars as 1000 independent trials, with each car having probability 1/200 of entering the lot during the given 30-second subinterval. From this point of view, the number of cars entering during the

30 seconds subinterval would be a binomial random variable with parameters n= 1000, p= 0.005. In that case the probability Pr(x=7) would be :

$$^{1000}C_7 \,(0.005)^7\,(0.995)^{93} \quad = \;0.104602.$$

It may be observed that this answer is quite close to the one obtained in example 2.9, where Poisson distribution was assumed instead. This can be considered as a numerical confirmation to the fact that: when n is large and p is small then the binomial distribution with parameters n and p is approximately equal to the Poisson distribution with parameter = n.p. That is why we call Poisson distribution as a bridge between discrete distributions and continuous distributions.

2.5 CONTINUOUS DISTRIBUTIONS

A random variable X and the corresponding distribution function F(x) are said to be continuous if the following condition is satisfied for any x:

$$F(x) = \int_{-\infty}^{x} f(y)\,dy \qquad (2.18)$$

The function f(x) is called the probability-density function and is piecewise continuous.

If the function F(x) is continuous, then its derivative is the density function,

i.e. $$f(x) = dF(x)/dx \qquad (2.19)$$

for every x for which f(x) is continuous.

It may be noted that this density function has the following properties:

1. $f(x) \geq 0$ for all x.
2. For any values a and b it satisfies the equation

$$Pr(a < x \leq b) = F(b)\text{-}F(a) = \int_{a}^{b} f(x)dx \qquad (2.20)$$

This means that the probability of the event $a< X \leq b$ equals the area under the curve of the density function f(x) between x=a and x=b.

3. $$\int_{-\infty}^{\infty} f(x)dx = 1 \qquad (2.21)$$

Example 2.10

Suppose x is a random variable having density function defined by $f(t) = 2t$ for $0<t<1$, and with $f(t) = 0$ otherwise. Find the distribution function.

Solution

Obviously $F(t) = 0$ for $t \leq 0$

For $0 < t < 1$

$$F(t) = \int_{-\infty}^{t} f(t)\, dt = \int_{0}^{t} 2t\, dt = t^2$$

Also $F(t) = 1$ for $t \geq 1$.

A plot of f(t) and F(t) for the example is shown in Fig 2.5

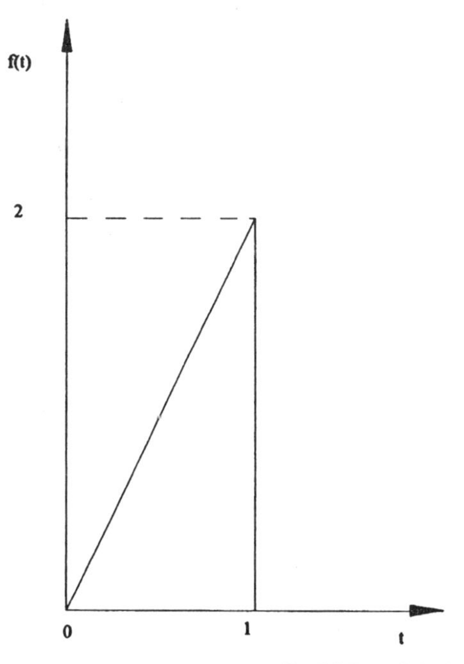

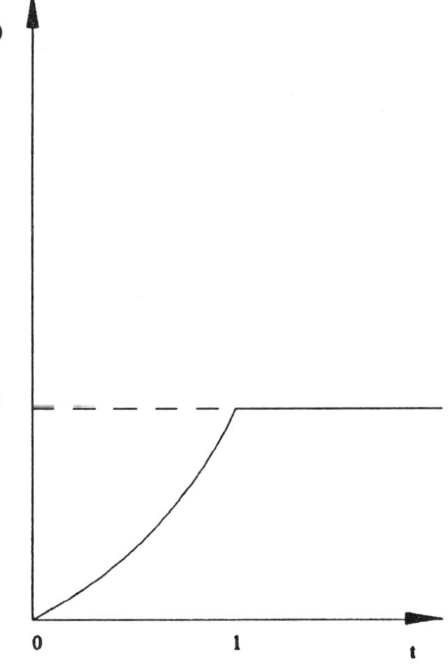

Fig. 2.5 Example 2.10

* * *

Example 2.11

Suppose $f(t) = c(4-t^2)$ for $-2< t <2$, with $f(t)=0$ otherwise. Determine the value that c must have in order for f to be a density function.

Solution

The total area under any density curve must be one. Hence,

$$\int_{-2}^{2} f(t)\, dt = 1$$

$$\text{or, } \int_{-2}^{2} c(4-t)^2\, dt = 1$$

$$\text{or, } c\,[4t - t^3/3] \Big|_{-2}^{2} = 1$$

$$\text{or, } c\,[8- 8/3 + 8 - 8/3] = 1$$

$$\text{or, } c = 3/32.$$

* * *

2.51 Uniform Distribution

A random variable X having the range of some finite interval $a < X \le b$ is said to have uniform distribution (Fig 2.6) if its probability density function is constant within the range, i.e.

$$f(x) = c, \qquad a < x \le b$$
$$= 0, \qquad \text{otherwise}$$

Since

$$\int_a^b f(x)dx = \int_a^b c\, dx = 1$$

It follows that $c = 1/(b-a)$ and therefore

$$f(x) = 1/(b-a), \quad a < x \le b \qquad (2.22)$$
$$= 0, \text{ otherwise}$$

The corresponding distribution function is

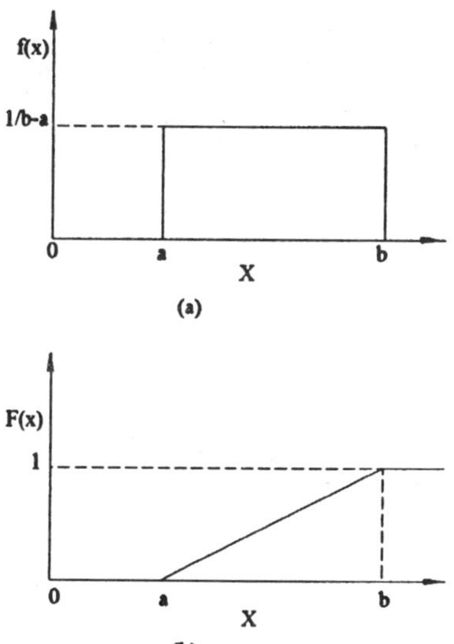

Fig. 2.6 f(x) and F(x) of a uniform distributed random variable x.

$$F(x) = \begin{cases} 0 & x \le a \\ (x-a)/(b-a) & a < x \le b \\ 1 & x > b \end{cases} \quad (2.23)$$

2.52 Exponential Distribution

A continuous random variable having the range $0<x<\infty$ is said to have an exponential distribution (Fig 2.7) if it has the probability-density function of the form

$$f(x) = \lambda \exp(-\lambda x), \quad 0 \le x < \infty \quad (2.24)$$

where λ is a positive constant. The corresponding distribution function is

$$F(x) = 1-\exp(-\lambda x), \quad 0 \le x < \infty \quad (2.25)$$

Exponential distribution plays an important role in reliability studies. In most cases, the time to failure of components obeys an exponential-distribution. This is particularly so in the case of most of the electronic components.

2.53 Rayleigh Distribution

A continuous random variable X having the density function

$$f(x) = \begin{cases} a\, x \exp[-(b\, x^2/2)], & 0 \le x < \infty \\ 0, & \text{otherwise} \end{cases} \quad (2.26)$$

is said to have Rayleigh distribution, where a and b are positive constants. Further, we know that

$$\int_{-\infty}^{\infty} f(x)dx = 1 \qquad (2.27)$$

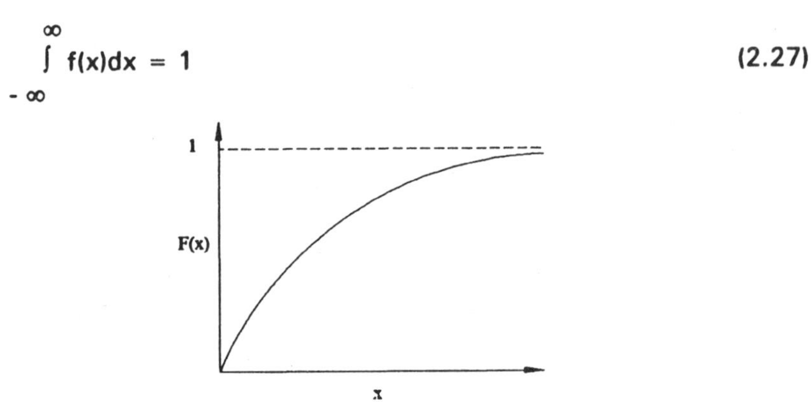

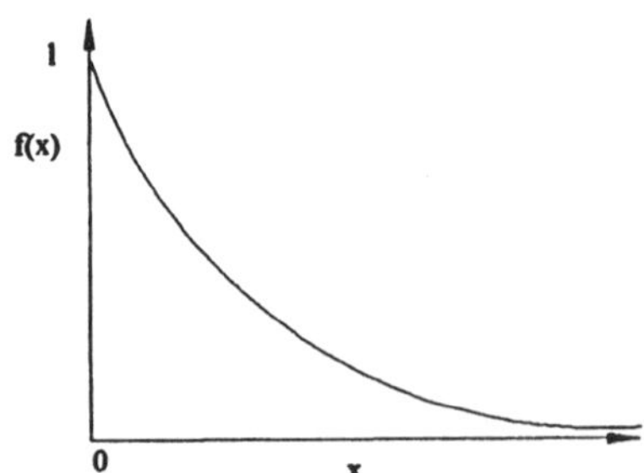

Fig.2.7 F(x) and f(x) of an exponential distribution.

and therefore

$$\int_{0}^{\infty} a\, x \exp[-(b\, x^2/2)]dx = a/b = 1 \qquad (2.28)$$

Thus, the Rayleigh density becomes:

$$f(x) = b\, x \exp[-(b\, x^2/2)], \quad 0 < x < \infty \qquad (2.29)$$

and the corresponding distribution function is

$$F(x) = 1 - \exp[-bx^2/2)], \quad 0 \leq x < \infty \qquad (2.30)$$

This distribution finds application in reliability when system components are characterized by linearly increasing failure rates such as Rubber components.

2.54 Weibull Distribution

A Weibull distribution has the density function defined by

$$f(x) = ax^b \exp[-a\, x^{(b+1)}/(b+1)], \quad x \geq 0 \tag{2.31}$$

and is shown in Fig 2.8.

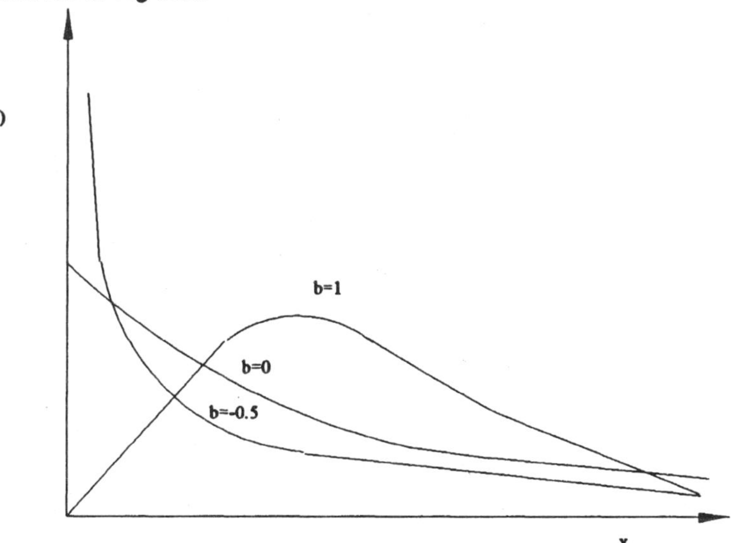

Fig. 2.8 The Weibull density function.

Then the distribution function is

$$F(x) = 1\text{-}\exp[-a\, x^{(b+1)}/(b+1)], \quad x > 0 \tag{2.32}$$

where a and b are positive constants and are known as *scale* and *shape* parameters respectively.

It is evident that the exponential and Rayleigh distributions are the special cases of the two-parameter Weibull distribution when $b=0$ and $b=1$ respectively. Weibull distribution is useful whenever failure is caused by the stress exceeding the strength at the weakest point of the item and is widely applicable for Mechanical components.

2.55 Gamma Distribution

A distribution of a continuous random variable X whose density function is given by

$$f(x) = c\, x^{a-1} \exp(-b\, x), \quad x \geq 0 \tag{2.33}$$

is known as Gamma distribution, where a and b are positive constants and the constant c can be obtained in terms of a and b from the equation

$$c \int_0^\infty x^{a-1} \exp(-bx)dx = 1 \qquad (2.34)$$

i.e.

$$c = b^a/\Gamma a$$

where

$$\Gamma a = \int_0^\infty u^{a-1} \exp(-u)du \qquad (2.35)$$

is termed as the gamma function.

It can be noted that the exponential distribution is a special case of the gamma distribution with a=1.

The time-to-failure of a stand-by system whose components are characterized by a constant failure rate is gamma-distributed.

2.56 Normal Distribution

Normal distribution is a two-parameter distribution of a continuous random variable whose probability has the form:

$$f(x) = \frac{\exp[-(x-\mu)^2/2\sigma^2]}{\sigma (2\pi)^{1/2}}; \qquad -\infty < x < \infty \qquad (2.36)$$

The constants μ and $\sigma > 0$ are arbitrary and represent the mean and standard deviation of the random variable. This function and the corresponding distribution function are shown in Fig 2.9. This is the most important probability distribution for use in statistics. It also has applications in Reliability engineering, for example in the failure of Ball- bearings.

2.6 STOCHASTIC PROCESSES

A *stochastic process* is a family of random variables $\{X(t) \mid t \in T\}$, defined on a given probability space, indexed by the parameter t, where t varies over an index set T.

The values assumed by the random variable X(t) are called *states*, and the set of all possible values forms the *state space* of the process. The state space is generally denoted by I.

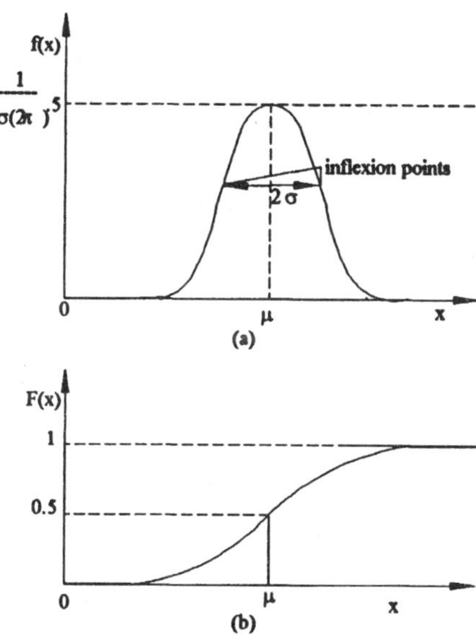

Fig. 2.9 The normal distribution.

Recall that a random variable is a function defined on the sample space S of the underlying experiment. Thus the above family of random variables is a family of functions $\{X(t,,s) | s \in S, t \in T\}$. For a fixed $t = t_1$, $X(t_1,s)$ is a random variable [denoted by $X(t_1)$] as s varies over the sample space S. At some other fixed instant of time t_2, we have another random variable $X(t_2,s)$. For a fixed sample point $s_1 \in S$, $X(t,s_1)$ is a single function of time t, called a *sample function* or a *realization* of the process. When both s and t are varied, we have the family of random variables constituting a stochastic process.

If the state space of a stochastic process is discrete, then it is called a *discrete-state process*, often referred to as a ***chain***. In this case, the state space is often assumed to be {0, 1, 2,...}. Alternatively, if the state space is continuous, then we have a *continuous-state process*. Similarly, if the index set T is discrete, then we have a *discrete (time)-parameter process;* otherwise we have a *continuous parameter process.*

2.7 MARKOV CHAINS

A *Markov process* is a stochastic process whose dynamic behaviour is such that probability distributions for its future development depend *only* on the present state and not on how the process arrived in that state. If we assume that the state space, I, is discrete (finite or countably infinite), then the Markov process is known as a ***Markov chain.***

In order to formulate a Markov model (to be more precise we are talking about continuous-time and discrete-state models) we must first define all the mutually exclusive states of the system. For example, in a system composed of a single non-repairable element x_1 there are two possible states: $s_o = x_1$, the element is good, and $s_1 = x'_1$, the element is bad. The states of the system at $t=0$ are called the *initial states*, and those representing a final or equilibrium state are called *final states.* The set of Markov state equations describes the probabilistic transitions from the initial to the final states.

The transition probabilities must obey the following two rules:

1. The probability of transition in time Δt from one state to another is given by $z(t)\ \Delta t$, where $z(t)$ is the hazard associated with two states in question. If all the $z_i(t)$'s are constant, $z_i(t) = \lambda_i$, and the model is called *homogeneous.* If any hazards are time functions, the model is called *nonhomogeneous.*

2. The probabilities of more than one transition in time Δt are infinitesimals of a higher order and can be neglected.

2.71 One Component System:

The probability of being in state s_o at time $t+\Delta t$ is written $P_o(t+\Delta t)$. This is given by the probability that the system is in state s_o at time t, $P_o(t)$, times the probability of no failure in time Δt, $1-z(t)\ \Delta t$, plus the probability of being in state s_1 at time t, $P_1(t)$, times the probability of repair in time Δt, which equals zero. (We are neglecting the possibility of repairs for the present).

The resulting equation is

$$P_o(t+\Delta t) = [1 - z(t)\ \Delta t]\ P_o(t) + OP_1(t) \tag{2.37}$$

Similarly, the probability of being in state s_1 at time $t+\Delta t$ is given by

$$P_1(t+\Delta t) = [z(t)\ \Delta t]\ P_o(t) + 1P_1(t) \tag{2.38}$$

The transition probability $z(t)\ \Delta t$ is the probability of failure (change from state s_o to s_1), and the probability of remaining in state s_1 is unity.

Rearrangement of the above equations yields

$$\frac{P_o(t+\Delta t) - P_o(t)}{\Delta t} = -z(t)\ P_o(t)$$

$$\frac{P_1(t+\Delta t) - P_1(t)}{\Delta t} = z(t)\, P_o(t)$$

Passing to a limit as Δt becomes small, we obtain

$$\frac{dP_o(t)}{dt} = -z(t)\, P_o(t) \qquad (2.39)$$

$$\frac{dP_1(t)}{dt} = z(t)\, P_o(t) \qquad (2.40)$$

These equations can be solved in conjunction with the appropriate initial conditions for $P_o(t)$ and $P_1(t)$. The most common initial condition is that the system is good at $t=0$, that is $P_o(t=0)=1$ and $P_1(t=0)=0$.

The solution of these equations is:

$$P_o(t) = \exp[-\int_0^t z(\tau)d\tau] \qquad (2.41)$$

and

$$P_1(t) = 1 - \exp[-\int_0^t z(\tau)d\tau] \qquad (2.42)$$

Ofcourse, a formal solution of the second equation is not necessary to obtain since it is possible to recognize at the outset that

$$P_o(t) + P_1(t) = 1 \qquad (2.43)$$

The role played by the initial conditions is clearly evident. If there is a fifty-fifty chance that the system is good at $t = 0$, then $P_o(0) = 1/2$, and

$$P_o(t) = (1/2) \exp[-\int_0^t z(\tau)d\tau] \qquad (2.44)$$

It is often easier to characterize Markov models by a graph composed of nodes representing system states and branches labeled with transition probabilities. Such a Markov graph for the problem described above is given in Fig 2.10. Note that the sum of transition probabilities for the branches leaving each node must be unity. Treating the nodes as signal sources and the transition probabilities as transmission coefficients, we can write difference equations by inspection. Thus, the probability of being at any node at time $t + \Delta t$ is the sum of all signals arriving at that node. All other nodes are considered probability sources at time t, and all transition probabilities serve as transmission gains. A simple algorithm for writing the differential equations by inspection is to equate the derivative of the probability at any node to the sum of the transmissions coming into the node. Any unity gain factors of the self-loops must first be set to zero, and the Δt factors are dropped from the branch gains.

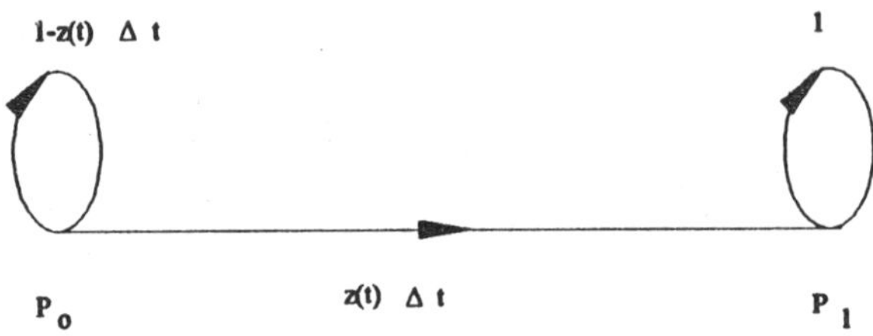

Fig. 2.10 Markov graph for a single nonrepairable element.

2.72 Two-element system

If a two element system consisting of elements x_1 and x_2 is considered, there are four system states: $s_o = x_1 x_2$, $s_1 = x'_1 x_2$, $s_2 = x_1 x'_2$ and $s_3 = x'_1 x'_2$. The Markov graph is shown in Fig 2.11. The probability expression for state s_o is given by

$$P_o(t + \Delta t) = \{1 - [z_{o1}(t) + z_{o2}(t)]\, \Delta t\}\, P_o(t) \tag{2.44}$$

where $[z_{o1}(t) + z_{o2}(t)]\, \Delta t$ is the probability of a transition in time Δt from s_o to s_1 or s_2. For state s_1,

$$P_1(t + \Delta t) = [z_{o1}(t)\, \Delta t]\, P_o(t) + [1 - z_{13}(t)\, \Delta t]\, P_1(t) \tag{2.45}$$

where $z_{13}(t)\, \Delta t$ is the probability of a transition from state s_1 to s_3. Similarly for state s_2,

$$P_2(t + \Delta t) = [z_{o2}(t)\, \Delta t]\, P_o(t) + [1 - z_{23}(t)\Delta t]\, P_2(t) \tag{2.46}$$

where $z_{23}(t)\ \Delta t$ is the probability of a transition from state s_2 to s_3.

For state s_3 the transition equation is

$$P_3(t + \Delta t) = [z_{13}(t)\ \Delta t]\, P_1(t) + [z_{23}(t)\, \Delta t]\, P_2(t) + 1 P_3(t) \tag{2.47}$$

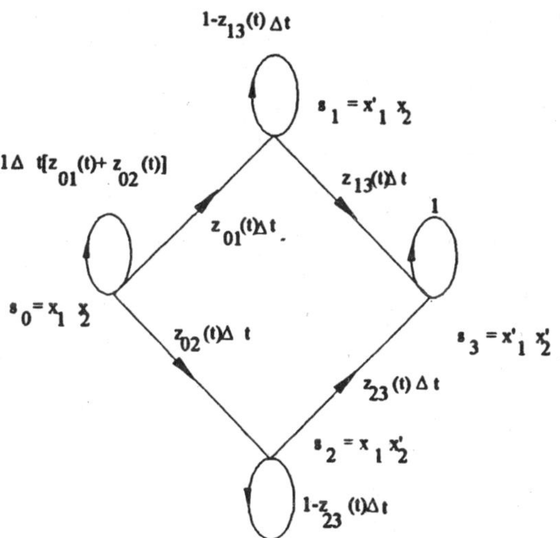

Fig. 2.11 Markov graph for two distinct nonrepairable elements.

Rearranging these equations and passing to a limit yields

$$\frac{dP_o(t)}{dt} = -[z_{o1}(t) + z_{o2}(t)]\ P_o(t) \tag{2.48a}$$

$$\frac{dP_1(t)}{dt} = -[z_{13}(t)]\ P_1(t) + [z_{o1}(t)]\ P_o(t) \tag{2.48b}$$

$$\frac{dP_2(t)}{dt} = -[z_{23}(t)]\ P_2(t) + [z_{o2}(t)]P_o(t) \tag{2.48c}$$

$$\frac{dP_3(t)}{dt} = [z_{13}(t)]P_1(t) + [z_{23}(t)]P_2(t) \tag{2.48d}$$

The initial conditions associated with this set of equations are $P_o(0)$, $P_1(0)$, $P_2(0)$, and $P_3(0)$. These equations, of course could have been written by inspection using the algorithm previously stated.

It is difficult to solve these equations for a general hazard function z(t), but if the hazards are specified, the solution is quite simple. If all the hazards are constant, $z_{o1}(t) = \lambda_1$, $z_{o2}(t) = \lambda_2$, $z_{13}(t) = \lambda_3$, and $z_{23}(t) = \lambda_4$.

The solutions are

$$P_o(t) = \exp[-(\lambda_1+\lambda_2)t] \tag{2.49a}$$

$$P_1(t) = \frac{\lambda_1[\exp(-\lambda_3 t) - \exp[-(\lambda_1+\lambda_2)t]}{\lambda_1+\lambda_2-\lambda_3} \tag{2.49b}$$

$$P_2(t) = \frac{\lambda_2[\exp(-\lambda_4 t) - \exp[-(\lambda_1+\lambda_2)t]}{\lambda_1+\lambda_2-\lambda_4} \tag{2.49c}$$

$$P_3(t) = 1 - [P_o(t)+P_1(t)+P_2(t)] \tag{2.49d}$$

where

$$P_o(0) = 1 \text{ and } P_1(0) = P_2(0) = P_3(0) = 0 \tag{2.50}$$

Note that we have not as yet had to say anything about the configuration of the system, but only have had to specify the number of elements and the transition probabilities. Thus, when we solve for P_o, P_1, P_2, we have essentially solved for all possible two element system configurations.

The complexity of a Markov model depends on the number of system states. In general we obtain for an m-state problem a system of m first order differential equations. The number of states is given in terms of the number of components n as

$$m = {}^nc_0 + {}^nc_1 + {}^nc_2 + \ldots + {}^nc_n = 2^n$$

Thus, our two-element model has four states, and a four-element model 16 states. This means that an n-component system may require a solution of as many as 2^n first-order differential equations. In many cases we are interested in fewer states. Suppose we want to know only how many failed items are present in each state and not which items have failed. This would mean a model with $n+1$ states rather than 2^n, which represents a tremendous saving. To illustrate how such simplifications affect the Markov graph we consider a collapsed flow graph shown in Fig 2.12 for the two element system. Collapsing the flow graph is equivalent to the restriction $P'_1(t) = P_1(t) + P_2(t)$. Note that this can collapse the flow graph only if $z_{13} = z_{23}$; however, z_{o1} and z_{o2} need not be equal.

Markov graphs for a system with repair are shown in Fig 2.13(a,b). The graph in Fig 2.13(a) is a general model, and that of Fig 2.13(b) is a collapsed model.

The system equations can be written for Fig 2.13(a) by inspection using the algorithm previously discussed.

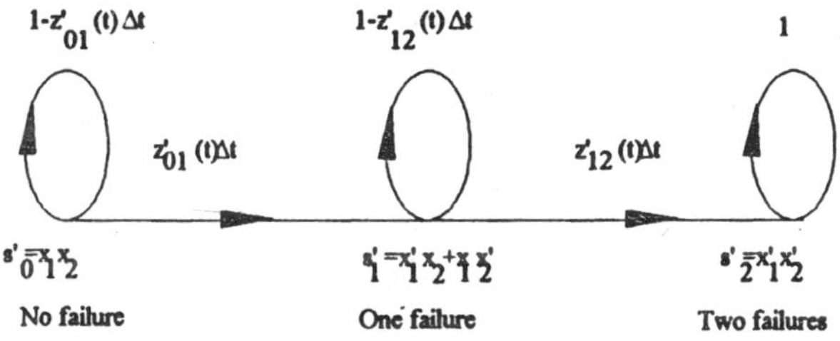

Fig. 2.12 Collapsed Markov graph without repair.

$$\dot{P}_o = -(z_{o1} + z_{o2})P_o + w_{1o}P_1 + w_{2o}P_2 \quad (2.51a)$$

$$\dot{P}_1 = -(z_{13}+w_{1o})P_1 + z_{o1}P_o \quad (2.51b)$$

$$\dot{P}_2 = -(z_{23} + w_{2o})P_2 + z_{o2}P_o \quad (2.51c)$$

$$\dot{P}_3 = z_{13}P_1 + z_{23}P_2 \quad (2.51d)$$

Similarly for Fig2.13(b)

$$\dot{P}'_o = - z'_{o1}P'_o + w'_{1o}P'_1 \quad (2.52a)$$

$$\dot{P}'_1 = -(z'_{12}+w'_{1o})P'_1 + z'_{o1}P'_o \quad (2.52b)$$

$$\dot{P}'_2 = z'_{12}P'_1 \quad (2.52c)$$

The probabilities in the general and the collapsed model are related by

$$P_{o'} = P_o \quad (2.53a)$$

$$P_{1'} = P_1 + P_2 \quad (2.53b)$$

$$P_{2'} = P_3 \quad (2.53c)$$

and the hazards must satisfy

$$z'o1 = zo1 + zo2 \quad (2.54a)$$
$$w'1o = w1o + w2o \quad (2.54b)$$
$$z'12 = z13 = z23 \quad (2.54c)$$

The solution to these equations for various values of the z's and w's can be obtained in a specific situation.

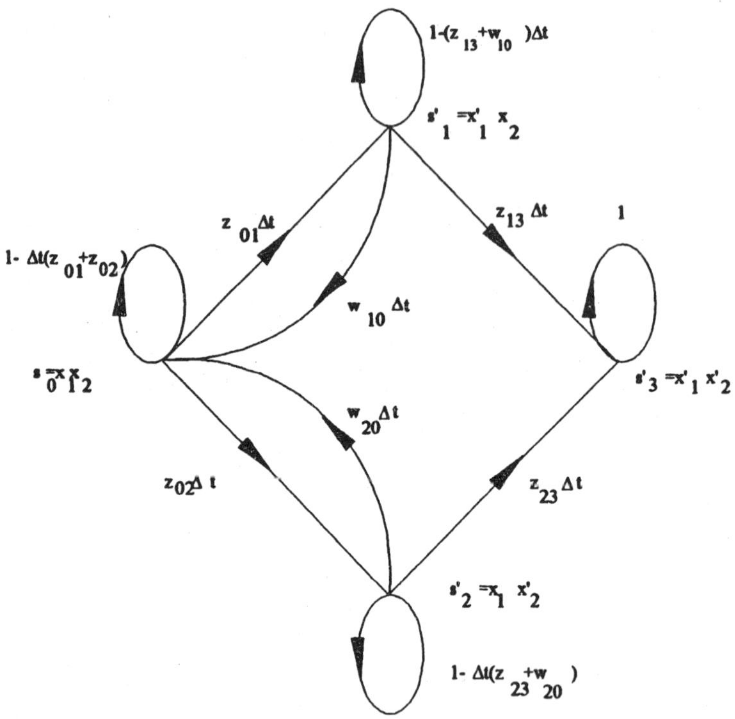

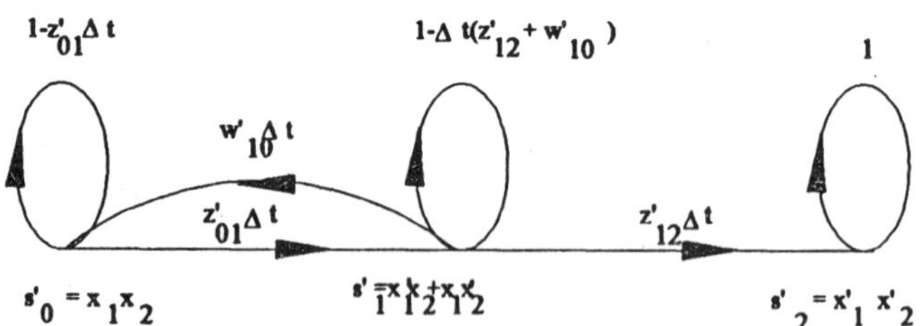

Fig. 2.13 Markov graph for a system with repair

(a) General Model (b) Collapsed Model

3

RELIABILITY ANALYSIS OF SERIES PARALLEL SYSTEMS

3.1 INTRODUCTION

Reliability is not confined to single components. We really want to evaluate the reliabilities of the systems, simple as well as extremely complex, and to use these evaluation techniques for designing reliable systems. System reliabilities are calculated by means of the calculus of probability. To apply this calculus to systems, we must have some knowledge of the probabilities of its components, since they affect the reliability of the system.

Component reliabilities are derived from tests which yield information about failure rates. The actual value of this failure rate can be obtained only by means of statistical procedures because of the two main factors which govern the probability of survival of a component:

1. The uncertainties of the production process.
2. The uncertainties of the stresses which component must withstand in operation.

In reliability tests we actually measure the failure rate of a component, which means we measure its instantaneous probability of failure at a given set of environmental and operating stress conditions. System reliability calculations are based on two important operations:

1. As precise as possible a measurement of the reliability of the components used in the system environment.

2. The calculation of the reliability of some complex combination of these components.

Once we have the right figures for the reliabilities of the components in a system, or good estimates of these figures, we can then perform very exact calculations of system reliability even when the system is the most complex combination of components conceivable. The exactness of our results does not hinge on the probability calculations because these are perfectly accurate; rather, it hinges on the exactness of the reliability data of the components. In system reliability calculations for Series-Parallel Systems we need use only the basic rules of the probability calculus.

The following *assumptions* are made:

1. The reliabilities of all constituent components of the system are known and these are constant during the time interval in which the reliability of the network is being examined.

2. All components are always operating except possibly in the case of redundancy.

3. There does not exist any correlation between failures of different links i.e. the states of all elements are s-independent.

4. The state of each element and of the entire network is either good (operating) or bad (failed).

5. The nodes of the network are perfect.

6. There is no limitation on the flow transmission capability of any component, i.e. each link/node can transmit the required amount of flow.

These assumptions are primarily made for mathematical practicability. Several of these assumptions are removed in the published work on Reliability Analysis.

3.2 RELIABILITY BLOCK DIAGRAMS

A block diagram which depicts the operational relationship of various elements in a physical system, as regards the success of the overall system, is called *Reliability Block Diagram or Reliability Logic Diagram.* While the system diagram depicts the physical relationship of the system elements, the reliability block diagram shows the functional relationship and indicates which elements must operate successfully for the system to accomplish its intended function. The function which is performed may be the simple action of a switch which opens or closes a circuit or may be a very complex activity such as the guidance of a spacecraft.

Two blocks in a block diagram are shown in *series* if the failure of either of them results in system failure. In a series block diagram of many blocks, such as Fig 3.1, it is imperative that all the blocks must operate successfully for system success. Similarly two blocks are shown in *parallel* in the block diagram, if the success of either of these results in system success. In a parallel block diagram of many blocks, such as Fig 3.2, successful operation of any one or more blocks ensures system success. A block diagram, in which both the above connections are used is termed as *Series-Parallel Block Diagram.*

A closely related structure is a *k-out-of-m* structure. Such a block diagram represents a system of m components in which any k must be good for system to operate successfully. A simple example of such a type of system is a piece of stranded wire with m strands in which at least k are necessary

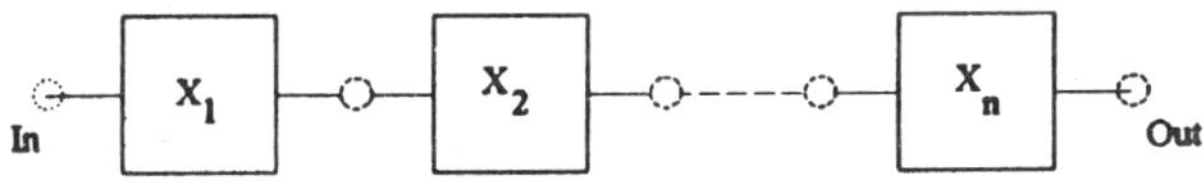

Fig. 3.1 A Series Block Diagram

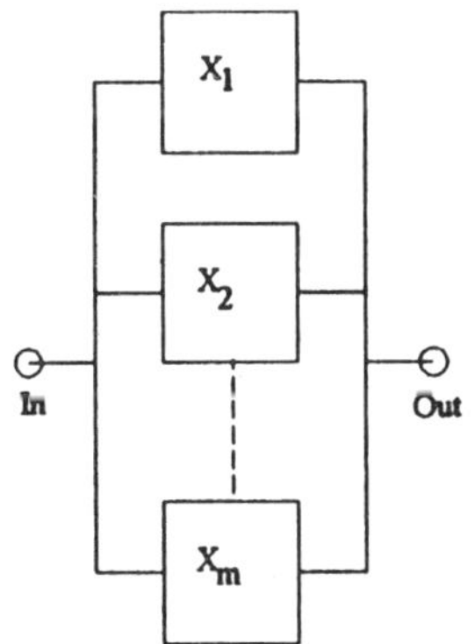

Fig. 3.2 A Parallel Block Diagram

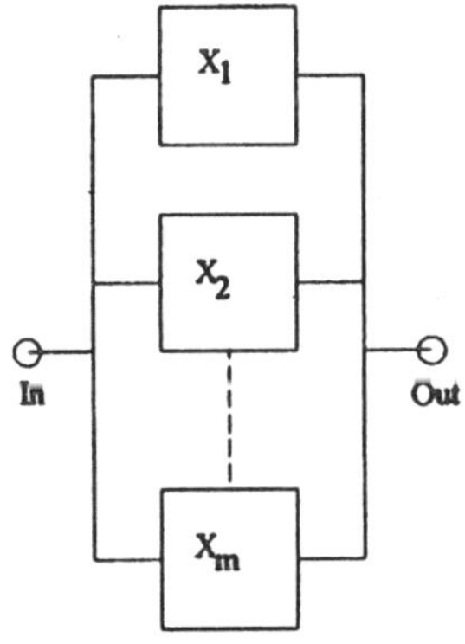

(atleast k needed)

Fig. 3.3 A k-out-of-m Block Diagram

to pass the required current. Such a block diagram can not be recognised without a description inscribed on it, as in Fig 3.3. Series and Parallel reliability block diagrams can be described as special cases of this type with k equal to m and unity respectively.

A block diagram which can not be completely described through series or parallel operational relationships, is called a non-series parallel block diagram. The analysis methods for such systems are discussed in the next chapter.

3.3 SERIES SYSTEMS

Many complex systems are series systems as per reliability logic. The block diagram of a series system was shown in Fig 3.1. If E_i and E_i' denote the events of satisfactory and unsatisfactory operation of the component i, the event representing system success is the logical intersection of $E_1, E_2, \ldots, E_n$. Reliability of the system is the probability of success of this event and is given by

$$R = Pr(E_1 \cap E_2 \cap \ldots\ldots \cap E_n) \quad (3.1)$$

$$= Pr(E_1)\, Pr(E_2/E_1)\, Pr(E_3/E_2E_1) \ldots \quad (3.2)$$

where $Pr(E_2/E_1)$ is the probability of event E_2 provided E_1 has occurred. For *independent* components

$$R = Pr(E_1)Pr(E_2)\ldots\ldots Pr(E_n) \quad (3.3)$$

If $Pr(E_i) = P_i(t)$; the time dependent reliability function is

$$R(t) = \prod_{i=1}^{n} P_i(t) \quad (3.4)$$

The above equation is commonly known as *product-law of reliabilities.*

In the case of exponential distributions, if λ_i is the failure rate of component i,

$$P_i(t) = \exp(-\lambda_i t)$$

and

$$R(t) = \exp\left[-t \sum_{i=1}^{n} \lambda_i\right] \quad (3.5)$$

Therefore, the reliability law for the whole system is still exponential. Also, for series systems with constant failure rate components the system failure rate is the sum of failure rates of individual components i.e.,

$$\lambda_s = \sum_{i=1}^{n} \lambda_i \tag{3.6}$$

and the MTBF of the system is related to the MTBF of individual components by

$$m_s = 1/\sum_{i=1}^{n} (1/T_i) \tag{3.7}$$

Example 3.1

An electronic circuit consists of 5 silicon transistors, 10 silicon diodes, 20 composition resisters, and 5 ceramic capacitors in continuous series operation and assume that under the actual stress conditions in the circuit the components have the following failure rates:

Silicon transistors	λ_t	$=0.000008$/hr
Silicon diodes	λ_d	$=0.000002$ /hr
Composition resistors	λ_r	$=0.000001$ /hr
Ceramic capacitors	λ_c	$=0.000004$ /hr

Estimate the reliability of this circuit for 10 hour operation.

Solution

Circuit failure rate is given as:

$$\lambda_s = \Sigma\, \lambda_i = 5\,\lambda_t + 10\,\lambda_d + 20\,\lambda_r + 5\,\lambda_c = 0.0001 \text{ /hr}$$

This sum is the expected hourly failure rate λ_s of the whole circuit. The estimated reliability of the circuit is then

$$R(t) = \exp(-0.0001t)$$

for an operating time t. For a 10 hour operation the reliability is

$$R(10) = 0.999 = 99.9\%$$

Also the expected mean time between failures is

$$m_s = 1/\lambda_s = 1/0.0001 = 10{,}000 \text{ hours}$$

This does not mean that the circuit could be expected to operate without failure for 10,000 hours. We know from the exponential function that its chance to survive for 10,000 hours is only about 37%.

* * *

It may be noted that the component failure rate figures apply to definite operating stress conditions-for instance, to an operation at rated voltage, current, temperature, and at a predicted level of mechanical stresses, such as shock and vibration. Failure rates usually change radically with changes in the stress levels. If a capacitor is operated at only half of its rated voltage, its failure rate may drop to $1/30^{th}$ of the failure rate at full rated voltage operation.

Thus, to upgrade the reliability of the circuit it becomes necessary to reduce the stresses acting on the components; that is, to use components of higher voltage and current ratings, and to make provisions for a reduction of the operating temperature levels. Using these techniques, component failure rate reductions by a factor of ten are often easily achieved.

Thus, when designing the circuits and their packaging, the circuit designer should always keep two things in mind:

1. *Do not overstress the components,* but operate them well below their rated values, including temperature. Provide good packaging against shock and vibration, but remember that in tightly packaged equipment without adequate heatsinks, extremely high operating temperatures may develop which can kill all reliability efforts.

2. *Design every equipment with as few components as possible.* Such simplification of the design increases reliability and also makes assembly and maintenance easier.

It may be observed that the time t used above is the system operating time. Only when a component operates continuously in the system will the component's operating time be equal to the system's operating time. In general, when a component operates on the average for t_1 hours in t system operating hours, it assumes in the system's time scale a failure rate of

$$\lambda = \lambda' t_1/t \tag{3.8}$$

Where λ' is the component's failure rate while in operation.

The above equation is based on the assumption that in the non-operating or de-energized condition the component has a zero failure rate even though the system is in operation. This is not always the case. Components may exhibit some failure rates even in their quiescent or idle condition while the system is operating. If the component has a failure rate of λ' when operating and λ'' when de-energized, and it operates for t_1 hours every t hours of system operation, the system will see this component behaving with an average failure rate of

$$\lambda = (\lambda' t_1 + \lambda''(t - t_1))/t \quad (3.9)$$

If the failure rate of a component is expressed in terms of operating cycles, and if the component performs on the average 'C' operations in t system hours, the system will see this component behave with a failure rate of

$$\lambda = C\,\lambda_c/t \quad (3.10)$$

But if this component also has a time dependent failure rate of λ' while energized, and a failure rate of λ'' when de-energized (with system still operating), the component assumes in the system time scale a failure rate of

$$\lambda = (C\,\lambda_c + t_1\,\lambda' + \lambda''(t-t_1))/t \quad (3.11)$$

Example 3.2

An electric bulb has a failure rate of 0.0002/hr when glowing and that of 0.00002/hr when not glowing. At the instant of switching -ON, the failure rate is estimated to be 0.0005/switching. What is the average failure rate of the bulb if on the average it is switched 6 times every day and it remains ON for a total of 8 hrs in the day on the average.

Solution

Here,

$$
\begin{aligned}
t &= 24 \text{ hrs}\\
t_1 &= 8 \text{ hrs}\\
\lambda' &= 0.0002/\text{hr}\\
\lambda'' &= 0.00002/\text{hr}\\
\lambda_c &= 0.0005/\text{switching}\\
C &= 6
\end{aligned}
$$

Therefore, using equation (3.11),

$$
\begin{aligned}
\lambda &= [6(0.0005) + 8(0.0002) + 16(0.00002)]/24\\
&= 0.00492/24 = 0.000205/\text{hr}.
\end{aligned}
$$

An interesting point to be made here is that purely from reliability considerations, it is better to keep the bulb on for the whole day rather than switching it off when not needed. (We have not discussed the question of energy consumption here -which may force the other decision on us).

* * *

In case the components in a series system are identical and independent each with reliability, p or unreliability, q

$$R = p^n = (1-q)^n \tag{3.12}$$

For high reliability region,

$$R \approx 1-nq \tag{3.13}$$

is a good approximation and can be used for fast calculation.

Example 3.3

A series system is composed of 10 identical independent components. If the desired value of system reliability is 0.99, how good must the components be from the reliability point of view?

Solution

Using relation (3.13),

$$\begin{aligned} R &\approx 1-nq \\ \text{or,}\quad 0.99 &= 1-10q \\ \text{or,}\quad q &= 0.001 \\ \text{Hence,}\ p &= 0.999 \end{aligned}$$

On the other hand, if we use the exact relationship,

$$\begin{aligned} R &= p^{10} \\ \text{or,}\ p^{10} &= 0.99 \\ p &= (0.99)^{0.1} = 0.99899. \end{aligned}$$

We can thus see that the difference between exact calculation and approximate calculation is negligible and hence the approximate realtion is frequently used in practical design which in simple words means that the system unreliability is the product of component unreliability by the number of components in the system.

* * *

3.4 PARALLEL SYSTEMS

When a system must be designed to a quantitatively specified reliability figure, it is generally not enough for the designer to simply reduce the number of components and the stresses acting on them. He must, during the various stages of the design, duplicate components, and sometimes whole circuits, to fulfill such requirements. In other words, he must use parallel systems, such as shown in Fig 3.2.

If E_i and E_i' are the events of satisfactory and unsatisfactory operation of the component i, the event for system success now is the union of E_1, $E_2, \ldots, E_m$. Reliability of the system is the probability of success of this event and is given by

$$R = \Pr(E_1 \cup E_2 \cup \ldots \cup E_m) \tag{3.14}$$

$$= 1 - \Pr(E_1' \cap E_2' \cap \ldots \cap E_m') \tag{3.15}$$

For independent components,

$$R = 1 - \Pr(E'_1)\Pr(E'_2) \ldots \Pr(E'_m) \tag{3.16}$$

If $\Pr(E_i') = q_i$ and $\Pr(E_i) = p_i$, the time dependent reliability function is

$$R(t) = 1 - \prod_{i=1}^{m} q_i(t) \tag{3.17}$$

$$= 1 - \prod_{i=1}^{m} (1 - p_i(t)) \tag{3.18}$$

In case of identical components,

$$R = 1 - [1 - p(t)]^m \tag{3.19}$$

and the unreliability

$$Q = q(t)^m \tag{3.20}$$

which is commonly called *product law of unreliabilities.* For designing a system having unreliability less than Q, the number of parallel components each with unreliability q can be determined easily using the above equation.

For constant failure rates,

$$R(t) = 1 - [1 - \exp(-\lambda t)]^m \tag{3.21}$$

and the MTBF for the system is given by

$$m_s = \int_0^{\infty} [1 - (1-\exp(-\lambda t))]^m \, dt \qquad (3.22)$$

It can be easily derived now that:

$$m_s = (1/\lambda) \sum_{i=1}^{m} (1/i) \qquad (3.23)$$

For large values of m, equation (3.23) can be reduced to:

$$m_s = (1/\lambda) \; [Ln(m) + 0.577 + 1/2m] \qquad (3.24)$$

Reliability improvement through redundancy is thus seen to be logarithimic.

It implies that although more number of components in parallel is advantageous from the reliability considerations, the incremental advantage keeps on reducing with every increase in the component used. A designer must study this vis-a-vis his observation that cost will be generally a linearly increasing function of the number of components. The above observation implies that while designer has the option of adding redundant components for improved reliability, this option should not be used indiscriminately.

When two components with the failure rates λ_1 and λ_2 operate in parallel, the reliability R_p of this parallel system is given by

$$R_p = \exp(-\lambda_1 t) + \exp(-\lambda_2 t) - \exp[-(\lambda_1 + \lambda_2)t] \qquad (3.25)$$

The mean time between failures in this case is

$$m_p = \int_0^{\infty} R_p \, dt = 1/\lambda_1 + 1/\lambda_2 - 1/(\lambda_1 + \lambda_2) \qquad (3.26)$$

When the failure rates of two parallel components are equal so that $\lambda_1 = \lambda_2 = \lambda$, the unreliability of this parallel combination of two identical components is

$$Q_p = Q_1 Q_2 = Q^2 = [1-\exp(-\lambda t)]^2$$

The reliability is

$$R_p = 1 - Q_p = 1 - [1-\exp(-\lambda t)]^2 = 2\exp(-\lambda t) - \exp(-2\lambda t) \qquad (3.27)$$

The mean time between failures now is

$$m_p = 2/\lambda - 1/(2\lambda) = 1/\lambda + 1/(2\lambda) = 3/(2\lambda) \qquad (3.28)$$

For three identical components in parallel, we have

$$R_p = 1\text{-}Q_p = 1\text{-}Q^3 = 1\text{-}[1\text{-}\exp(\text{-}\lambda t)]^3$$

$$= 3\exp(-\lambda t) - 3\exp(-2\lambda t) + \exp(-3\lambda t) \quad (3.29)$$

or, $m_p = 3/\lambda\text{-}3/2\lambda + 1/3\lambda = 11/6\lambda$ which can also be expressed as:

$$m_p = 1/\lambda + 1/(2\lambda) + 1/(3\lambda) = 11/(6\lambda) \quad (3.30)$$

When three components in parallel are not similar,

$$R_p = 1 - Q_1Q_2Q_3$$

$$= 1\text{-}[1\text{-}\exp(-\lambda_1 t)][1\text{-}\exp(-\lambda_2 t)][1\text{-}\exp(-\lambda_3 t)]$$

$$m_p = 1/\lambda_1 + 1/\lambda_2 + 1/\lambda_3 - 1/(\lambda_1+\lambda_2) - 1/(\lambda_1+\lambda_3) - 1/(\lambda_2+\lambda_3)$$

$$+ 1/(\lambda_1+\lambda_2+\lambda_3) \quad (3.31)$$

Finally, for n similar components in parallel, we obtain,

$$R_p = 1 - Q_p = 1\text{-}Q^n = 1\text{-}[1\text{-}\exp(-\lambda t)]^n$$

$$m_p = 1/\lambda + 1/(2\lambda) + 1/(3\lambda) + \ldots + 1/(n\lambda) \quad (3.32)$$

Although the improvement in reliability achieved by operating components in parallel is quite obvious, it must be remembered that not all components are suitable for what we have defined as parallel operation, i.e., continuous operation of two parallel sets for the sole purpose of having one to carry on the operation alone should the other fail. Resistors and capacitors are particularly unsuitable for this kind of operation because if one fails out of two parallel units, this changes the circuit constants. When high reliability requirements make redundant arrangements of such units a necessity, these arrangements must then be of the stand-by type where only one unit operates at a time and the second unit, which is standing by idly, is switched into the circuit if the first unit fails. Such systems are discussed in a subsequent section.

Example 3.4

A broadcast station has three active and independent transmitters. At least one of these must function for the system's success. Calculate the reliability of transmission if the reliabilities of individual transmitters are 0.92, 0.95, and 0.96 respectively.

Solution

$$R_p = 1-\prod_{i=1}^{m}(1-p_i)$$

$$=1-(0.08)(0.05)(0.04) = 0.99984 \text{ (or 99.98\%)}$$

* * *

3.5 SERIES PARALLEL SYSTEMS

In such systems, we have to apply the product law of reliability and product law of unreliability repeatedly for reliability analysis of the systems. This is best clarified with the help of some examples:

Example 3.5

A system consists of five components connected as shown in Fig 3.4 with given values of component reliabilities. Find the overall system reliability.

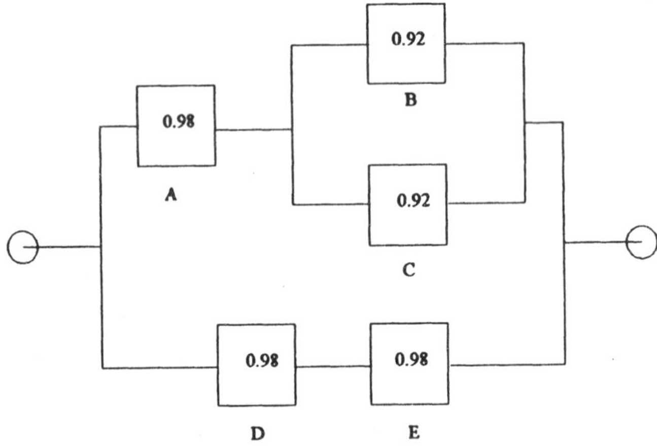

Fig. 3.4: System for Example 3.5

Solution

The reliability for series combination D-E is:

$$R_d R_e = 0.98 * 0.98 = 0.9604$$

The reliability for parallel combination B-C is:

$$R_b + R_c - R_bR_c = 0.92 + 0.92 - 0.92*0.92 = 0.9936$$

Hence, the reliability of ABC together is:

$$(0.98)(0.9936) = 0.9737$$

Therefore the overall system reliability is:

$$0.9737 + 0.9604 - (0.9737)(0.9604) = 0.99896$$

* * *

Example 3.6

Three generators, one with a capacity of 100 kw and the other two with a capacity of 50 kw each are connected in parallel. Draw the reliability logic diagram if the required load is:
(i) 100 kw (ii) 150 kw

Determine the reliability of both the arrangements if the reliability of each generator is 0.95.

Solution

The reliability logic diagram for case (i) is drawn as shown in Fig 3.5(a) because in this case either one 100 kw or two 50 kw generators must function. Similarly, the logic diagram for case (ii) is drawn as shown in Fig 3.5(b) as in this 100 kw generator must function and out of the remaining two any one is to function.

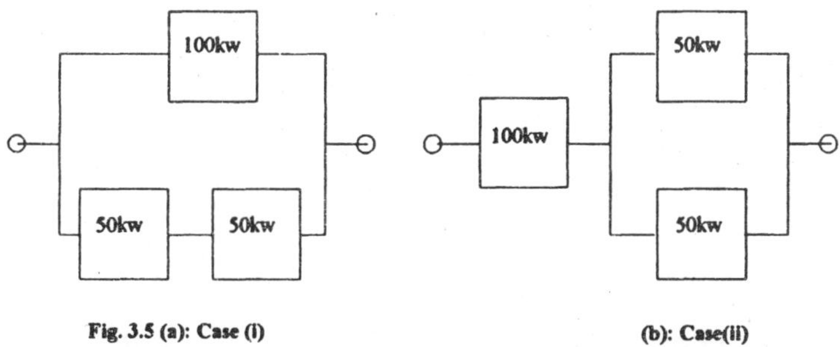

Fig. 3.5 (a): Case (i) **(b): Case(ii)**

If r is the reliability for each component, the system reliability R_1 and R_2 is respectively computed as:

$R_1 = r + r^2 - r^3$
$R_2 = r[2r - r^2]$

With $r = 0.95$, $R_1 = 0.995$ and $R_2 = 0.948$

* * *

3.51 Redundancy at Component Level

The pertinent question here is, at what level should the components be duplicated, i.e, at component level, subsystem level or system level?. We will explain this with the help of an example. Consider the two configurations as given in Fig 3.6.

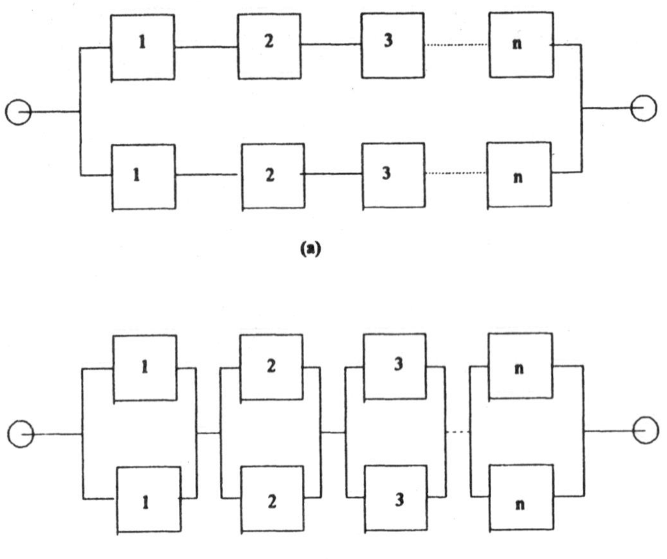

Fig 3.6: Redundancy at Component Level

In this configuration 3.6(a), there are n components connected in series, and the set of this n components, is placed in parallel with another set. In configuration 3.6(b), the components have been first placed in parallel, and in turn connected in series. Which configuration gives the better reliability, that is, the components duplicated at component level [Fig 3.6(b)], or at the subsystem level [Fig 3.6(a)] ?

Let the reliability of each component be r. The reliability of the system (R_s) in the case of configuration 3.6(a) can be expressed as

$$R_s = 1-(1-r^n)^2 = r^n(2-r^n)$$

The reliability of the system (R_s') in the case of configuration 3.6(b) is expressed as

$$R_s' = [1-(1-r)^2]^n = r^n(2-r)^n$$

The ratio of R_s' and R_s gives

$$\frac{R_s'}{R_s} = \frac{r^n(2-r)^n}{r^n(2-r^n)}$$

It can be shown that the ratio $R'_s:R_s$ is greater than unity for $r<1$. Hence, the configuration 3.6(b) would always provide higher reliability. Thus, as a generalisation, it can be said that the components if duplicated in the system at the component level give higher system reliability than if duplicted at the subsystem level (here each set is considered as a subsystem). In general, it should be borne in mind that the redundancy should be provided at the component level until and unless there are some overriding reasons or constraints from the design point of view.

3.6 K-OUT- OF-M SYSTEMS

In many practical systems more than one of the parallel components are required to work satisfactorily for successful operation of the system. For example, we can consider a power plant where two of its four generators are required to meet the customer's demand. In a 6-cylinder automobile, it may be possible to drive the car, if only four cylinders are firing. Such systems are known as *k-out-of-m systems.* For identical, independent components, with p as the reliability of each component, the probability that exactly x out of m components are successful is:

$$P = {}^mC_x \; p^x (1-p)^{m-x} \tag{3.33}$$

For a k-out-of-m system, the event of system success will be when k, k+1, k+2,... or m components function successfully. So the system reliability is the sum of probabilities for x varying from k to m i.e.

$$R = \sum_{i=k}^{m} {}^mC_i \, p^i (1-p)^{m-i} \tag{3.34}$$

For constant failure rates,

$$R(t) = \sum_{i=k}^{m} {}^mC_i \; \exp(-i\lambda t) \, [1-\exp(-\lambda t)]^{m-i} \tag{3.35}$$

and

$$m_s = (1/\lambda) \sum_{i=k}^{m} 1/i \tag{3.36}$$

In a k-out-of-m system, (m-k) components are redundant components and any increase in the value of k decreases the system reliability. For example let us suppose that there are four generators of 200 KW each in a power plant and the demand is 400 KW. This demand can be met by any two of the generators and this becomes a 2-out-of-4 system, leaving 2 generators as redundant. In case the demand increases to 600 KW, this can be met by 3 generators and this would become a 3-out-of-4 system leaving only one generator as redundant with a decreased system reliability.

If the components are not identical but have different reliabilities, the calculations become more complicated.

Assume three components with the reliabilities R_1, R_2 and R_3 operating simultaneously and in parallel. Then,

$$\begin{aligned}(R_1+Q_1)(R_2+Q_2)(R_3+Q_3) &= R_1R_2R_3 + (R_1R_2Q_3 + R_1R_3Q_2+R_2R_3Q_1) \\ &\quad + (R_1Q_2Q_3 + R_2Q_1Q_3 + R_3Q_1Q_2) + Q_1Q_2Q_3 \\ &= 1\end{aligned}$$

To obtain system reliability for 1-out-of-3 system, we will discard the last term only, i.e., $Q_1Q_2Q_3$ and for 2-out-of-3 system, the last four terms are to be discarded.

Example 3.7

An electrical system consists of four active, identical, and independent units whose failure rates are constant. For the system's success atleast three units must function normally. Each unit has a constant failure rate equal to 0.0005 failures/hr. Calculate the system mean time to failure.

Solution

Now, $m=4$, $k=3$ and $\lambda=0.0005$ failures/hr

Using equation (3.35),

$$R(t) = \sum_{i=3}^{4} {}^4C_i\, e^{-i\lambda t}\,(1-e^{-\lambda t})^{4-i} = 4e^{-3\lambda t} - 3e^{-4\lambda t}$$

Also using equation (3.36),

$$m_s = (1/\lambda)(1/3+1/4) = 7/12\lambda = 1{,}167 \text{ hr}$$

The above result for MTBF could also be derived by integrating the expression for R(t) from 0 to ∞ .

* * *

3.7 OPEN-AND-SHORT-CIRCUIT FAILURES

The previous redundant models were based on the assumption that individual element or path failure has no effect on the operation of the surviving paths. Consider a simple parallel unit composed of two elements, A and B, each of which can fail in either of two ways-open failure or short-circuit failure. Since a short in either of the two elements will result in unit failure, the assumption that individual path failure does not result in unit failure is not always true. The conditional probabilities of short and open failures are then used to represent element failure probabilities.

It may be noted that there are several elements which can fail open or short. The familiar examples are diodes and electrolytic capacitors in electronic circuits. Several other elements having two modes of failures can be similarly treated. For example, a valve fails to open when required or it fails to close when needed has two modes of failure. The analysis given below is applicable to such situations also.

Let, q_o' = conditional probability of an open = $Pr(O|F) = q_o/q$
and q_s' = conditional probability of a short = $Pr(S|F) = q_s/q$

Then the following relationships hold true:

$$q_o' + q_s' = 1.0$$
$$q_o'q = q_o$$
$$q_s'q = q_s$$

For two elements A and B in the active-parallel redundant configuration, the unit will fail if

1. Either A or B shorts, or
2. Both A and B open.

The respective probabilities of these two events are

1. $P_a(S)$ or $P_b(S)$ $= P_a(S) + P_b(S) - P_a(S)P_b(S)$
$= 1 - [1-P_a(S)]\,[1-P_b(S)]$
$= 1 - (1-q_{sa})\,(1-q_{sb})$

2. $P_{ab}(O)$ $= P_a(O)\,P_b(O)$
$= q_{oa}\,q_{ob}$

Where $P_i(O)$ is the probability that element i opens and $P_i(S)$ is the probability that element i shorts. Since events (1) and (2) are mutually exclusive, the probability of unit failure is the sum of the two event probabilities, or,

$$\begin{aligned} P(F) &= Q = P_a(S) \text{ or } P_b(S) + P_{ab}(O) \\ &= 1 - (1-q_{sa})(1-q_{sb}) + q_{oa}q_{ob} \end{aligned} \tag{3.37}$$

In general, if there are m parallel elements,

$$Q = 1 - \prod_{i=1}^{m} (1-q_{si}) + \prod_{i=1}^{m} q_{oi} \tag{3.38}$$

and the reliability is, of course equal to

$$R = \prod_{i=1}^{m} (1-q_{si}) - \prod_{i=1}^{m} q_{oi} \tag{3.39}$$

If all elements are identical, the reliability of the unit is

$$R = (1-q_s)^m - q_o^m \tag{3.40}$$

It is apparent that by introducing the possibility of short-circuit failures, unit reliability may be significantly decreased by adding parallel elements.

For any range of q_o and q_s, the optimum number of parallel elements is one if $q_s > q_o$. For most practical values of q_o and q_s, the optimum number turns out to be two. In general, for a given q_s and q_o, the reliability as a function of m would have the form shown in Fig.3.7.

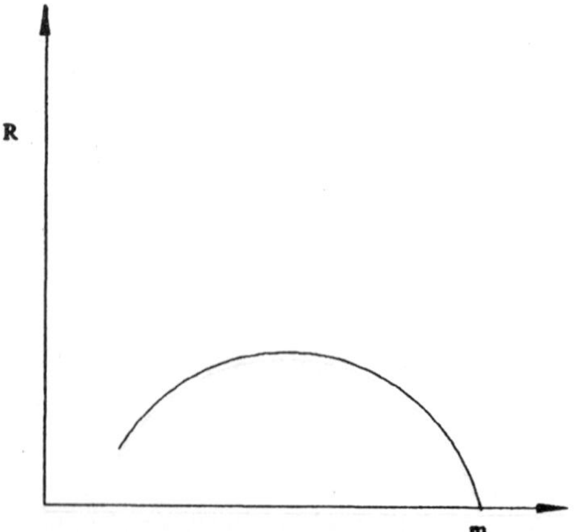

Fig. 3.7: Reliability versus number of elements

Therefore, by taking the derivative of R with respect to m, we can find the optimum number of parallel elements for maximizing reliability. Now, equating $\partial R/\partial m = 0$, we have

$$\frac{\partial R}{\partial m} = \frac{\partial[(1-q_s)^m - q_o^m]}{\partial m} = 0 \qquad (3.41)$$

Or,

$$(1-q_s)^m \ln(1-q_s) - q_o^m \ln(q_o) = 0$$

Solving for m gives

$$m = \frac{\ln[\ln(q_o)/\ln(1-q_s)]}{\ln[(1-q_s)/q_o]} \qquad (3.42)$$

A chart for giving optimum values of m for given q_s and q_o is shown in Fig.3.8.

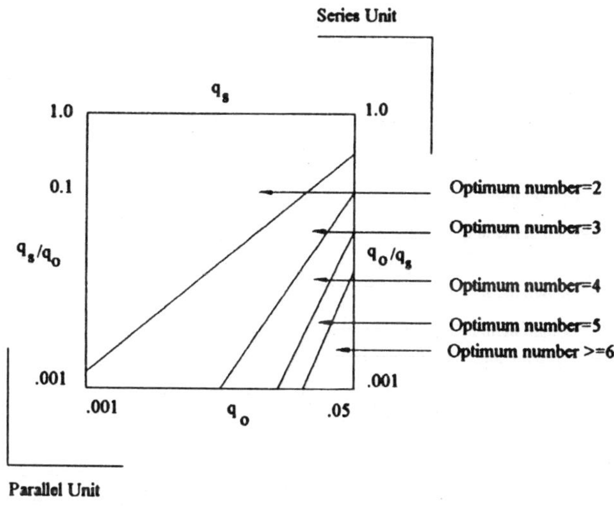

Fig. 3.8: Optimum number of elements for parallel or series units whose elements can be short & open

The result given above indicates that if $q_s > q_o$, the optimum number of parallel paths is one. However, addition of an element in series will result in an increase in reliability if q_s is much greater than q_o.

The reliability of a series system in which both short-circuit and open-circuit failures are possible is estimated below, with a two- element series unit discussed for illustration:

The unit will fail if

1. Both A and B short.
2. Either A or B opens.

The probabilities of these two events are

1. $P_{ab}(S) = P_a(S)\,P_b(S) = q_{sa}\,q_{sb}$

2. $P_a(O)$ or $P_b(O)$ $= P_a(O) + P_b(O) - P_a(O)P_b(O)$

$$= 1 - [1-P_a(O)]\,[1-P_b(O)]$$

$$= 1 - (1-q_{oa})\,(1-q_{ob})$$

Since events (1) and (2) are mutually exclusive,

$$P(F) = Q = P_a(O) \text{ or } P_b(O) + P_{ab}(S)$$

$$= 1 - (1-q_{oa})(1-q_{ob}) + q_{sa}q_{sb}$$

In general, if there are n series elements,

$$Q = 1 - \prod_{i=1}^{n} (1-q_{oi}) + \prod_{i=1}^{n} q_{si}$$

and the reliability is, of course equal to

$$R = \prod_{i=1}^{n} (1-q_{oi}) - \prod_{i=1}^{n} q_{si} \tag{3.43}$$

If all elements are identical, the reliability of the n-element series unit is

$$R = (1-q_o)^n - q_s^n \tag{3.44}$$

Using the same approach as that for the parallel configuration case, it is easily shown that the optimum number of series elements for a given q_o and q_s is

$$n = \frac{\ln[\ln(q_s)/\ln(1-q_o)]}{\ln((1-q_o)/q_s)} \tag{3.45}$$

The optimum value for n can also be read from Fig.3.8.

Example 3.8

The estimated failure probability for an element that can short or open is 0.15. The ratio of short to open failure probabilities is known to be 0.25. What is the optimum number of parallel elements to use ?.

Solution

Here,

$$q_o + q_s = 0.15 \text{ and } q_s/q_o = 0.25$$

Hence, $q_o = 0.12$ and $q_s = 0.03$
Using equation (3.42),

$$m_{opt} = \frac{\ln[\ln(0.12)/\ln(1-0.03)]}{\ln[(1-0.03)/0.12]} \approx 2$$

It may be pertinent to point out here that if the numerical value of the optimum number does not come out to be close to an integer, we should determine the reliability by considering integers on both sides of the real value and then choose the optimum one.

* * *

3.71 Fail-Safe and Fail-to-Danger

When we are determining the reliability, there are situations in which different modes of failure can have very different effects. Consider an alarm system, or for that matter any safety-related system. The alarm may fail in one of two ways. It may fail to function even though the danger is present or it may give a spurious or false alarm even though no danger is present. The first of these is referred to as fail-to-danger and the second as fail-safe. Generally, the probability of fail-to-danger is made much smaller than the fail-safe probability. Even then, small fail-safe probabilities are also required. If too many spurious alarms are sounded, they will tend to be ignored. Then, when the real danger is present, the alarm is also likely to be ignored. This difficulty can be circumvented by automating the safety actions, but then each spurious alarm may lead to a significant economic loss. This would certainly be the case were a chemical plant, a nuclear reactor, or any other industrial installation shut down frequently by the spurious operation of safety systems.

The distinction between fail-safe and fail-to-danger has at least two important implications for reliability engineering. First, many design alterations that may be made to decrease the fail-to-danger probability are likely to increase the fail-safe probability. An obvious example is that of power supply failures, which are often a primary cause of faliure of crudely designed safety systems. Often, the system can be redesigned so that if the power supply fails, the system will fail-safe instead of to-danger. Specifically, instead of leaving the system unprotected following the failure, the power supply failure will cause the system to function spuriously. Ofcourse, if no change is made in the probability of power supply failure, the reduction in the probability for system fail-to -danger will be compensated for by the increased number of spurious operations.

A second implication for reliability engineering is that the more redundancy is used to reduce the probability of fail-to-danger, the more fail-safe incidents are likely to occur. To demonstrate this, consider a parallel system with which are associated two failure probabilities p_d and p_s, for fail to-danger and fail-to-safe, respectively. The fail-to-danger unreliability for the system is found by noting that all units must fail. Hence

$$Q_d = (p_d)^N \tag{3.46}$$

However, the system fail-safe unreliability is calculated by noting that any one-unit failure with probability p_s, will cause the system to fail-safe. Thus

$$Q_s = 1-(1-p_s)^N \tag{3.47}$$

Using the approximation $p_s << 1$, we see that the fail-safe probability grows linearly with the number of units in parallel,

$$Q_s \approx N\, p_s \tag{3.48}$$

The k-out-of-m configuration has been extensively used in electronic and other protection systems to limit the number of spurious operations at the same time that the redundancy provides high reliability. In such systems the fail-to-danger unreliability is given by

$$Q_d = \sum_{j=m-k+1}^{m} {}^mC_j\, (p_d)^j (1-p_d)^{m-j} \tag{3.49}$$

With the rare-event approximation this reduces to

$$Q_d = {}^mC_{m-k+1}\, p_d^{\,m-k+1} \tag{3.50}$$

Conversely, atleast k spurious signals must be generated for the system to fail-safe. Assuming independent failures with probability p_s, we have

$$Q_s = \Pr\{j \geq k\} = \sum_{j=k}^{m} {}^mC_j \, (p_s)^j \, (1-p_s)^{m-j} \qquad (3.51)$$

Again using the rare-event approximation that $p_s << 1$, we may approximate this expression by

$$Q_s \approx {}^mC_k \, p_s^k \qquad (3.52)$$

From Eqs.(3.50) and (3.52) the trade-off between fail-to-danger and spurious operation is seen. The fail-safe unreliability is decreased by increasing k and the fail-to-danger unreliability is decreased by increasing m-k.

3.8 STAND-BY SYSTEMS

Often it is not feasible or practical to operate components or units in parallel and so called *Stand-by* arrangements must be applied; that is, when a component or unit is operating, one or more components or units are standing by to take over the operation when the first fails.

Stand-by arrangements normally require failure sensing and switchover devices to put the next unit into operation. Let us first assume that the sensing and switchover devices are 100 percent reliable and that the operating component and the stand-by components have the same constant failure rate.

We can regard such a group of stand-by components as being a single unit or system which is allowed to fail a number of times before it definitely stops performing its function. If n components are standing by to support one operating component, we have (n + 1) components in the system, and n failures can occur without causing the system to fail. Only the (n + 1)th failure would cause system failure.

Since $\exp(-\lambda t)\exp(\lambda t) = 1$

We have,

$$\exp(-\lambda t)[1 + \lambda t + (\lambda t)^2/2! + (\lambda t)^3/3! + \text{--------}] = 1$$

In this expression the term $\exp(-\lambda t)*1$ represents the probability that no failure will occur, the term $\exp(-\lambda t)*(\lambda t)$ represents the probability that exactly one failure will occur, $\exp(-\lambda t)(\lambda t)^2/2!$ represents the probability that exactly two failures will occur, etc. Therefore, the probability that two or one or no failure will occur or the probability that not more than two failures will occur equals:

$$\exp(-\lambda t) + \exp(-\lambda t)\,\lambda t + \exp(-\lambda t)\,(\lambda t)^2/2!$$

If we denote by R_s and Q_s the reliability and the unreliability of the system, and because $R_s + Q_s = 1$ we can write

$$\begin{aligned} R_s + Q_s &= \exp(-\lambda t)[1 + \lambda t + (\lambda t)^2/2! + (\lambda t)^3/3! + \text{---------}] \\ &= \exp(-\lambda t) + \exp(-\lambda t)\,\lambda t + \exp(-\lambda t)\,(\lambda t)^2/2! + \text{-----} \\ &= 1 \end{aligned}$$

If in this expanded form, we allow one failure, then the reliability of a stand-by system composed of one operating component and another standing by idly to take over if the first fails is given by:

$$R_s = \exp(-\lambda t)[1 + \lambda t] \qquad (3.53)$$

The mean time between failures for a two-component system is:

$$m_s = \int_0^\infty R_s dt = 1/\lambda + \lambda/\lambda^2 = 2/\lambda \qquad (3.54)$$

For a stand-by system of three units which have the same failure rate and where one unit is operating and other two are standing by to take over the operation in succession, we have

$$R_s = \exp(-\lambda t)[1 + \lambda t + \lambda^2 t^2/2!] \qquad (3.55)$$

and

$$m_s = (1/\lambda) + (1/\lambda) + (1/\lambda) = 3/\lambda \qquad (3.56)$$

In general, when n identical components or units are standing by to support one which operates,

$$R_s = \exp(-\lambda t) \sum_{i=0}^{n} (\lambda t)^i/i! \qquad (3.57)$$

$$m_s = (n+1)/\lambda \qquad (3.58)$$

The stand-by arrangements are slightly more reliable than parallel operating units, although they have a considerably longer mean time between failures. However, these advantages are easily lost when the reliability of the sensing-switching device R_{ss} is less than 100%, which is more often the case. Taking this into consideration and when the circuits are arranged so that the reliability of the operating unit is not affected by the unreliability of the sensing-switching device, we obtain for a system in which one stand-by unit is backing up one operating unit:

$$R_s = \exp(-\lambda t) + R_{ss} \exp(-\lambda t)\, \lambda t \qquad (3.59)$$

It is the exception rather than the rule that the failure rates of the stand-by units are equal to those of the operating unit. For instance, a hydraulic actuator will be backed up by an electrical actuator, and there may be even a third stand-by unit, pneumatic or mechanical. In such cases, the failure rates of the stand-by units will not be equal and the formulae which we derived above will no longer apply.

If the system contains two different elements, A and B, the reliability functions can be found directly as follows:

The system will be successful at time t if either of the following two condtions holds (letting A be the primary element).

1. A succeeds up to time t or
2. A fails at time $t_1 < t$ and B operates from t_1 to t.

Translation of these two condtions to the time dependent probabilities gives

$$R(t) = \int_t^{\infty} f_a(t)dt + \int_0^t [f_a(t_1) \int_{t-t_1}^{\infty} f_b(t)dt]\, dt_1 \qquad (3.60)$$

where f(t) is the time-to-failure density function of an element.

The first term of this equation represents the probability that element A will succeed until time t. The second term excluding the outside integral, is the density function for A failing exactly at t_1 and B succeeding for the remaining $(t-t_1)$ hours. Since t_1 can range from 0 to t, t_1 is integrated over that range.

For the exponential case where the element failure rates are λ_a and λ_b

$$R(t) = \int_t^{\infty} \lambda_a \exp(-\lambda_a t)\, dt + \int_0^t [\lambda_a \exp(-\lambda_a t_1) \int_{t-t_1}^{\infty} \lambda_b \exp(-\lambda_b t)\, dt]\, dt_1$$

$$= \exp(-\lambda_a t) + \int_0^t \lambda_a \exp(-\lambda_a t_1) \exp[-\lambda_b(t-t_1)]\, dt_1$$

$$= \exp(-\lambda_a t) + \lambda_a \exp(-\lambda_b t) \int_0^t \exp[-(\lambda_a-\lambda_b)t_1]\, dt_1$$

or, $$R(t) = [\lambda_b \exp(-\lambda_a t) - \lambda_a \exp(-\lambda_b t)]/(\lambda_b - \lambda_a) \qquad (3.61)$$

and $$m_s = 1/\lambda_a + 1/\lambda_b \qquad (3.62)$$

It can be shown that it does not matter whether the more reliable element is used as the primary or the stand-by element.

Example 3.9

One generator is placed in standby redundancy to the main generator. The faliure rate of each generator is estimated to be λ = 0.05/hr. Compute the reliability of the system for 10hrs and its MTBF assuming that the sensing and switching device is 100% reliable. If the reliability of this device is only 80%, how are the results modified?

Solution

When sensing and switching device is 100% reliable,

$$R_s = (1+\lambda t)\exp(-\lambda t) = (1+(0.05)(10))\exp(-(0.05)(10))$$
$$= 0.9098.$$

Also, $\text{MTBF} = 2/\lambda = 2/0.05 = 40$ hrs.

When sensing and switching device is 80% reliable,

$$R_s = (1+0.80\lambda t)\exp(-\lambda t) = 0.8491$$

and,

$$\text{MTBF} = (1+0.80)/\lambda = 1.80/0.05 = 36 \text{ hrs}$$

The appreciable decrease in the values of reliability and MTBF may please be observed by the reader because of the imperfect nature of sensing and switching over device.

* * *

3.81 Types of Standby Redundancy

There could be several variations of the standby arrangements in actual practice some of these are discussed in the section below;

1. *Cold Standby*

The standby configuration discussed earlier having perfect or imperfect sensing and switching over devices, is known as cold standby, as in this case, the primary component operates and one or more secondary components are placed in as standbys. It is assumed that the secondary components in the standby mode do not fail.

2. *Tepid Standby*

In this case, the value of the standby component changes progressively. For example, components having rubber parts deteriorate over time and ultimately affect the reliability of standby component.

3. *Hot Standby*

The standby component in this case, fails without being operated because of a limited shelf life. For example, batteries will fail even in standby due to some chemical reactions.

4. *Sliding Standby*

Consider a system consisting of N components connected in series. To this system, a sliding standby component is attached which will function when any of the components of the system fails. This is shown in Fig 3.9.

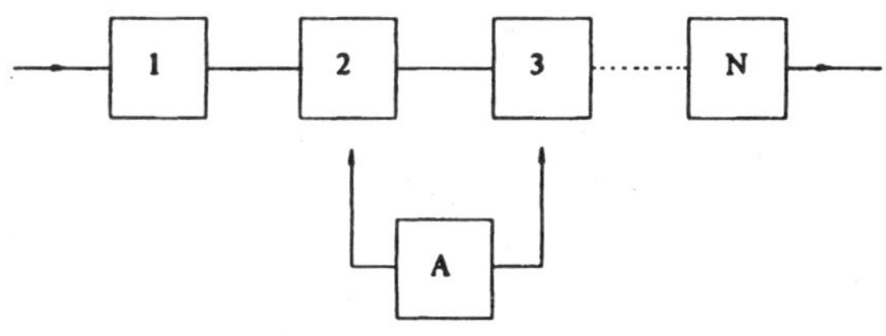

Fig 3.9: Sliding Standby

It may be noted that sliding standby components may have more than one component in standby depending upon the reliability requirement.

5. *Sliding Standby with AFL*

In this case, an Automatic Fault Locator (AFL) is provided with the main system which accomplishes the function of locating the faulty component, disconnecting it and connecting the standby component. AFL's are generally provided in automatic and highly complex systems. The sliding standby redundancy having AFL is shown in Fig 3.10.

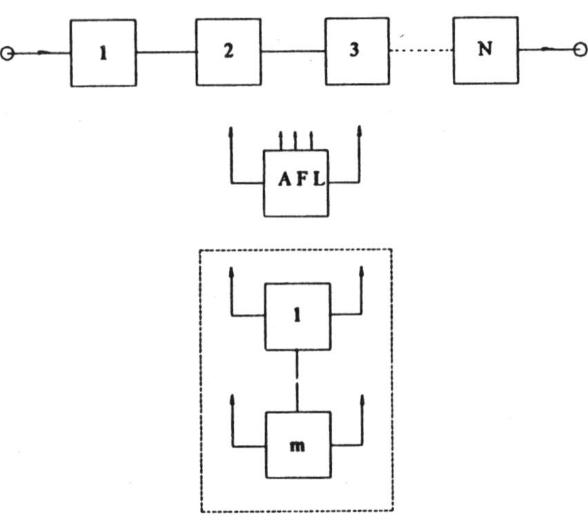

Fig 3.10: Sliding Standby with AFL

4

RELIABILITY ANALYSIS OF NONSERIES PARALLEL SYSTEMS

4.1 INTRODUCTION

System reliability evaluation is a basic step in all reliability studies. Therefore, derivation of the symbolic reliability expression in a simplified and compact form for a general system is very desirable.

In system reliability analysis, it is customary to represent the system by a probabilistic graph in which each node and each branch has a probability of being operative. The techniques for reliability evaluation depend on the logic diagram of the system. For a series- parallel or partial redundant structure, the reliability evaluation is relatively straight forward as has been discussed in the last chapter. Unfortunately, several practical systems lead to non-series-parallel reliability logic diagrams, where reliability evaluation is quite involved.

An example of a general system which leads to nonseries-parallel logic diagram is a high pressure oxygen supply system for a spacecraft as shown in Fig.4.1. The high-pressure oxygen in the cabin is supplied through a system of regulators and valves from a high-pressure oxygen tank. There are two pairs of the sub-systems of check valves, shut-off valves and non-return automatic shut-off valves in the system. The function of these valves is to stop the reverse flow of air from the cabin to the gas tank in the case of pressure drop and to close the line supply if there is some sudden pressure drop in header line or cabin in order to avoid the waste of the gas. Each pair of the valve systems consists of two alternative branches. One consists of a non-return automatic emergency shut-off valve, and the other consists of a check valve and a shut-off valve in series. Any branch of the two pairs is capable of supplying sufficient gas to the cabin. There are three alternative paths between the oxygen tank and the pair of valves. Oxygen can be transmitted to the cabin through either of the two regulators and the pair of valves connected to the regulator. It can also be transmitted to the cabin through a selector valve and either of the two pairs of valves.

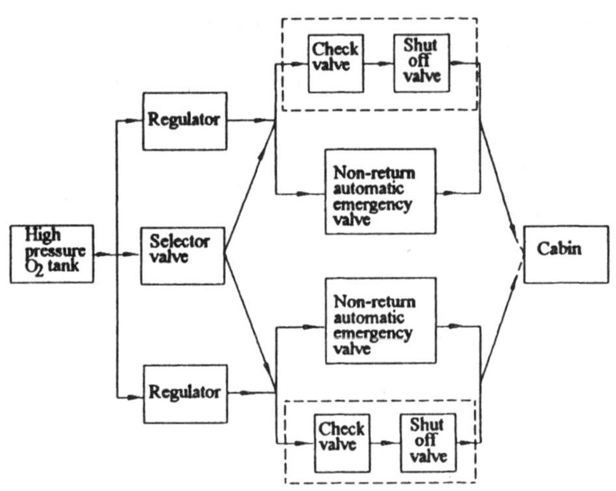

Fig. 4.1: High pressure Oxygen supply system of a spacecraft life support system.

Another very common example of the non-series parallel systems is the computer communication networks. A subset of the ARPA computer network is shown in Fig.4.2.

The most common problem which arises in the analysis of such a network is to compute in an efficient and systematic manner the source to terminal reliability between a given pair of nodes, namely, the probability that there exists at least one path between these two nodes. Although not necessary, it is generally convenient to simplify the diagram by removing purely series, purely parallel, self-loops and dead-end connections before applying any of these general algorithms.

The algorithms for the reliability analysis of general systems can be divided into two groups. The methods in the first group, to which majority of the reported algorithms belong, require a prior knowledge of all minimal paths (or minimal cutsets) of the network. These methods make use of AND-OR expressed system success (or failure) function as the starting point. Boolean algebra is then used to transform this function into another equivalent function in which all the terms are disjoint with respect to each other. The derivation of reliability expression is then straight-forward as simpler probability laws are applicable. A method of this group is discussed in the next section in detail. Most of the other methods have only minor variations with respect to each other and several of those are reported to be more efficient also.

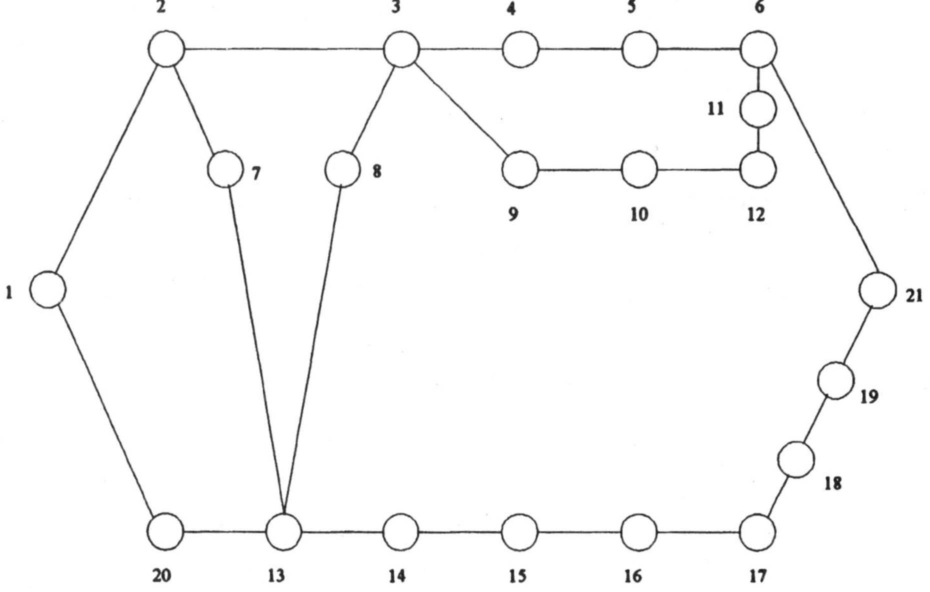

Fig. 4.2: Topology of subset of ARPA computer network.

The methods in the second group do not require a prior knowledge of all paths of the network. These methods are also important as the computer time needed to determine all minimal paths is sometimes comparable to the time required for making the terms of the success function disjoint. Three such methods viz. Delta-Star Method, Logical Signal Relations method and Baye's Theorem Method are also discussed.

An example has been solved by all the methods discussed below. This allows the reader to easily compare the algorithms and also ensures correctness of calculations by all methods.

4.2 PATH DETERMINATION

As already discussed, in using several methods to evaluate system reliability, determination of all m paths is necessary. In a simple network this may be possible by inspection; but in a general network some systematic method has to be used. One such method is discussed below which is based on the use of the connection matrix.

A connection matrix is an analytic correspondence of the system graph and has a size (n x n) where n is the number of nodes in the graph. Although not necessary, it is convenient to number the source node as 1 and terminal node as 2. In this matrix,

$$C_{ij} = \begin{cases} 0; & \text{if there is no branch from i to j} \\ X; & \text{if there is a branch from i to j (X denotes the event of this branch being successful)} \end{cases}$$

For the bridge network of Fig.4.3; the connection matrix [C] is written as:

$$[C] = \begin{bmatrix} 0 & 0 & A & C \\ 0 & 0 & 0 & 0 \\ 0 & B & 0 & E \\ 0 & D & E & 0 \end{bmatrix}$$

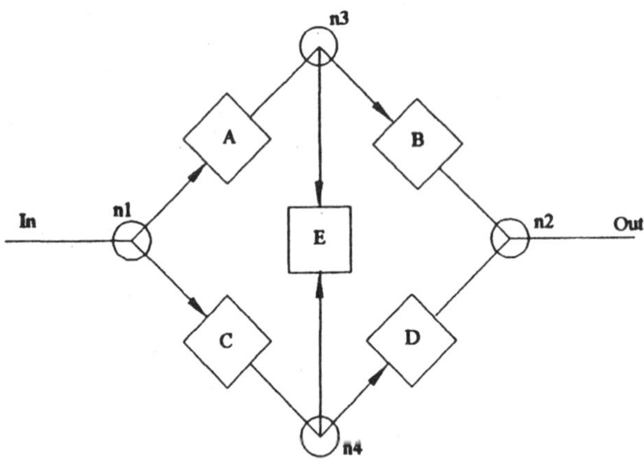

Fig. 4.3: Non-Series Parallel Network.

The method requires *removal* of the last row and last column after modifying the remaining entires of [C] as:

$$C_{ij,\ new} = C_{ij,\ old} + C_{in}\, C_{nj} \qquad i,j = 1,2,....,(n-1) \qquad (4.1)$$

where n^{th} row (column) is the last row (column) in the matrix. This operation will lead to all required paths from i to j through n. Thus, a reduced connection matrix of size (n-1) is built. The above steps are successively repeated till a matrix of size 2 is obtained. Element C_{12} of this matrix corresponds to all the paths. Removing nodes 4 and 3 respectively from the connection matrix,

$$C(4) = \begin{bmatrix} 0 & CD(4) & A+CE(4) \\ 0 & 0 & 0 \\ 0 & B+ED(4) & 0 \end{bmatrix}$$

$$C(4,3) = \begin{bmatrix} 0 & CD(4) + AB(3) + CEB(4,3) + AED(4,3) \\ 0 & 0 \end{bmatrix}$$

Hence, the minimal paths are: CD, AB, CEB and AED. The number(s) in parenthesis denotes the node which has been traversed and is recorded to avoid going over that node again. The algorithm is attractive as it does not require matrix multiplications and the size of the matrix reduces in every step.

4.3 BOOLEAN ALGEBRA METHODS

In this section, we will briefly demonstrate by means of an example, the ideas contained in the Boolean algebra methods that have been developed for computing the terminal reliability of a probabilistic network.

As already stated, we first derive the s-o-p expression for the system success function as a pure Boolean algebraic statement. If it is to be interpreted as a probability expression, certain modifications may be necessary. The modifications are necessary because the following relation for expressing the probability of the union of n events is true *only if* the events are mutually exclusive

$$Pr(E_1 \cup E_2 \cup \ldots \cup E_n) = Pr(E_1) + Pr(E_2) + \ldots + Pr(E_n) \qquad (4.2)$$

To appreciate the effect of this, consider the Boolean expression,

$$Z = AB + ACD \qquad (4.3)$$

This function is plotted on a Karnaugh map in Fig.4.4. This map can be re-interpretted as a probability map where A, B, C, D represent four primary events with individual probabilities of occurrence p_a, p_b, etc. and individual probabilities of non-occurrence as q_a, q_b etc. On this basis, therefore, the probability of event Z is the algebraic sum of five events defined by the five locations containing a 1 in the Karnaugh map, i.e.,

$$Pr(Z) = Pr(E_1) + Pr(E_2) + Pr(E_3) + Pr(E_4) + Pr(E_5) \tag{4.4}$$

where,

$$Pr(E_1) = p_a p_b q_c q_d$$
$$Pr(E_2) = p_a p_b q_c p_d$$
$$Pr(E_3) = p_a p_b p_c p_d$$
$$Pr(E_4) = p_a p_b p_c q_d$$
$$Pr(E_5) = p_a q_b p_c p_d$$

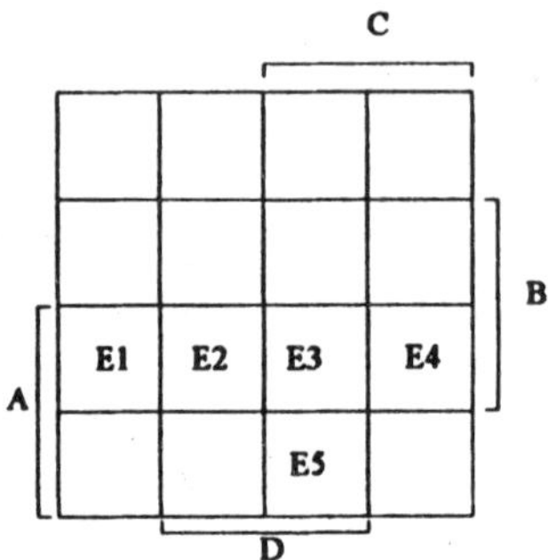

Fig. 4.4: Karnaugh Map for Z= AB+ACD.

This equation could have been obtained directly from the orginal Boolean expression by converting the same into its canonical form as:

$$Z = AB + ACD = AB(C+C')(D+D') + ACD(B+B')$$

or

$$Z = ABC'D' + ABC'D + ABCD + ABCD' + AB'CD \tag{4.5}$$

There is now a one-to-one correspondence between the terms of the two equations. However, it must be realised that one can not use equation (4.3) to derive Pr(Z) directly, as

$$Pr(Z) \neq p_a p_b + p_a p_c p_d \tag{4.6}$$

In terms of the probability map interpretation, the modification is necessary to compensate for the fact that the groupings of the AB and ACD terms are not disjoint. In this particular case, ABCD is common to both. An alternative solution therefore would be to modify the Boolean terms until they do represent a disjoint grouping and one possibility in this case is:

$$Z = AB + AB'CD \tag{4.7}$$

which leads directly to

$$Pr(Z) = p_a p_b + p_a q_b p_c p_d \tag{4.8}$$

The above Boolean expression thus represents a valid alternative to full canonical form and can still be interpreted as a probability expression.

The key problem of all Boolean algebra methods thus is to rewrite the Boolean statement of system success /failure function in a form (as concise as possible) such that all terms are mutually disjoint. It may be observed that two conjunctive terms T_1 and T_2 will represent disjoint groupings if there exists at least one literal in T_1 such that the same literal occurs in its complemented form in T_2.

4.4 A PARTICULAR METHOD

The algorithm described in this section gives quite a simplified reliability expression derived in a relatively straight forward manner.

From a knowledge of the paths, we find

$$S = P_1 \cup P_2 \ldots \cup P_m \tag{4.9}$$

This expression is required to be rewritten in another equivalent form in which all the terms are mutually disjoint. The method for making P's disjoint is easier if paths are enumerated in such a way that the path having minimum number of branches is listed first and so on. To select $P_{2,dis}$ from P_2, expand P_2 about a variable set K_1 (corresponding to a branch k_1) which is contained in P_1 but not in P_2.

$$P_2 = (P_2 \cap K_1) \cup (P_2 \cap K'_1) \tag{4.10}$$

Now if (P_2K_1) is contained in P_1, it is *dropped* from further considerations (because it is already included); otherwise, it is further expanded about K_2 and so on. If $(P_2K'_1)$ is disjoint with P_1, it is *retained;* otherwise it is also further expanded about K_2 and so on.

Ultimately, we shall find all subsets of P_2 which are disjoint with P_1. Union of all these subsets is $P_{2,dis}$. Similarly we find $P_{j,dis}$ for all j such that $P_{j,dis} \cap P_i = 0$ for all $i<j$. This step is fastest if we first expand P_j about a branch which has occurred in P_i's most often. Then

$$S_{dis} = \bigcup_{i=1}^{m} P_{i,dis} \tag{4.11}$$

where $$P_{1,dis} = P_1 \tag{4.12}$$

The reliability then is:

$$R = Pr(S_{dis}) = Pr(\bigcup_{i=1}^{m} P_{i,dis})$$

or,

$$R = \sum_{i=1}^{m} Pr(P_{i,dis}) \tag{4.13}$$

A formalization of the above method is represented in the form of the following steps of an algorithm;

1. Define a b-dimensional vector E_i ($i=1,2, \ldots ,m$) corresponding to P_i such that element k of this vector is 1 if the branch k is included in path P_i and 0 otherwise.

2. Define $T_j = \sum_{i \le j} E_i \,, \ j=1,2,\ldots,m$ (4.14)

3. $P_{1,dis} = P_1; \quad j = 1$ (4.15)

4. Let $j = j+1$

   (A) If there are any non-zero entries in T_j corresponding to zero entries in E_j, record their positions in order of their descending magnitude in T_j. Let these be $k_1,k_2,\ldots,k_r$. This ordering helps in getting the minimal expression fast.

   (B) Decompose E_j in two components $E_j(K_1)$ and $E_j(K'_1)$ corresponding to expanding P_j about K_1. $E_j(K_1)$ and $E_j(K'_1)$ are formed by replacing 0 in k_1th position of E_j by 1 and -1 respectively. If $E_j(K_1)$ contains 1's in ALL the positions where there have been 1's in ANY E_j ($i < j$); then $E_j(K_1)$ is *DROPPED* from further analysis because it is already included in a previous path. If $E_j(K'_1)$ contains -1 in ANY position where there is 1 in E_j for ALL $i<j$; then $E_j(K'_1)$ is *RETAINED* as a disjoint subset. If $E_j(K_1)$ is not dropped and/or $E_j(K'_1)$ is not retained; then these are further decomposed about K_2 and so on, carrying out the dropping and retaining tests at each step. Union of the retained components of E_j is $P_{j,dis}$.

5. If $j < m$; go to step 4.

6. Apply Probability Addition Rule to derive the reliability expression.

Example 4.1

The above steps of the algorithm are illustrated with the help of the non-series-parallel reliability logic diagram in Fig.4.3.

The sets associated with the paths of the above network, properly arranged, are:

$$P_1 = AB,\ P_2 = CD,\ P_3 = ADE,\ P_4 = BCE$$

Corresponding E_j's and T_j's are:

$$E_1 = [\,1\ \ 1\ \ 0\ \ 0\ \ 0\,]$$
$$E_2 = [\,0\ \ 0\ \ 1\ \ 1\ \ 0\,]$$
$$E_3 = [\,1\ \ 0\ \ 0\ \ 1\ \ 1\,]$$
$$E_4 = [\,0\ \ 1\ \ 1\ \ 0\ \ 1\,]$$

$$T_1 = [\,1\ \ 1\ \ 0\ \ 0\ \ 0\,]$$
$$T_2 = [\,1\ \ 1\ \ 1\ \ 1\ \ 0\,]$$
$$T_3 = [\,2\ \ 1\ \ 1\ \ 2\ \ 1\,]$$
$$T_4 = [\,2\ \ 2\ \ 2\ \ 2\ \ 2\,]$$

$$P_{1,dis} = P_1 = AB$$

Considering E_2 and T_2, $K_1 = A$, $K_2 = B$

$E_2(A)$	$= [\,1\ \ 0\ \ 1\ \ 1\ \ 0\,]$	CONTINUE
$E_2(A')$	$= [-1\ \ 0\ \ 1\ \ 1\ \ 0\,]$	RETAIN
$E_2(A)(B)$	$= [\,1\ \ 1\ \ 1\ \ 1\ \ 0\,]$	DROP
$E_2(A)(B')$	$= [\,1\ \ -1\ \ 1\ \ 1\ \ 0\,]$	RETAIN

Hence, $P_{2,dis} = A'CD + AB'CD$

Similarly,

$$P_{3,dis} = AB'C'DE$$
$$P_{4,dis} = A'BCD'E$$

Therefore,

$$S_{dis} = AB + CD\,(A' + AB') + AB'C'DE + A'BCD'E \tag{4.16}$$

Hence,

$$R = p_ap_b + p_cp_d(q_a + p_aq_b) + p_aq_bq_cp_dp_e + q_ap_bp_cq_dp_e \tag{4.17}$$

* * *

4.5 CUT SET APPROACH

An alternative approach for reliability analysis is to first derive the unreliability expression using a knowledge of the s-t cutsets of the system rather than paths. An s-t cutset is defined as a minimal set of branches which if cut renders the graph in two separate parts such that source node is in one part and the terminal node in the other. This approach is preferable in the case of systems where the number of cutsets may be smaller than the number of paths. It has been observed that this is the case if the average number of branches incident on a node in the graph is more than four.

The method for finding the unreliability expression using this approach is just the dual of the method for finding the reliability expression using a knowledge of paths. The basic philosophy remaining same, all the reported methods for the reliability analysis using paths can be easily transformed for the dual analysis. The method described in section 4.4 is shown applied in the following example using cutset approach.

Example 4.2

Derive the reliability expression for the graph shown in fig.4.3 using cutset approach.

Solution:

It can be seen easily that s-t cutsets are AC, BD, ADE and BCE.

We can then write system failure function as:

$$S' = A'C' \cup B'D' \cup A'D'E' \cup B'C'E' \qquad (4.18)$$

We now proceed to first make the second term disjoint with respect to the first as follows:

$$S' = A'C' \cup B'D' (A \cup A') \cup A'D'E' \cup B'C'E'$$

$$= A'C' \cup AB'D' \cup A'B'D' \cup A'D'E' \cup B'C'E'$$

Now AB'D' is disjoint with respect to the first term but not with A'B'D'. Hence expanding A'B'D' further, we have:

$$S' = A'C' \cup AB'D' \cup A'B'CD' \cup A'B'C'D' \cup A'D'E' \cup B'C'E'$$

Now A'B'CD' is disjoint with respect to the first two terms and A'B'C'D' can be dropped because it is completely contained in the first term. Therefore,

$$S' = A'C' \cup AB'D' \cup A'B'CD' \cup A'D'E' \cup B'C'E'$$

Proceeding similarly for making third and fourth terms also disjoint, we have finally the following expression for S' in which all terms are mutually disjoint.

$$S' = A'C' \cup B'D' (A \cup A'C) \cup A'BCD'E' \cup AB'C'DE' \tag{4.19}$$

As all the terms are mutually disjoint, probability calculations are relatively straight forward and we have the following expression for Q i.e. Pr{S'}:

$$Q = q_a q_c + q_b q_d (p_a + q_a p_c) + q_a p_b p_c q_d q_e + p_a q_b q_c p_d q_e \tag{4.20}$$

Hence, system reliability expression can be written as:

$$R = 1 - q_a q_c - q_b q_d (p_a + q_a p_c) - q_a p_b p_c q_d q_e - p_a q_b q_c p_d q_e \tag{4.21}$$

It can be shown that this expression is exactly equivalent to the system reliability expression (4.17) derived by following the method based on the knowledge of all paths of the system.

* * *

4.6 DELTA-STAR METHOD

In the delta star method, three independent conditions are to be imposed as there are three elements to be determined. Consider the block diagram shown in Fig.4.5. It is assumed that the branches of one (two) set out of the sets S^1, S^2, S^3 have flow into the corresponding node and the branches of

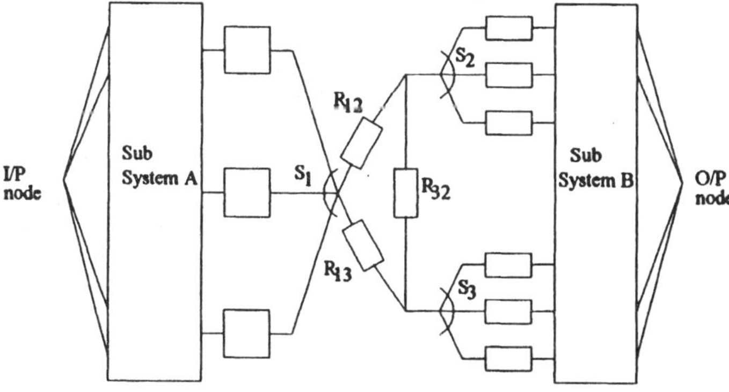

Fig. 4.5: Reliability diagram for Delta-Star method.

the remaining two (one) sets have flows coming out of the corresponding nodes.

The reliability before and after the transformation is the same:

(a) Between node 1 and nodes 2 and 3 (Fig.4.6a) when all three sets are present.

(b) Between node 1 and node 2 (Fig.4.6b) when S^3 is a null set.

(c) Between node 1 and node 3 (Fig.4.6c) when S^2 is a null set.

For example, three components of a system with reliabilities R_{13}, R_{12}, R_{32} connected to form the delta configuration shown in Figs.4.5 & 4.6 can be transformed into star equivalent with reliabilities R_{10}, R_{20}, R_{30}.

Applying s-independent event probability laws to series and parallel components results in

$$R_{10}[1 - (1-R_{20})(1-R_{30})] = 1 - (1-R_{12})(1-R_{13}) \quad (4.22a)$$

$$R_{10}R_{20} = 1 - (1-R_{12})(1-R_{13}R_{32}) \quad (4.22b)$$

$$R_{10}R_{30} = 1 - (1-R_{13})(1-R_{12}R_{32}) \quad (4.22c)$$

Solving the above equations for R_{10}, R_{20}, R_{30} results in:

$$R_{10} = M_1M_2/M_3 \quad (4.23a)$$

$$R_{20} = M_3/M_2 \quad (4.23b)$$

$$R_{30} = M_3/M_1 \quad (4.23c)$$

Where,

$$M_1 = R_{12} + R_{13}R_{32} - R_{12}R_{32}R_{13} \quad (4.24a)$$

$$M_2 = R_{13} + R_{12}R_{32} - R_{12}R_{32}R_{13} \quad (4.24b)$$

$$M_3 = R_{12}R_{32} + R_{12}R_{13} + R_{32}R_{13} - 2R_{12}R_{32}R_{13} \quad (4.24c)$$

Example 4.3

Again for the bridge structure of Fig.4.3, with

$p_a = R_{12}$, $p_c = R_{13}$ and $p_e = R_{32}$

$$M_1 = p_a + p_c p_e - p_a p_c p_e \tag{4.25a}$$

$$M_2 = p_c + p_a p_e - p_a p_c p_e \tag{4.25b}$$

$$M_3 = p_a p_e + p_a p_c + p_c p_e - 2p_a p_c p_e \tag{4.25c}$$

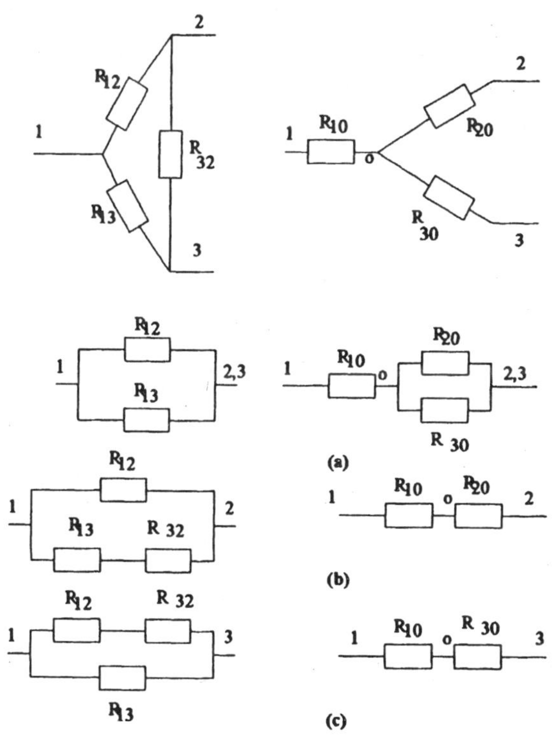

Fig. 4.6: (a), (b) & (c): Delta to Star equivalents.

Solving for R_{10}, R_{20} and R_{30} from the above equations, we have

$$R_{10} = \frac{(p_a + p_c p_e - p_a p_c p_e)(p_c + p_a p_e - p_a p_c p_e)}{p_a p_e + p_a p_c + p_c p_e - 2p_a p_c p_e} \tag{4.26a}$$

$$R_{20} = \frac{p_a p_e + p_a p_c + p_c p_e - 2p_a p_c p_e}{p_c + p_a p_e - p_a p_c p_e} \tag{4.26b}$$

$$R_{30} = \frac{p_a p_e + p_a p_c + p_c p_e - 2p_a p_c p_e}{p_a + p_c p_e - p_a p_c p_e} \tag{4.26c}$$

System reliability R is given by

$$R = R_{10}[1 - (1-R_{20}p_b)(1-R_{30}p_d)] \tag{4.27}$$

After lot of algebraic manipulations, we can verify that the system reliability expression is equal to the one obtained earlier in the last section. It is seen however that the method involves complicated and large expressions and extensive manipulations.

* * *

4.7 LOGICAL SIGNAL RELATIONS METHOD

In the application of this method, the numbering of the nodes of the reliability logic diagram begins from the source and continues in such a way that the output terminal of each branch is assigned a number greater than the number used for its input, taking further care that each node is assigned a different number. Thus, the previous network is redrawn as Fig.4.7.

A pair of nodes n_i and n_j are *fused* if the two nodes are replaced by a single new node such that all branches that were incident on either n_i or n_j or on both are now incident on the new node. We denote the fusion of n_i and n_j as $\overline{n_i n_j}$. More than two nodes are fused by taking them two at a time until all are fused.

The logical signal relations for some common sub-networks are given in Fig.4.8. Each relation is expressed so that its terms are always mutually disjoint. Sub-networks at serial number 4 and 5 refer to 2 and 3 branches, respectively, incident on a node. This concept can easily be extended for b branches incident on a node by observing the recursive nature of relations.

The steps of the algorithm are:

1. (a) Write the logical signal relation for the sink node.

   (b) Successively proceed towards the source node using the required relations. Repeat until the source node is reached.

   Substitute,

$$S(n_1) = S(\overline{n_1....}) = 1$$

Where $(\overline{n_1...})$ indicates the fusion of any number of nodes, one of which is source node n_1. The above equation signifies that the signal is assumed to be present at the source node.

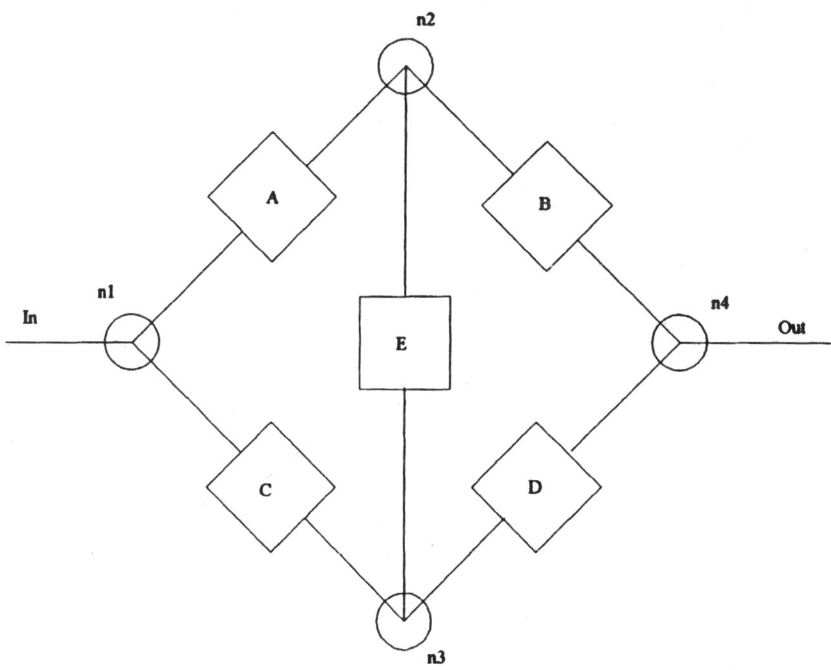

Fig. 4.7: Bridge Network (Redrawn).

2. In the expression thus obtained for the logical signal at the output node, replace the logical variables by the corresponding probability variables to obtain the reliability expression.

Example 4.4

We intend evaluating the reliability of same bridge network (Fig.4.7) with the above method. Relevant sub-networks to be used are given at Sr. No.4 and 6 of Fig.4.8.

Proceeding from the output node,

$$S(n_4) = BD'S(n_2) \cup DB'S(n_3) \cup BDS(\overline{n_2 n_3}) \tag{4.28}$$

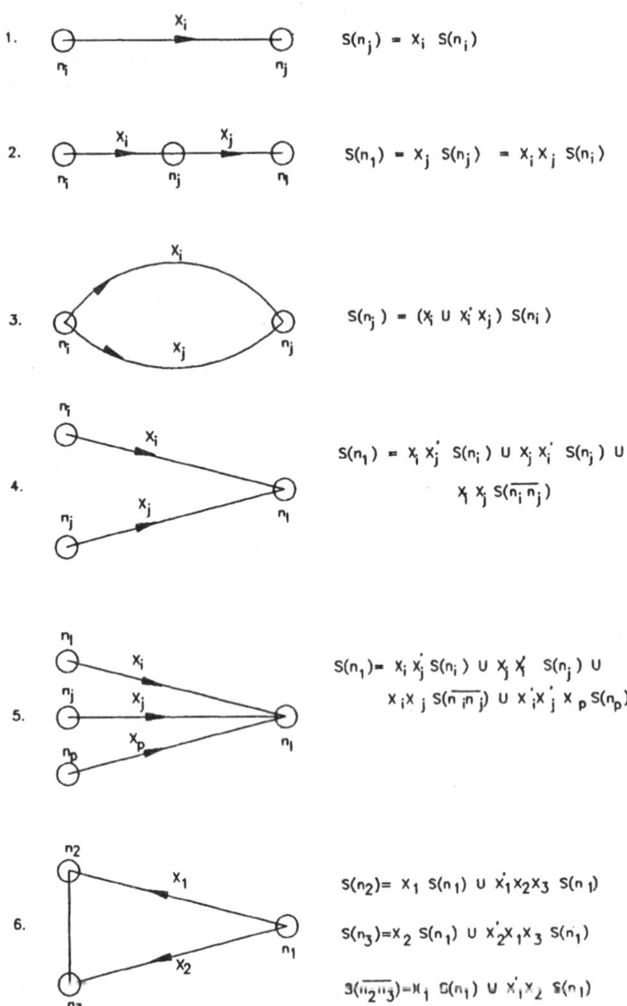

Fig. 4.8: Some common subnetworks.

or,

$$S(n_4) = BD'(A \cup A'CE)\ S(n_1) \cup B'D(C \cup AC'E)\ S(n_1) \cup BD(A \cup A'C)\ S(n_1) \quad ..(4.29)$$

Substituting $S(n_1) = 1$

$$S(n_4) = ABD' \cup A'BCD'E \cup B'CD \cup AB'C'DE \cup ABD \cup A'BCD \quad (4.30)$$

Therefore,

$$R = p_a p_b q_d + q_a p_b p_c q_d p_e + q_b p_c p_d + p_a q_b q_c p_d p_e + p_a p_b p_d + q_a p_b p_c p_d \quad ..(4.31)$$

After a few algebraic manipulations, this expression can of course be shown equal to the one obtained earlier. This method is slightly less economical as compared to Boolean algebra method as far as simplicity of the derived symbolic expression is concerned.

4.8 BAYE'S THEOREM METHOD

In this method a *keystone* component is chosen. This keystone element can have only two states viz, good or bad. When keystone is good it is shorted and a new reduced system is obtained and when keystone is bad then it is opened and again a reduced system is obtained. This process is repeated until the reduced system is series parallel system. Then the reliability of reduced systems is obtained by methods of series parallel reliability evaluation and is multiplied by the probability of proper keystone states and added together to get system reliability.

Example 4.5

Consider the bridge system of Fig.4.7. Let E be the keystone component. Then two reduced systems are, as shown in Fig.4.9. As these reduced systems are series parallel systems, the overall system reliability is obtained as

(a) When E is good

$$R_{s1} = [1-P(A')P(C')]\ [1-P(B')P(D')] \quad (4.32)$$
$$= (1-q_a q_c)(1-q_b q_d)$$

(b) When E is bad

$$R_{s2} = [1-P(A)P(C)]\ [1-P(B)P(D)] \quad (4.33)$$

$$= (1-p_a p_c)(1-p_b p_d)$$

Therefore , system reliability is

$$R = P(E)\, R_{s1} + P(E')\, R_{s2} \tag{4.34}$$
$$= p_e R_{s1} + q_e R_{s2}$$

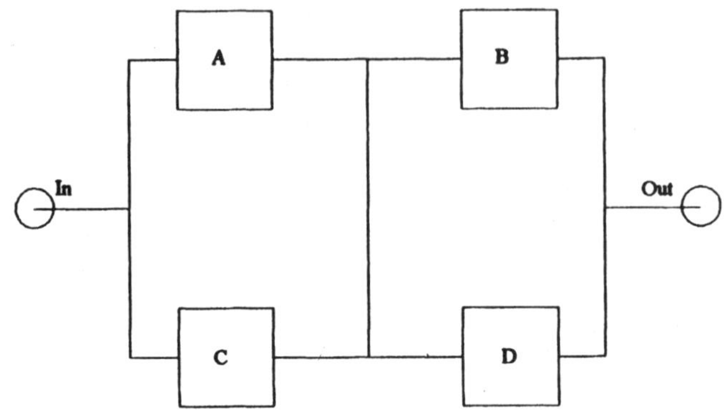

(a) E- Good

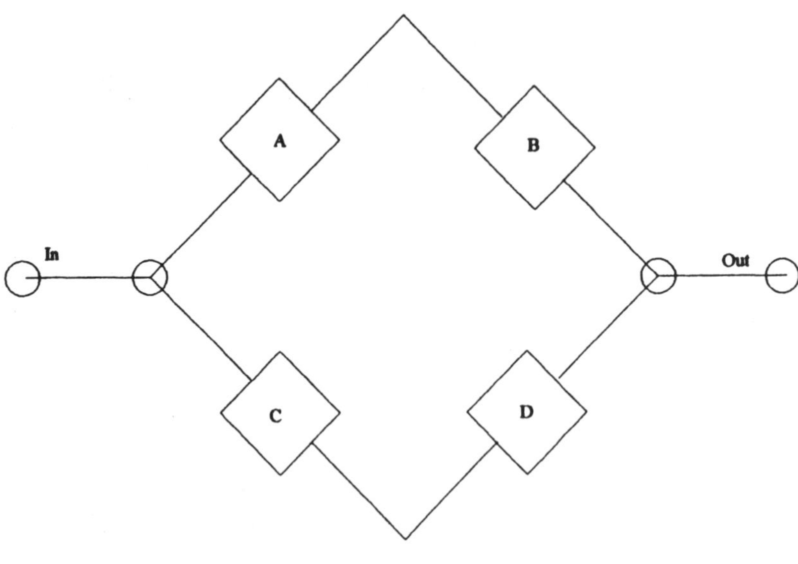

(b) E- Bad

Fig. 4.9: Reduced Networks.

After a few algebraic manipulations, this expression can of course be shown equal to the one obtained by the earlier method.

* * *

Example 4.6

Derive an expression for s-t reliability of the network shown in fig.4.10.

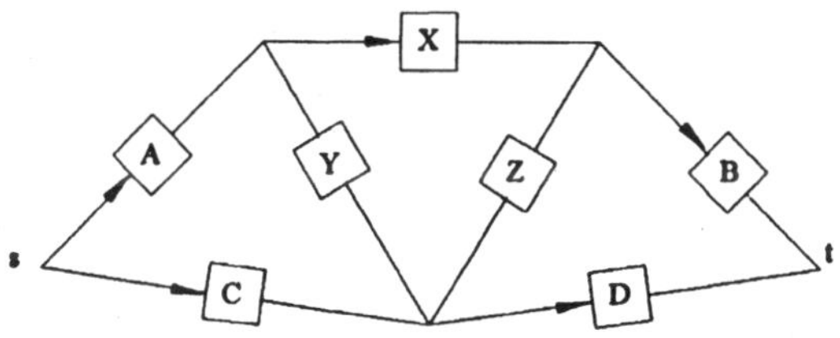

Fig.4.10 Network for Example 4.6.

Solution:

Let us choose element X to be keystone element, in this case. Two reduced networks by considering X-bad and X-good are shown in fig.4.11(a) and (b) respectively.

Fig.4.11(a) is a series parallel network whose reliability is easily seen as:

$$R_1 = (p_c + p_a p_y - p_c p_a p_y)(p_d + p_z p_b - p_d p_z p_b) \qquad (4.35)$$

Fig.4.11(b) is a bridge network which is identical to fig.4.7 provided branch E in fig.4.7 is considered as a group of two parallel branches Y & Z in fig.4.11(b). Hence, R_2 can be written following expression (4.34) as:

$$R_2 = p_e Rs_1 + (1-p_e) Rs_2 \qquad (4.36)$$

where Rs_1 and Rs_2 are given in equations (4.32) and (4.33) respectively and p_e is given as:

$$p_e = p_y + p_z - p_y p_z \qquad (4.37)$$

After derivation of R_1 and R_2, system reliability can be expresed as:

$$R = p_x R_2 + (1-p_x) R_1 \qquad (4.38)$$

* * *

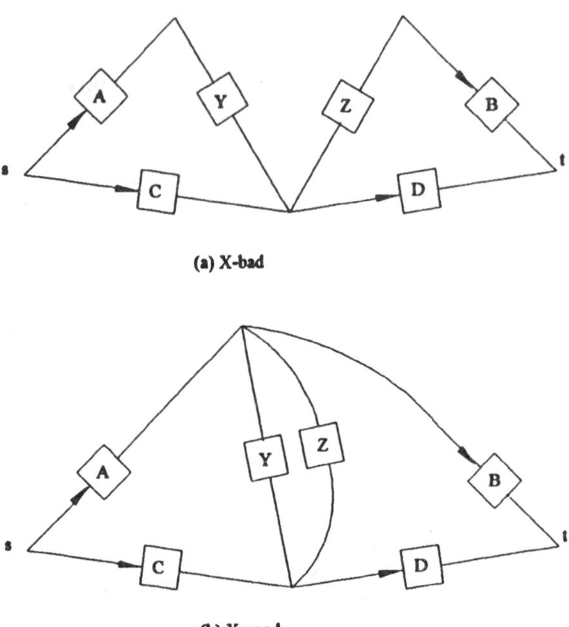

Fig.4.11 Reduced Networks for Example 4.6.

5
RELIABILITY PREDICTION

5.1 INTRODUCTION

Reliability prediction is an essential function in evaluating a system design from its conceptual stage through development and manufacture and also in assisting in controlling changes during the production. Prediction provides a rational basis for design decisions, involving choice between alternative concepts, variations in part quality levels, appropriate application of derating factors and use of proven vs state-of-art methods and other related factors.

An accurate prediction of the reliability of a new product before it is manufactured or marketed, is highly desirable, since with the advance knowledge of its reliability accurate forecasts of support costs, spares requirements, warranty costs, marketability, etc. could be made with reasonable certainity. When the design of an electronic system to perform a complex and demanding job is proposed, it is assumed that the required investment will be justified according to the perfection by which the job is performed for a large number of times by the system. The assumption cannot be justified, when the system fails to perform the job upon demand or fails to perform repeatedly. Hence, in the design of a complex electronic system, sufficient effort is made to obtain reliable system performance.

Amongst the various evolving technologies, electronics evolution is particularly rapid, it is sometimes referred to as an exploding technology. As there is very little time for an orderly evolution of systems, applications suffer most from unreliability. The ratio of the new to tried portions of electronic systems is relatively high and till the new becomes proven and true, its reliability must be suspected.

5.2 PURPOSE

Reliability prediction should be used in formulating design decisions. It should begin early in the design phase and continue during design effort. Early predictions may be based primarily on part counts of known reliability of similar components. As design information becomes available predictions can be updated using stress data on specific parts and reflecting the actual components utilized in design. A flow diagram of the different inputs, interactions and outcome for Reliability Prediction Activity to be purposeful is shown in Fig.5.1.

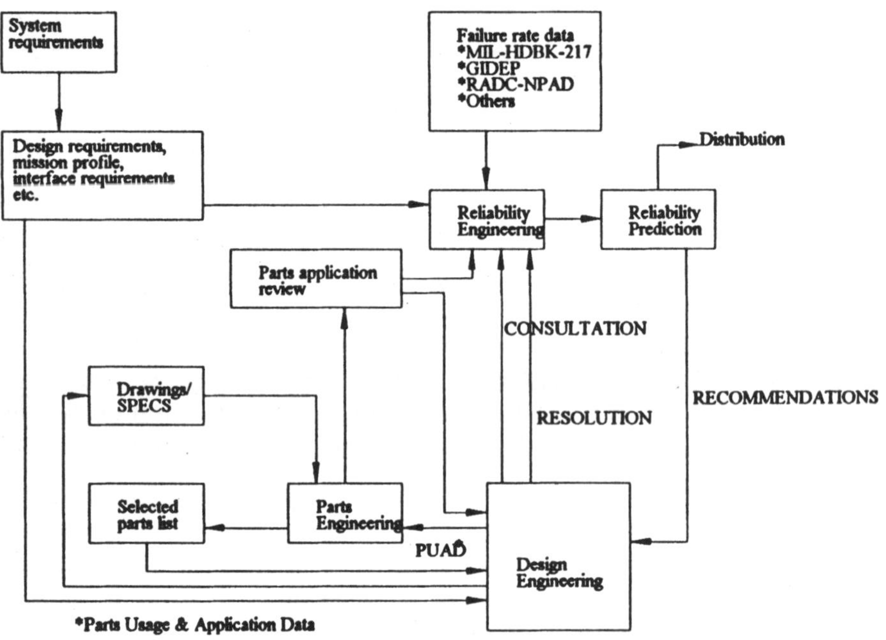

Fig. 5.1:Reliability Prediction Activity

Reliability Prediction has many purposes as under:

1. Basis for selection among competing designs (prediction must use some assumptions and data sources).
2. Disclose critical or reliability limiting items in the design.
3. Sensitivity of design to electrical stress, thermal stress and parts quality.
4. Basis for reliability trade-offs among system components.
5. Describe numerically the inherent reliability of the design.
6. Provide inputs to design review, failure mode effects and criticality analysis (FMECA), mantainability analysis, safety analysis, logistic

support and thermal design.

5.3 CLASSIFICATION

Reliability predictions, as defined herein, are classified as follows:

Type I - Feasibility prediction
Type II - Preliminary design prediction
Type III - Detailed design prediction

5.31 Feasibility Prediction

Feasibility prediction is intended for use in the conceptual phase of item development. During this phase the level of detailed design information is generally restricted to overall aspects of the item. Detailed configuration data generally are limited to that which may be derived from existing items having functional and operational requirements similar to those of the item being developed. Feasibility prediction methods include similar item method, similar circuit method and Active Element Group Method.

5.32 Preliminary Design Prediction

Preliminary design prediction is intended for use in the early detailed design phase. During this phase design configuration data are documented by engineering sketches and preliminary drawings. The level of detailed information available may be restricted to part listings. Stress analysis data are not generally available. Parts Count Method is one such preliminary design prediction method.

5.33 Detailed Design Prediction

Detailed design prediction is intended for use in and subsequent to the detailed design phase. This phase is characterized by drawings which identify all parts, materials, and processes needed to produce the item. Operating stress and temperature analysis data are necessary for each part in the item. The analysis data shall be based on acceptable design analysis and measurement techniques. Parts Stress Method is a detailed design prediction method.

5.4 INFORMATION SOURCES FOR FAILURE RATE DATA

Reliability Prediction is accomplished by solving the reliability model using appropriate failure rates at part or component levels. The sources for these failure rates can be had from MIL-HDBK-217, Non-electronic Parts Reliability Data (NPRD), Government Industry Data Exchange Programme (GIDEP) or derivatives from test on products or data from a large body of devices which are in use in service. Some sources are given below:

* MIL-STD-1670 Environmental Criteria and Guidelines for Air Launched Weapons

* MIL-HDBK-217 Reliability Prediction of Electronic Equipment

* MIL-HDBK-251 Reliability/Design Thermal Applications

* RADC-TR-73-248 Dormancy and Power On-Off Cycling Effects on Electronic Equipment and Part Reliability

* RADC-TR-74-269 Effects of Dormancy on Non-electronic Components and Materials

* LC-78-1Storage Reliability of Missile Material Program, Missile Material Reliability Handbook

* GIDEP Government Industry Data Exchange Program, Summaries of Failure Rates

* NPRD-1 Non-electronic Parts Reliability Data

The failure rates should be corrected for applied and induced stress levels with duty cycles determined by Mission Analysis.

Sufficient Data has been generated by these agencies through a coordinated effort regarding the laboratory and field generated data through inhouse tests and field performance feedback in a systematic manner. These are regularly analysed and updated information is available through the RADC documents like Micro circuits Reliability Data (MDR), Non-electronic Parts Reliability Data (NPRD), etc. The GIDEP consists of different Data banks for Engineering Data, Reliability and Maintainability Data, Meterology Data and Failure experience Data which help in the intended configuration and environment. The Exchange of Authenticated Component Test Data (EXACT) mostly operate amongst the European Countries and some other outside member countries and through this scheme, duplication of testing efforts is minimized. Further, it provides through data updating and analysis, scope to know about the Quality level of components. There are many other types of data generated by the major component and equipment industries who are manufacturing sophisticated electronic parts, equipments and systems for various professional applications, as it is helpful to them to keep track of their performance in various environments and give an indication of their credibility.

5.5 GENERAL REQUIREMENTS

The general requirements for reliability prediction are:

1. *Part Description:* Part and application descriptions shall be provided for any prediction based upon part failure rates.

2. *Environmental Data:* Environmental data affecting part failure rates must be defined. These data include the associated natural and induced environments (Refer to Table 5.1 for typical environments).

TABLE 5.1

Environmental Symbol Identification and Description

Ground, Benign G_B: Nearly zero environmental stress.

Space, Flight S_F: Earth orbital. Approaches Ground Benign conditions. Vehicle neither under powered flight nor in atmospheric reentry.

Ground, Fixed G_F: Conditions less than ideal to include installation in permanent racks with adequate cooling air and possible installation in unheated buildings.

Ground, Mobile G_M: Conditions more severe than those for G_F, mostly for vibration and shock. Cooling air supply may also be more limited.

Naval, Sheltered N_S: Surface ship conditions similar to G_F but subject to occasional high shock and vibration.

Naval, Unsheltered N_U: Nominal surface shipborne conditions but with repetitive high levels of shock and vibration.

Airborne, Inhabited Transport A_{IT}: Typical conditions in transport or bomber compartments occupied by air crew without environmental extremes of pressure, temperature, shock and vibration, and installed on long mission aircraft such as transports and bombers.

Airborne, Inhabited Fighter A_{IF}: Same as A_{IT} but installed on high performance aircraft such as fighters and interceptors.

Airborne, Inhabited Helicopter A_{IH}: Same as A_{IT} but installed on rotary wing aircraft such as helicopters.

Airborne, Uninhabited Transport A_{UT}: Bomb bay, equipment bay, tail, or wing installations where extreme pressure, vibration and temperature cycling may be aggravated by contamination from oil, hydraulic fluid and engine exhaust. Installed on long mission aircrafts such as transports and bombers.

Airborne, Uninhabited Fighter A_{UF}: Same as A_{UT} but installed on high performance aircrafts such as fighters and interceptors.

Airborne, Uninhabited Helicopter A_{UH}: Same as A_{UT} but installed on rotary wing aircraft such as helicopters.

Missile, Launch M_L: Severe conditions of noise, vibration and other environments related to missile launch and space vehicle boost into orbit, vehicle reentry and landing by parachute. Conditions may also apply to installation near main rocket engines during launch operations.

Missile, Captive Carry M_C: Same as A_{UT}, A_{UF} or A_{UH} depending on the applicable aircraft platform.

Missile, Free Flight M_F: Typical conditions of pressure, vibration and temperature experienced in atmospheric flight to target.

---

3. *Part Operating Temperature:* Part temperatures used for prediction purposes shall include the item internal temperature rise as determined by thermal analysis or test data.

4. *Stress Analysis:* Analysis shall be performed to determine the operating stresses to be experienced by each part commensurate with the prediction classification and the design details available. Failure rates shall be modified by appropriate factors to account for the effect of applied stress.

5. *Failure Rates:* Failure rates for all electronic, electrical, electro-mechanical and mechanical items are required for each significant event and environment. Basic failure rates from data sources must be modified with appropriate factors to account for the specific item application under consideration. Operating failure rates, non-operating failure rates and storage failure rates for parts shall be derived from the available data sources.

5.6 PREDICTION METHODOLOGIES

There are different methods of predicting the reliability of the electronic equipment or system depending on the period when the information is required and to what level information/data is available that is authentic. The predictions are mostly based on experience, data from similar items or those produced in the same manner. Extreme caution must be exercised in ascertaining the similarity of items and degree of similarity in the conditions of use. It is essential to establish common ground rules for techniques and data sources on the formulation of reliability models and prediction so that there is uniform application and interpretation.

Reliability Prediction can be made by using the following methods:

5.61 Similar Item Method

This prediction method utilizes specific experience on similar items. The most rapid way of estimating reliability is to compare the item under consideration with a similar item whose reliability has previously been determined by some means and has undergone field evaluation. This method has a continuing and meaningful application for items undergoing orderly evolution. Not only is the contemplated new design similar to the old design, but small differences can be easily isolated and evaluated. In addition, difficulties encountered in the old design are signposts to improvements in the new design.

5.62 Similar Circuit Method

The similar circuit method should be considered if a similar item comparison cannot be made. This method utilizes specific experience on similar circuits such as oscillators, discriminators, amplifiers, modulators, pulse transforming networks, etc. This method is employed either when only a circuit is being considered or the similar item method cannot be utilized. The most rapid way of estimating reliability is to compare the circuits of the item under consideration with similar circuits whose reliability has previously been determined by some means and has undergone field evaluation. Individual circuit reliabilities can be combined into an item reliability prediction. This method has a continuing and meaningful application for circuits undergoing orderly evolution.

5.63 Active Element Group Method

The Active Element Group (AEG) method is termed as feasibility estimating procedure because it is useful for gross estimates of a design in the concept formulation and preliminary design stages. Only an estimate of the number of series AEG's required to perform the design function is needed. The AEG method relates item functional complexity (active element groups) and application environment to failure rates experienced in fleet usage.

5.64 Parts Count Method

The parts count method is a prediction method used in the preliminary design stage when the number of parts in each generic type class such as capacitors, resistors, etc., are reasonably fixed and the overall design complexity is not expected to change appreciably during later stages of development and production. The parts count method assumes the time to failure of the parts as exponentially distributed(i.e. a constant failure rate).

5.641 Procedure

The item failure rate can be determined directly by the summation of part failure rates if all elements of the item reliability model are in series or can be assumed in series for purposes of an approximation. In the event the item reliability model consists of non-series elements (e.g. redundancies, alternate modes of operation), item reliability can be determined by summing part failure rates for the individual elements and calculating an equivalent series failure rate for the non-series elements of the model.

The information needed to support the parts count method includes:

(a) Generic part types (including complexity for microelectronics)
(b) Part quantity
(c) Part quality levels (when known or can be assumed)
(d) Item environment

The general expression for item failure rate with this method is:

$$\lambda_{item} = \sum_{i=1}^{i=n} N_i \lambda_{Gi} \Pi_{Qi} \qquad (5.1)$$

for a given item environment.
Where

λ_{item} = total failure rate
λ_{Gi} = generic failure rate for the ith generic part
Π_{Qi} = quality factor for the ith generic part
N_i = quantity of the ith generic part
n = number of different generic part categories

The above equation applies to an entire item being used in one environment.

If the item comprises several units operating in different environments (such as avionics with units in airborne, inhabited, fighter (A_{IF}) and uninhabited, fighter (A_{UF}) environment, then this equation should be applied to the portions of the item in each environment. These 'environment item' failure rates should be added to determine total item failure rate.

Quality factors are to be applied to each part type where quality level data exists or can be reasonably assumed. Multi-quality levels and data exist for parts, such as microelectronics, discrete semiconductors, and for established reliability(ER) resistors and capacitors. For other parts such as non-electronics, $\Pi_Q = 1$ provided that parts are procured in accordance with applicable parts specifications.

Table 5.2 shows typical parts count method prediction of a transmitter unit.

5.65 Part Stress Analysis

This is a detailed design prediction method. This is characterized by drawings which identify all parts, materials, processes needed to produce the equipment or system. This method is applicable when most of the design is complete and a detailed parts list including part stresses are available. It is also used during later design phases for reliability tradeoffs vs. part selection and stresses. Normally there are no assumptions necessary and details about the parts used, stress derating, their quality factors and the operating environment are all fully known. The failure rate models for a broad variety of parts used in electronic equipment are utilized for arriving at the failure rate figures. The Parts Stress Method is an accurate method of Reliability Prediction prior to the measurement of reliability under actual or simulated conditions.

Major parts that are used in electronic equipment which have an influence on the reliability of the system and their behaviour is dependent on the stresses are:

* Microelectronics
* Discrete Semiconductors
* Electron Tubes
* Lasers
* Resistors
* Capacitors
* Inductive Components
* Rotary Components
* Relays
* Switches
* Connectors

* Wires & Printed Wiring boards
* Connections
* Miscellaneous

TABLE 5.2

100 W Transmitter Unit

Component	Total No	Failure rate λ_px10⁶	Product x10⁶
Resistos(Fixed)			
Carbon	4	0.033	0.132
Metal film	30	0.046	1.380
Resistors(variable)			
Non Wire wound	6	0.900	5.400
Capacitors(fixed)			
Ceramic	40	0.054	2.160
Tantalum	4	0.042	0.168
Electrolytic	8	0.660	5.280
Capacitors(variable)			
Air Dielectric	4	5.70	22.800
Diodes			
General purpose	2	0.031	0.062
Transistors(NPN)	8	0.160	1.280
ICs	1	1.085	1.085
RF Coils	8	0.011	0.088
Connectors			
Multipin	6	0.051	0.306
RF Coaxial	4	0.051	0.204
Microstriplines	10	0.072	0.720
Miscellaneous			
Lamps	4	1.000	4.000
Air movers	2	2.400	4.800
Circulator	1	0.240	0.240
Meters	4	10.000	40.000
Connections	350	0.027	9.450
		Total :	99.56

Normally there are three major factors that influence the failure rate of the part.

I Part Quality
II The use environment
III The Thermal Aspect

The quality factor of the part has a direct effect on the Part failure rate and appears on the Part Models as Π_Q.

A small Regulated Power Supply Circuit as given in Fig.5.2 is taken as an example for demonstrating the methodology of approach using Part Stress Analysis to arrive at the failure rate or MTBF figure. As per the procedure, the types of components used in the circuit are listed with their values, ratings and associated specifications. Based on the type of component the appropriate model expression from MIL-HDBK-217 is utilized.

The applicable model parameters based on the stress and other related factors are obtained from the relevant tables and substituted in the corresponding expressions. The failure rate for each part is obtained and considering all parts as a series system (because the absence of any part

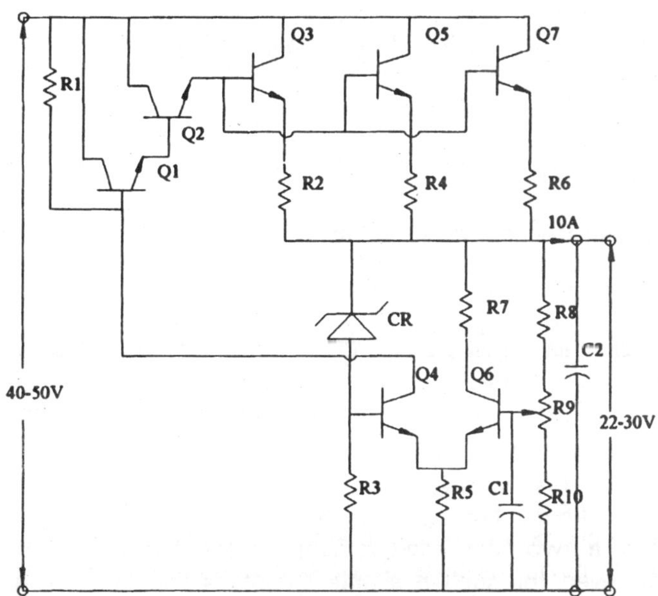

Fig. 5.2: Regulated power supply.

will not make the circuit functional) the total failure rate (or MTBF) is obtained as a summation taking into account the interconnections and printed wiring board configuration.

The different types of components used in the circuit are

RF Power Transistor, NPN, Silicon	:	4
Medium Power Transistor, NPN, Silicon	:	2
Low Power Transistor, NPN	:	1
Zener Diode, Silicon	:	1
Resistor, Power 0.1	:	3
Resistor, film	:	5
Capacitor-electrolytic	:	2
Variable resistor	:	1

The model expressions are as under:

Transistor $\lambda_p = \lambda_b(\Pi_E \Pi_A \Pi_Q \Pi_R \Pi_S \Pi_C)$ failures/10^6 hrs (5.2)

Zener $\lambda_p = \lambda_b(\Pi_E \Pi_A \Pi_Q)$ failures/10^6 hrs (5.3)

Resistor $\lambda_p = \lambda_b(\Pi_E \Pi_R \Pi_Q)$ failures/10^6 hrs (5.4)

Variable Potentiometer

$\lambda_p = \lambda_b(\Pi_{taps} \Pi_Q \Pi_R \Pi_V \Pi_C \Pi_E)$ failures/10^6 hrs (5.5)

Capacitor $\lambda_p = \lambda_b(\Pi_E \Pi_Q \Pi_{cv})$ failures/10^6 hrs (5.6)

PWB $\lambda_p = \lambda_b(N \Pi_E)$ failures/10^6 hrs (5.7)

Connections $\lambda_p = \lambda_b(\Pi_E \Pi_Q)$ failures/10^6 hrs (5.8)

The details of the components with their applicable stresses based on Circuit Analysis are given in Table 5.3. Substituting the appropriate values from MIL-HDBK-217, in the corresponding model expressions the failure rate values are calculated taking into consideration the number of identical components with similar stress factors etc., the details of which are given in Table 5.4.

The total failure rate for the circuit using Part Stress Analysis works out to 0.606 x 10^{-6} hrs, whereas that by Parts count method it is calculated as 1.45 x 10^{-6} hours. From this, it can be observed that in this case there is more than a two fold improvement on the failure rate or MTBF figure. However, even for such a simple circuit as the one given in Fig.5.2 the manual work associated with circuit analysis and calculation of values of failure rates refering to the appropriate MIL-HDBK-217 tables with applicable Π factors requires about one full man day as compared to less than an hour for calculations by the Parts Count Method. This is the price to be paid for Parts Stress Method which is more refined and leads to better and accurate prediction.

Table 5.3

Details of Circuit Parts with Actual and Rated Stresses

S.No	Code	Type	Applied Stress	Max.Ratings
1	Q1	2N 1479	1.00W	5.00 W
2	Q2	2N 3055	10.00W	117.00W
3	Q3	2N 3055	66.00W	117.00W
4	Q4	2N 3053	0.50W	5.00 W
5	Q5	2N 3055	66.00W	117.00W
6	Q6	2N 3053	0.50W	5.00 W
7	Q7	2N 3055	66.00W	117.00W
8	R1	1.2K	0.39W	1.00 W
9	R2	0.1K	1.16W	2.50 W
10	R3	2.0K	0.16W	0.25 W
11	R4	.1K	1.16W	2.50 W
12	R5	570	0.50W	1.00 W
13	R6	.1K	1.16W	2.50W
14	R7	270	0.06W	0.25W
15	R8	1 K	0.10W	0.25W
16	R9	1 K Pot	0.10W	0.25W
17	R10	1 K	0.10W	0.25W
18	C1	1 MF	18.00V	50.00V
19	C2	100 MF	30.00V	63.00V
20	CR	BZV 58 C12	0.10W	0.40 W

TABLE 5.4

Failure Rate Calculation by Parts Stress Analysis

Part Ref.	Failure rate $\lambda_p \times 10^6$	No. of similar Parts	Total failure rate $\lambda_p \times 10^6$
Q3, Q5, Q7	0.04200	3	0.126
Q2	0.00430	1	0.0043
Q1	0.05600	1	0.056
Q4, Q6	0.00315	2	0.0063
CZ	0.01100	1	0.011
R2, R4, R6	0.03230	3	0.097
R8, R10	0.01150	2	0.023
R1, R5	0.00280	2	0.0056
R3, R7	0.00840	2	0.0168
R9	0.036	1	0.036
PWB	0.000576	1	0.000576
Connections	0.0055	40	0.2244
		Total :	0.606

5.7 SOFTWARE PREDICTION PACKAGES

Presently there has been significant change in the approach for Reliability Prediction Methods. A microcomputer revolution has taken place, and personal computer systems like, PC, PC/XT, PC/AT have flooded the market. Manual calculations and data generation have become time consuming and the present day computer having multi- tasking, multi-user features with interactive facility and powerful software packages have helped to unburden the design and reliability engineer. Most of the softwares have been developed on Microcomputer Systems having a 256 KB memory, 2 floppies, 10 MB Hard disk, Monitor (Colour Display) and printer with the cost of software being nominal. The use of the computer as a tool for all these and availability of many sources for software on 5.25"/ 3.50" floppies assure portability and easy access etc. The language mostly used is dBase III/ IV.

Some of the commonly available software packages relating to Reliability Prediction are:

* Predictor
* 217 Predict
* HARP, (Parts Count using Standard failure rate lists other than MIL-HDBK-217)
* RELECALC 217
* IRAS

These are available from different non-defence US vendors.

A software package (for performing reliability predictions) named ORACLE was developed to mechanise the implementation of MIL-HDBK-217. It is interactive in nature and structured. A few days of training for a reliability engineer would help in productively using the package. The program provides queries to the user, guides in program execution and development of proper data inputs. The original version of ORACLE had its genesis under an Army Project. It was modified and updated by Airforce at RADC and was subsequently called RADC-ORACLE.

The main features of the software is to help the reliability engineer in using it as a tool for reliability prediction of electronic equipment and systems and the factors considered for MIL-HDBK-217 implementation are:

(a) Piece parts making up the system and their breakdown into modules
(b) Part dependent parameters for each piece part
(c) Failure rate models and failure rate confirmation covered by it for each piece part

(d) Part application dependant parameters for each part
(e) Contingency parameters (treatment of default values, trade-off analysis, redundancy)
(f) Forms of prediction results
(g) Structuring of ORACLE outputs to meet the data item description

5.8 ROLE AND LIMITATION OF RELIABILITY PREDICTION

Reliability Prediction should be timely, if it is to be of value. However, lot of difficulties will be encountered if it is needed earlier. The early predictions made on the basis of little knowledge can form a rationale for changing the method of production. Reliability cannot be known with certainty, but a lot of knowledge about it can be accumulated during a short early period of its life. Prediction can become a process for designing for future. The process of prediction, action, measurement and repetition of the cycle should be a continuous programme of development. The two trends in prediction are:

(a) To gain better record of class characteristics in more usable and realistic forms
(b) To develop improved techniques for applying consequent knowledge to prediction in appropriate confidence settings

The current state-of-art in prediction rests at the level of development of these data and techniques.

Practical limitations depend on data gathering and technique complexity which are due to difficulty of accumulation of performance field data on newer class of devices in actual use environment and the derivation of their failure rates with valid confidence values.

Applicability of failure rate data is based on past system usage and relevance in newer environments and future concepts. Relevance of similarity of system use in one environment compared to that in another environment is questionable. Variants may be different users, operators, maintenance practices, measurement techniques and detection of failure.

The one Fundamental Limitation is the accumulation of data with known validity for new application while the other is complexity of prediction technique.

6

RELIABILITY ALLOCATION

6.1 INTRODUCTION

In a complex system, it is necessary to translate overall system characteristics, including reliability, into detailed specifications, for the numerous units that make up the system. The process of assigning reliability requirements to individual units to attain the desired system reliability is known as reliability allocation. The allocation of system reliability involves solving the basic inequality.

$$f(R_1^*, R_2^*, \ldots\ldots, R_n^*) \geq R^* \tag{6.1}$$

where,

R^*: system reliability requirement

R_i^*: i^{th} subsystem reliability requirement

For a series system, the above equation is simplified as

$$R_1^* . R_2^* \ldots\ldots R_n^* \geq R^* \tag{6.2}$$

Theoretically, this equation has an infinite number of solutions, assuming no restrictions on the allocation. The problem is to establish a procedure that yields a solution by which consistent and reasonable reliabilities may be allocated.

Reliability requirements determined through an allocation procedure would be more realistic, consistent, and economically attained than those obtained through subjective or haphazard methods, or those resulting from crash programs initiated after unfortunate field experiences.

Some of the advantages of the reliability allocation program are:

1. The reliability allocation program forces system design and development personnel to understand and develop the relationships between component, subsystem, and system reliabilities. This leads to an understanding of the basic reliability problems inherent in the design.

2. The design engineer is obliged to consider reliability equally with other system parameters such as weight, cost, and performance characteristics.

3. Reliability allocation program ensures adequate design, manufacturing methods, and testing procedures.

The allocation process is approximate and the system effectiveness parameters, such as reliability and maintainability apportioned to the subsystems, are used as guidelines to determine design feasibility. If the allocated parameters for a system cannot be achieved using the current technology, then the system must be modified and the allocations reassigned. This procedure is repeated until an allocation is achieved that satisfies the system requirements (Fig.6.1).

Apportionment has its greatest value at the first level of breakdown of a system into its major subsystems. It is also rather necessary at this level for, frequently each of the major subsystems is produced by a separate division or agency. The physical make up of the system plays an important role in determining how far down into the assembly we would approtion our system requirements.

6.2 SUBSYSTEMS RELIABILITY IMPROVEMENT

In any complex system there may always be some subsystems whose reliability is known to us a-priori. On the other hand in an era of fast technological innovations, a sophisticated system will often have several units which are to be used first time and no reliability predictions are possible for these units. In the former category there may be several units in which reliability improvement is possible, while there may be some units which we like to use as such without putting in any effort for their reliability improvement. This may be because we do not intend altering the design of these subsystems for several reasons such as cost of redesign, non-availability of alternatives, restricted time schedule for mission completion, etc. It is obviously desirable not to include such subsystems in the purview of reliability allocation as we will not be in a position to incorporate the allocated values in any manner whatsoever. All such subsystems are, therefore, identified and the required system reliability goal is divided by the product of the reliabilities for such units. The new goal is thus established which is to be met by the remaining units. Because of the simplicity in these calculations, without any loss of generality, we assume henceforth that there is no such unit in the reliability allocation program.

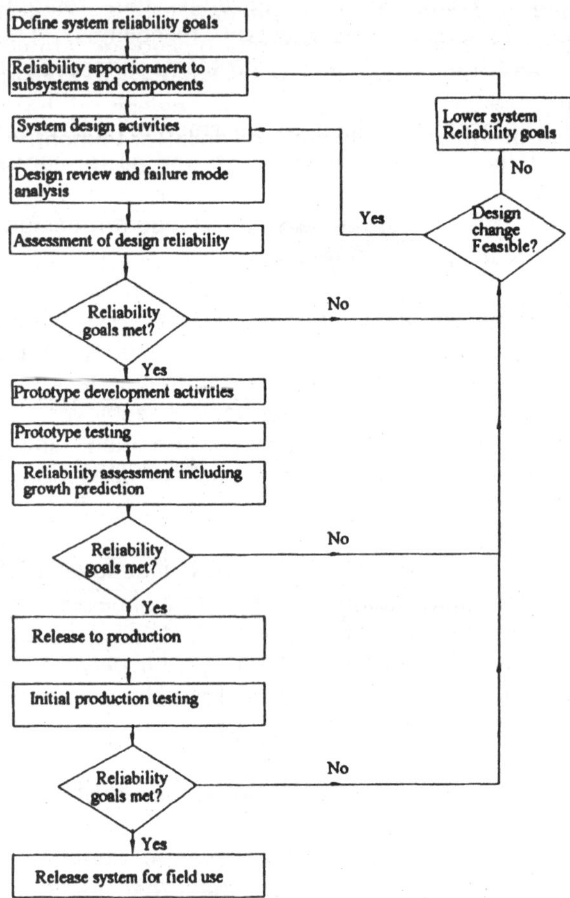

Fig. 6.1: Reliability allocation process

Let there be N subsystems in the system whose reliability goal is R^*. Out of these N subsystems, let there be $m(\leq N)$ subsystems whose estimated or predicted reliabilities are known and reliability improvements are considered feasible. Let $n(=N-m)$ be the remaining subsystems whose estimated or predicted reliabilities are not known and we have to allocate reliabilities to these subsystems considering parameters such as cost, complexity, state of art, etc. These n units are beyond the purview of this section and the problem of reliability allocation for this group is discussed in the next section.

As we are planning to decompose the problem of reliability allocation to two independent sub-problems involving m and n(=N-m) units respectively, it is necessary to partition the reliability goal into two sub-goals. For the first m components, the goal is taken as $(R^*)^{m/N}$ and for the remaining n components in the second category, the goal is taken as $(R^*)^{n/N}$. For simplicity of notation

Let

$$R' = (R^*)^{m/N} \tag{6.3}$$

and

$$R'' = (R^*)^{n/N} \tag{6.4}$$

For the purpose of this section, therefore, the statement of the problem is:

A system has m components with predicted reliabilities R_1, R_2, ...,R_m. The desired system reliability is R'. Allocate new reliability values as R_1^*, R_2^*, ...,R_m^*.

We discuss below two methods for the solution of this problem.

6.21 Basic Allocation Method

In this method, reliability of every constituent subsystem is improved so as to achieve the reliability goal. The basic philosophy of the method is to decrease the failure rate of each subsystem by the same factor.

Let

λ_s^* : system failure rate
λ_j : predicted failure rate for j^{th} subsystem
λ_j^* : allocated failure rate for j^{th} subsystem

The steps of this method are:

(i) If λ_s^* is the system failure rate requirement, allocated unit failure rates λ_j^* must be chosen so that

$$\lambda_1^* + \lambda_2^* + \ldots + \lambda_n^* \leq \lambda_s^* \tag{6.5}$$

(ii) Relative unit weights are computed from observed failure rates as:

$$w_j = \lambda_j \Big/ \sum_{j=1}^{m} \lambda_j \tag{6.6}$$

(iii) Since w_j represents the relative failure vulnerability of jth unit and $\Sigma w_j = 1$,

$$\lambda_j^* = w_j \lambda_s^* \tag{6.7}$$

(iv) If reliability values are to be allocated,

$$R_j^* = (R')^{w_j} \tag{6.8}$$

Example 6.1

A system has four serial units with predicted failure rates of 0.002, 0.003, 0.004 and 0.007/hr. If system failure rate is desired to be 0.010, allocate failure rates to four units.

Solution

$$\Sigma\lambda_j = 0.002 + 0.003 + 0.004 + 0.007 = 0.016$$

Therefore,

$$w_1 = 0.002/\, 0.016 = 0.1250$$
$$w_2 = 0.003/\, 0.016 = 0.1875$$
$$w_3 = 0.004/\, 0.016 = 0.2500$$
$$w_4 = 0.007/\, 0.016 = 0.4375$$

Hence,

$$\lambda_1^* = 0.1250 \times 0.010 = 0.001250$$
$$\lambda_2^* = 0.1875 \times 0.010 = 0.001875$$
$$\lambda_3^* = 0.2500 \times 0.010 = 0.002500$$
$$\lambda_4^* = 0.4375 \times 0.010 = 0.004375$$

* * *

Example 6.2

If the system reliablity requirement for the system considered in example 6.1 is 0.90, allocate relaibilities to four serial units.

Solution

Unit weights have already been computed in example 6.1. Hence, allocated relaibilities are directly computed as:

$R_1^* = (R')^{w_1} = (0.90)^{0.1250} = 0.987$

Similarly, $R_2^* = 0.980$, $R_3^* = 0.974$ and $R_4^* = 0.955$

* * *

6.22 Minimum Effort Method

In the previous method, reliability of each subsystem was expected to be improved to achieve the system reliability goal. This implies even high reliability components have to be improved further. It is well known that cost of improving reliability for such components is prohibitive. We therefore, describe a method below which has been proved to require minimum effort under certain assumptions.

To increase R_i to R_i^* obviously needs an effort which would be some function of the number of tests, amount of manpower applied to the task and so forth. Let the effort function $G(R_i, R_i^*)$ be a measure of the amount of effort needed to increase the reliability of ith subsystem from R_i to R_i^*. Assuming that the nature of the effort function is the same for all subsystems, the suggested method is outlined in the following steps:

1. The estimated or predicted reliabilities are arranged in an ascending order. Having done this, assume that

   $$R_1 \le R_2 \le \ldots \le R_m \qquad (6.9)$$

2. The reliabilities of first k components are increased to the same value R_o. The reliabilities R_{k+1}, R_{k+2}, ..., R_m of the remaining (m-k) subsystems are left unchanged.

3. The number k is the maximum value of j, such that

   $$R_j < [R' / \prod_{i=j+1}^{m+1} R_i]^{1/j} = r_j \text{ (say)} \qquad (6.10)$$

where $R_{m+1} = 1$ (by definition) (6.11)

For illustration, we first calculate value of $[R' /(R_2R_3...R_m)]^{1/1} = r_1$. If $R_1 < r_1$, then R_1 will have to be increased to the value of R_o. Then we calculate the value of $[R' /(R_3R_4.....R_m)]^{1/2} = r_2$. Now if $R_2 < r_2$, R_1 and R_2 both will have to be increased to a common value. The procedure is continued as long as $R_j < r_j$.

4. The value of R_o, the allocated reliability for first k units, is determined as:

$$R_o = [R' / \prod_{i=k+1}^{m+1} R_i]^{1/k} \tag{6.12}$$

where $R_{m+1} = 1$

Example 6.3

A system consists of three units connected in series with reliabilities $R_1=0.70$, $R_2=0.80$ and $R_3=0.92$. It is desired that the reliability of the system be 0.65. How this is apportioned among the three units, using minimum effort method?

Solution

The component reliabilities are correctly arranged in this case.

$$r_1 = 0.65/(0.80)(0.92) = 0.883 > 0.7$$
$$r_2 = (0.65/0.92)^{1/2} = 0.841 > 0.8$$
$$r_3 = (0.65)^{1/3} = 0.866 < 0.92$$

Hence, minimum effort method suggests not to improve the reliability of third component, but improve the first two component reliabilities to a value R_o which is given as:

$$R_o = (0.65/0.92)^{1/2} = 0.841$$

Hence, $R_1^* = R_2^* = 0.841$ and $R_3^* = R_3 = 0.92$

* * *

6.3 APPORTIONMENT FOR NEW UNITS

Very often reliability is to be allocated to subsystems for which no estimated or predicted reliability values are known. It can be logically done keeping in view factors like complexity, cost, state of art, maintenance, time of operation. In this section, a series logic diagram is assumed. If the modules are connected in parallel in any subsystem to provide redundancy; the parallel unit is considered as one entity for the purpose of reliability allocation. This approach allows us to retain the validity of series structure.

The problem of reliability allocation can thus be to apportion the reliability goal R" to n units such that:

$$\prod_{j=1}^{n} R_j^* \geq R" \qquad (6.13)$$

If λ_j^* is the allocated failure rate for j^{th} subsystem and $\lambda^{\wedge}$ is the required failure rate for the system, the above equation is equivalent to

$$\sum_{j=1} \lambda_j^* \leq \lambda^{\wedge} \qquad (6.14)$$

As λ_j^* is obviously a fraction of the total failure rate,

let $\lambda_j^* = w_j \lambda^{\wedge}$ (6.15)

where w_j is the weightage factor for j^{th} subsystem. These weightage factors have obviously to be defined in such a manner so as to have

$$\Sigma w_j = 1 \qquad (6.16)$$

To make sure that the above equation is satisfied, we define w_j, in terms of proportionality factors Z_j's as

$$w_j = Z_j / \Sigma Z_j \qquad (6.17)$$

Also, R_j^* can be expressed as:

$$R_j^* = (R")^{w_j} \qquad (6.18)$$

The proportionality factor Z_j for j^{th} subsystem is defined in terms of various factors based on which reliability is desired to be allocated. A discussion of these factors follows. It may, however, be kept in mind that higher Z_j implies higher w_j which implies higher allocated failure rate and hence lower allocated reliability.

6.31 Reliability Allocation Factors

1. *Complexity*

In any system, different subsystems in general have widely varying complexity. As reliability of any module heavily depends upon the number of components comprising it, reliability allocation should have a strong dependence on complexity. It is known that the failure rate of any module is the sum of the failure rates of the constituent components. As a first approximation, therefore, it is logical to have allocated failure rate of any module proportional to the number of components comprising it.

Hence, $Z_j \alpha K_j$ (6.19)

where, K_j is the complexity factor for j^{th} subsystem. These complexity factors are generally measured as the number of Active Element Groups, AEG's.

2. *Cost*

Consideration of cost factor in the reliability allocation program is important as the cost increment for reliability improvement for a relatively costly subsystem is often very large.

The apportioned reliabilities are supposed to be demonstrated also and demonstration of a high reliability value for a costly system may be extremely uneconomical. From this point of view also, a relatively lower value of reliability is desired to be allocated to a costlier subsystem. As higher failure rate is desired to be allocated to a costlier system,

$Z_j \alpha C_j$ (6.20)

where C_j is the cost for j^{th} subsystem.

3. *State of Art*

If a component has been available for a long time and has experienced an extensive development program including failure analysis and corrective action of deficiencies, it may be quite difficult to further improve its reliability even if the reliability is considerably lower than desired. Other components which have initially high reliabilities may be further improved relatively rather economically.

As state of art is the possibility of achieving improvement, the value of this factor is an engineering judgement. When no improvement can be achieved, a factor of 1 is chosen and for the subsystem where lot of improvement is possible, a larger value S_j is chosen. Obviously a larger S_j means higher reliability and hence a lower failure rate. Therefore,

$Z_j \alpha 1/S_j$ (6.21)

4. *Redundancy Introduction*

It is suggested that the possibility of redundancy introduction should also find a place in reliability allocation. A stage where it is feasible to use redundant modules can offer itself for higher reliability allocation rather easily. It is known that when two components are connected in active parallel redundancy, the failure rate of the combination reduces to 2/3 of its value for a single component. This figure can reduce further for more than two units in parallel. Also the exact number, even if it is more than two, cannot be known at the design stage. Therefore,

$$Z_j \alpha \ F_j \tag{6.22}$$

The value of factor F_j is taken as 2/3 if j^{th} subsystem can have units connected in parallel and is taken as 1 otherwise.

5. *Maintenance*

A component which is periodically maintained or one which is regularly monitored or checked and repaired as necessary will have, on an average, a higher availability than one which is not maintained. The quantification of this factor is also an engineering judgement. For non-maintainable subsystems a factor of 1 is chosen and for the subsystems where maintenance is possible, a larger value, M_j is chosen. As stipulated, larger M_j implies higher availability enabling us to apportion relatively lower reliability corresponding to a higher failure rate. Therefore,

$$Z_j \alpha \ M_j \tag{6.23}$$

6. *Time of Operation*

If T is the mission time and also the operating time of all subsystems, time of operation need not be considered in reliability allocation. However, for a sophisticated mission, it is probable that some subsystems are required to operate for periods less than the mission time.

It is logical to apportion relatively lower reliability to subsystems whose operating time is less than the mission time. Hence,

$$Z_j \alpha \ 1/d_j \tag{6.24}$$

where d_j can be defined as the duty ratio for j^{th} subsystem i.e. the fraction of the mission time for which j^{th} subsystem operates. So,

$$d_j = t_j/T \tag{6.25}$$

6.32 Reliability Allocation Formula

After a consideration of various factors, we are now in a position to suggest the following formula for reliability allocation:

$$Z_j = \frac{K_jC_jF_jM_j}{S_jd_j} \tag{6.26}$$

The proportionality sign has been replaced by equality without any loss of generality as any constant will cancel out during computation of weight factors.

The discussion on the use of various factors now follows:

1. K_j is the complexity factor for j^{th} subsystem. The value of K_j is higher for a more complex system. It is logical to choose this factor proportional to the number of AEG's, if possible. Otherwise, of course, relative values have to be assigned using engineering judgement.

2. C_j is the relative cost for j^{th} subsystem. A knowledge of the subsystems generally enables us to guess their costs and hence relative costs. Otherwise, a scale can be used for judgement of the system designer.

3. F_j is the factor which considers whether redundancy is possible to be incorporated in j^{th} subsystem. As already discussed, $F_j = 2/3$ if redundancy is possible and $F_j = 1$ if no redundancy is possible.

4. M_j is the maintenance factor. It is taken to be 1 in case of non-maintainable systems and a larger value is assigned for other systems depending upon their maintenance capabilities. A scale of 1 : 10 may be appropriate for most situations.

5. S_j is the state of art factor which is chosen to be 1 if no improvement is expected in the unit and a larger value if lot of improvement is possible. The state of art factor should also consider our inherent expectations of reliability. A class of components is at times known to be more reliable than the other.

   It may not be possible to determine the state of art factors in an absolute sense and only a relative scale has to be used. As unreliability comes out to be directly proportional to this factor, a high relative scale may not be appropriate particularly when the reliability goal itself is not very high.

6. d_j is the duty cycle for j^{th} subsystem. A knowledge of the functional relationship of various subsystems to the system should usually help us to write the values of these factors without any difficulty.

All the factors are included in the suggested relation for generalization. If any factor is considered insignificant for a particular system or if any factor is not valid in a particular case, this can be easily incorporated in the formula by taking unity as the value for such factor(s) for all the subsystems. For example, if it is not possible to use redundancy for any subsystems, F_j 's $=1$; if it is not possible to carry out the maintenance work for any subsystem, M_j 's $=1$, if all the subsystems are operating for the complete mission time, d_j 's $=1$; and so on.

After calculating Z_j 's for all stages, it is a simple exercise to calculate the weightage factors and hence the values of apportioned reliabilities.

Example 6.4

Consider a system having 10 subsystems whose reliability is desired to be 0.70. The estimated reliabilities of the first five subsystems are 0.95, 0.90, 0.94, 0.96, and 0.98 respectively. The reliabilities of the first two of these cannot be improved while the remaining three are available for possible improvement. The estimated reliabilities of the last five subsystems are not known but the following data are known about these:

1. Subsystems 7 and 8 operate for 75% and 50% of the mission time respectively. All other subsystems operate for complete mission time.
2. Redundancy can be used at subsystems 6 and 10 only.
3. Maintenance is not possible for any of the subsystems.
4. The values of complexity factor, cost factor and state of art factor for these subsystems are:

j	K_j	C_j	S_j
6	6	2	1.0
7	5	3	4.0
8	3	2	3.0
9	7	4	5.0
10	2	6	2.0

Reliability is to be allocated for the subsystems of this system.

Solution

Subsystems 1 and 2 have their estimated reliabilities knwon which cannot be improved further. Therefore, we can take these two subsystems out of the purview of reliability allocation by modifying the reliability goal as:

$R^* = 0.70/(0.95)(0.90) = 0.819$

The reliability goals for subsystems 3-5 and for subsystems 6-10 are established as:

$R' = (R^*)^{m/N} = (0.819)3/8 = 0.928$

$R'' = (R^*)^{n/N} = (0.819)5/8 = 0.883$

First consider the problem of reliability allocation to the first group of subsystems. We decide to use minimum effort method. Hence,

$r_3 = [0.928/(0.96)(0.98)] = 0.986 > 0.94$

$r_4 = [0.928/0.98]^{1/2} = 0.973 > 0.96$

$r_5 = [0.928/1]^{1/3} = 0.978 < 0.98$

Therefore subsystem 5 need not be improved while subsystems 3 and 4 are to be improved to R_o where,

$R_o = [0.928/0.98]^{1/2} = 0.973$

Hence, $R_3^* = R_4^* = 0.973$ and $R_5^* = 0.980$ (Unchanged)

For reliability allocation to subsystems 6-10, we first calculate the proportionality factors using

$$Z_j = \frac{K_j C_j F_j M_j}{S_j\, d_j}$$

$M_j = 1$ for all j
$F_6 = F_{10} = 2/3;\ F_7 = F_8 = F_9 = 1$
$d_7 = 0.75,\ d_8 = 0.50,\ d_6 = d_9 = d_{10} = 1$

Using the above and the table of data given,

$Z_6 = 8,\ Z_7 = 5,\ Z_8 = 4,\ Z_9 = 5.6,\ Z_{10} = 4$

The weightage factors are now calculated as

$w_j = Z_j / s\ Z_j$
$w_6 = 0.3007,$
$w_7 = 0.1880,$

$w_8 = 0.1504,$
$w_9 = 0.2105,$
and $w_{10} = 0.1504$

Hence, the allocated reliabilities to these subsystems are:

$R_6^* = 0.963$, $R_7^* = 0.977$, $R_8^* = 0.981$, $R_9^* = 0.974$ and $R_{10}^* = 0.981$

* * *

6.4 CRITICALITY

In the above allocation procedure, an important factor *Criticality* has not been considered intentionally. In this section, we introduce this factor and in the next section discuss its role in reliability allocation.

It may so happen that failure of a component in the system (although non-redundant) may still result in system success with a finite non-zero probability. This is particularly so for mass produced consumer products which are to be used by several users with varying operating conditions and using the products in varying environmental conditions.

For a quantitative treatment of this aspect, criticality of a component is defined as the probability of system failure if that particular component fails. A value of 1 (or 100%) for the criticality implies certainity of the system failure consequent upon the failure of the component. Thus, the commonly used series model assumes all components to have criticality 1- an assumption which is not always valid.

Several practical situations can be cited necessitating the incorporation of this concept. A few examples are:

1. Does the failure of an audio channel imply the failure of TV receiver? Is it not that some users will interpret it as system failure while others may ignore it? Does it not depend upon the situation? Does the failure of an audio channel during the transmission of a football match or a musical concert lead to the same conclusions?

2. An automobile's head-lights or dynamo fails. Is it a failure of the vehicle? Does the answer not depend upon the time of driving (day or night)? When we are carrying out the reliability analysis of the automobile, it has to be considered driven at any time.

3. A radar system is designed for the detection of fixed as well as moving targets. Therefore, it has a unit known as MTI (Moving Target Indicator). If this unit fails, does it mean failure of the radar system? Many targets

to be detected may be fixed only.

4. A stabilizer is connected to a sophisticated equipment for the adjustment of voltage on the line. If the stabilizer fails, does it mean failure of the equipment? The answer to this question obviously depends upon the line voltage at that time. We may keep in mind a large multitude of electric supply corporations in various countries with all kinds of standards and practices set for themselves.

5. A meteriological satellite is to monitor several parameters and transmit the relevant information. Suppose it fails to monitor one of the parameters. Does it imply system failure? Does it not depend upon the particular user and particular time of the year?

6. The dial of a telephone instrument becomes defective. Will we call it the failure of the instrument? We can still receive incoming calls.

7. One of the units in a channel of a stereo-system fails, say a loudspeaker or an amplifier. Does it imply the failure of the system? The answer will depend upon the sensitivity of the user to the music quality as also on the music program being relayed at that time.

To quantify the role of criticality in reliability evaluation, consider a system having just two components A and B. There can be 4 possible states(00), (01), (10) and (11). Assume component A to be 100% critical (its failure definitely fails the system). However, assume component B to be 100 x_b% critical implying that if component B fails, the probability of system failure is x_b only and there is a 100(1-x_b)% chance that the system still does not fail. Therefore, states (00), (01) will always be failure states, state (11) will always be a success state but state (10) will result in failure 100 x_b% of the time and in success [100-100x_b]% of the time. Hence, system reliability in this case is not $p_a p_b$, but will have another additional term arising out of the state (10). Now,

$$R = p_a p_b + [1-x_b]\, p_a q_b$$

$$= p_a[1-x_b + x_b p_b] \qquad (6.27)$$

The above relation suggests that in a system reliability expression a component with reliability p_b and criticality x_b can be replaced by an equivalent fully critical component with reliability R_b where R_b is given by:

$$R_b = 1 - x_b + x_b p_b$$

or, $$R_b = 1 + x_b[p_b-1] \qquad (6.28)$$

Obviously, $R_b = p_b$ for $x_b = 1$

and $R_b = 1$ for $x_b = 0$

A graph of the relationship between these two parameters is shown in Fig.6.2. As a numerical example, if $p_b = 0.90$ and $x_b = 0.70$, $R_b = 0.93$, i.e. a fully critical component with 93% reliability will have the same contribution to the system reliability as a 70% critical component with 90% reliability.

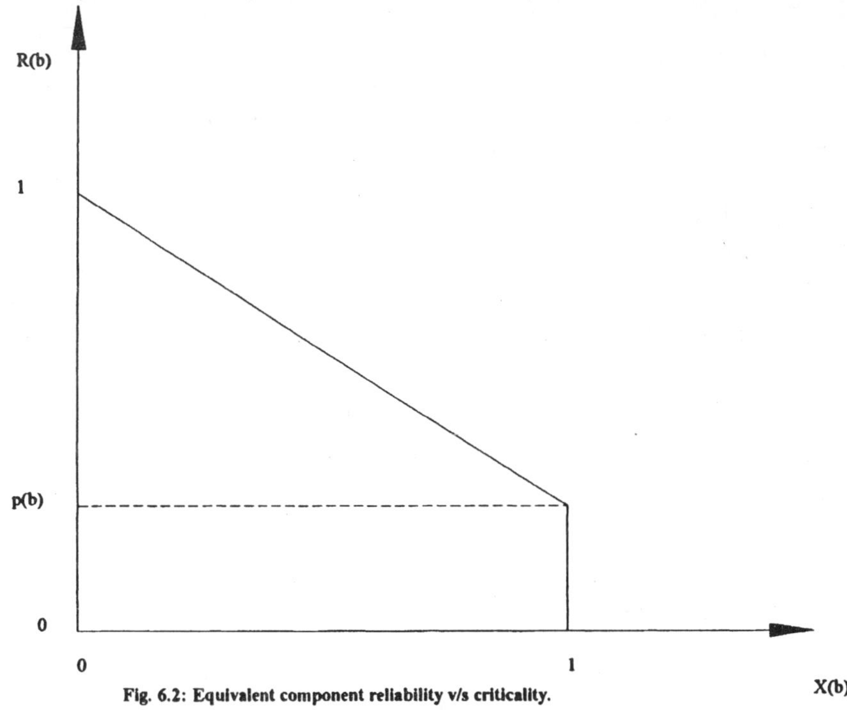

Fig. 6.2: Equivalent component reliability v/s criticality.

6.41 Role of Criticality in Reliability Allocation

After having established that a component with reliability p_b and criticality x_b can be replaced by a fully critical component with reliability R_b, it is suggested that criticality be ignored in the initial reliability apportionment. This approach permits us to use product law of reliabilities which is so simple and a tempting approach for work. The problem of reliability allocation can thus be to consider all other factors, except criticality, and apportion the system reliability goal R^* to n-units such that:

$$\prod_{k=1}^{n} R_j \geq R^* \tag{6.29}$$

As a result, if reliability R_j is allocated in the apportionment (assuming all components fully critical) to a component whose criticality is x_j; then its actual reliability allocation can be $R_j^*[R_j^* < R_j]$. The relationship between R_j^* and R_j is given as follows:

$$R_j = 1 + x_j[R_j^* - 1]$$

or,

$$R_j^* = [R_j + x_j - 1]/x_j \tag{6.30}$$

This approach thus makes the reliability allocation for partially critical components also a relatively simple exercise.

A plot of R_j^* versus x_j for an obtained value of R_j is given in Fig.6.3 which leads to an interesting observation. Mathematically, for a very low criticality component actually allocated reliability can even be negative i.e. in a practical sense we need not bother about the reliability values of such components. The transition occurs at a value of x_j^* given by :

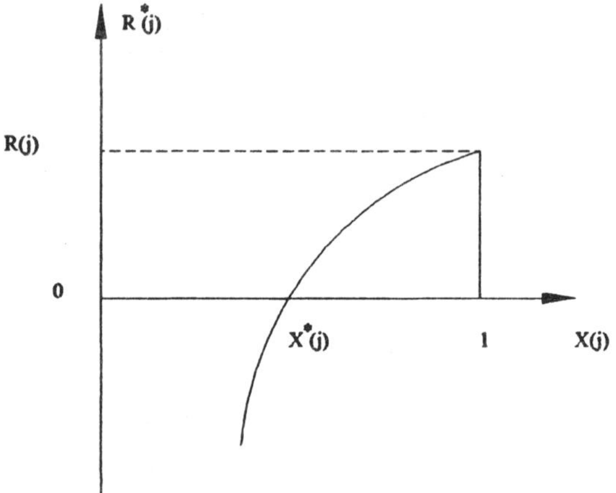

Fig. 6.3: Actual allocated reliability v/s criticality.

$$R_j + x_j^* - 1 = 0 \quad \text{or} \quad x_j^* = 1 - R_j \tag{6.31}$$

Hence, if the allocated reliabilty of a component (ignoring criticality) is R_j and if the criticality of the component is x_j which is less than x_j^*, we need not bother about the actual reliability allocation for this component. This is pictorially shown in Fig.6.4.

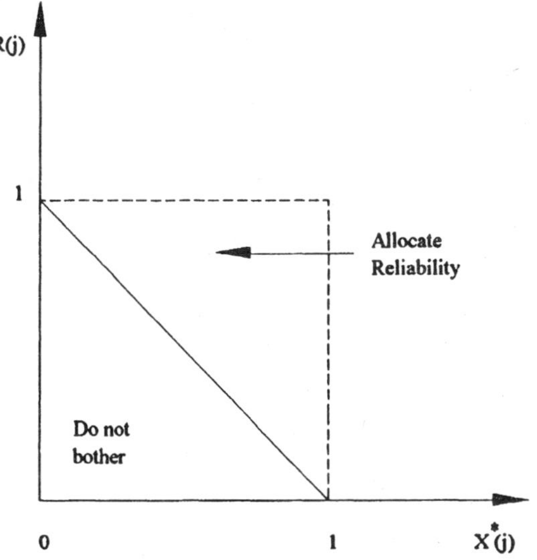

Fig. 6.4: Applicable range for actual reliability allocation.

7

REDUNDANCY TECHNIQUES FOR RELIABILITY OPTIMIZATION

7.1 INTRODUCTION

One of the major functions of a reliability engineer is to achieve the desired level of system reliability or improve the system reliability as far as possible. Several methods of improving system reliability exist. These methods approach the problem by :

1. Using a large safety factor
2. Reducing the complexity of the system
3. Increasing the reliability of constituent components:

   (a) Through a product improvement programme, or

   (b) By derating
4. Practising a planned maintenance and repair schedule
5. Using structural redundancy.

Of these, the last method is most effective and most commonly used. The other methods are generally limited by the level of improvement which can be achieved. For example, it is well known that system reliability can be improved by using superior components, i.e., highly reliable components with low failure rates. But it is not always possible to produce such highly reliable components with reasonable effort and/or cost. We describe commonly used Redundancy Techniques in this chapter.

What is redundancy? In simple terms redundancy is addition of information, resources, or time in excess of what is needed for normal system operation for the purpose of tolerating and/or detecting failures. The redundancy may take several forms, including :

(a) Signal redundancy
(b) Time redundancy
(c) Software redundancy
(d) Hardware redundancy

7.2 SIGNAL REDUNDANCY

Signal redundancy is commonly used in digital systems and is the addition of information beyond that required to implement a function e.g, Hamming error correcting codes. These codes are used to detect double errors and to correct single errors. The basic principle is to extend the M information bits with k-parity-checking bits to form a code word of M+k bits. The position of each of the M+k bits within a code word is given a decimal value between 1 for the most significant bit and M+k for the least significant bit. Then k parity checks are performed and recorded as bits C_1, C_2,.....C_k. The decimal value of the word $C_1C_2....C_k$ will be equal to the decimal value given to the faulty bit, if any; otherwise its value is zero.

k must be large enough to allow the location of any of the M+k possible single errors and a fault free case. So k must satisfy the inequality $2^k \geq M+k+1$. For example, a four bit message requires k=3. However, as the number of message bits increases, the efficiency of the code also increases, e.g., k=6 for M=32.

Table 7.1
Hamming Code for BCD

No.	p_1 1	p_2 2	d_3 3	p_3 4	d_2 5	d_1 6	d_0 7
0	0	0	0	0	0	0	0
1	1	1	0	1	0	0	1
2	0	1	0	1	0	1	0
3	1	0	0	0	0	1	1
4	1	0	0	1	1	0	0
5	0	1	0	0	1	0	1
6	1	1	0	1	1	1	0
7	0	0	0	1	1	1	1
8	1	1	1	0	0	0	0
9	0	0	1	1	0	0	1

Table 7.1 shows Hamming code corresponding to BCD code. Each parity bit

when combined with selected data bits, produces even parity. Parity check bit p_1 is associated with data bits d_3, d_2, d_0 and gives C_3; p_2 with d_3, d_1, d_0 and gives C_2; and p_3 with d_2, d_1, d_0 and gives C_1. Error detection and location are performed by checking the code words at receiving end to form word $C_1C_2C_3$.

7.3 TIME REDUNDANCY

Time redundancy is used to provide protection against transient or intermittent failures. In this redundancy the processor performs the computations one or more times after detecting the first error; if the error condition clears, the processor proceeds further, otherwise it rolls back to the last check point and if fault still persists it is taken as a permanent fault and the processor is shut down for repairs. The main difficulty with time redundancy is to assume that the processor has the same data to manipulate each time it redundantly performs a computation. If a transient error has truly occurred, the processor's data may be completely scrambled, making it difficult to obtain agreement on results computed at two points in time. The correct identification of the failing instruction is a necessary condition for the success of the instruction retry step. A major reason for incorrect identification of the failing instruction is a latency period between the occurrence of the fault and the manifestation of the consequent error. An additional latency period may appear between the occurrence of the error and its detection. The system success can be achieved in any of the following three manners:

1. The instruction is completed successfully (without retry) when first executed and there is no undetected fault in the system.
2. The instruction fails, is correctly identified and the jth retry is the first successful one.
3. The instruction fails and is correctly identified, all k retries fail but the instruction is completed successfully after program rolls back.

7.4 SOFTWARE REDUNDANCY

Software redundancy is simply addition of extra software to provide some fault tolerance. Probably the most common form of software redundancy is the validity check. Here additional software is added to verify that the results being produced are within certain acceptable ranges as is the case in several practical cases.

A second type of software redundancy is the periodic self test. Often, a large percentage of faults can be detected by allowing software to periodically exercise the hardware and set a 'watch dog' timer if the test is passed. The timer, in such cases, is designed to generate an error interrupt if it is not appropriately reset. The technique often provides very good protection against a variety of faults. It is infact a mixture of software and time redundancy, it requires additional software and uses additional processor time during its implementation.

A third example of software redundancy is the use of multiple copies of programs. The programs are prepared by different teams and one runs simultaneously in multiple processors or frequently on a single processor. The results are compared to provide a means of fault detection. This is referred to as 'N-version programming'.

7.5 HARDWARE REDUNDANCY

We rather concentrate here on Hardware Redundancy which is the physical replication of system for the purpose of detecting and tolerating failures. In this method, we improve the system reliability by connecting several identical components in parallel to the given component or creation of new parallel paths in a system structure.

There is a basic conflict in increasing the reliability of a system. The improvement of reliability is causative of increasing the consumed amounts of resources; e.g. cost, weight, volume, area, etc. This conflict cannot be circumvented, but it can be minimized through optimum design. The conflict between quality and the outlay of resources is present everywhere. It is prominent, for example, in the design of complex electronic equipment for space use. There are constraints on some of the resources. In the case of space systems, the pay-load weight is limited by the capability of the launch vehicle.

No doubt, use of redundancies will increase the overall system reliability; but this cannot be done without looking into the availability of resources. The object of redundancy allocation, therefore, is to maximise the system reliability with certain constraints imposed such as cost, weight, power requirements, etc.

The optimization techniques which have been applied in the improvement of system reliability are:

1. Dynamic programming
2. The discrete maximum principle
3. The sequential unconstrained minimization technique (SUMT)
4. Method of Lagrange multipliers and the Kuhn-Tucker conditions
5. Geometric programming
6. Integer programming
7. Heuristic approaches

The above techniques can be classified as Exact and Approximate Methods. Exact methods give us optimum solution but require large amount of computer time and memory. Approximate methods are faster but may not result into the optimum solution.

Most of these optimization techniques, (except the heuristic approaches) are very time consuming. Heuristic methods are approximate methods for the optimum allocation of redundancies. These methods, do not have any guarantee for optimum solution, although they may provide an optimum solution in many cases.

The main advantages of these methods are as follows:

1. These methods are extremely easy and computationally economical. The methods can be conveniently applied to any problem of large size.
2. Any number of linear or non-linear constraints can be taken care of without much increase in the computational work.
3. In addition to other constraints, availability constraint of any component can be very conveniently taken care of in these methods.
4. These methods provide the true optimum solution in many cases. Otherwise in most of the situations they provide near optimum solution which may be acceptable in most of the cases.

Let there be n number of stages in a system connected in series where stage-i is a parallel configuration of X_i components each with reliability P_i. It is assumed that all elements are working simultaneously and for a stage to fail all these elements must fail. For the whole system to be operating all the stages must be operating. In such a case, the system reliability is expressed by:

$$R = \prod_{i=1}^{n} 1 - (1-P_i)^{Xi} \tag{7.1}$$

As an aid to understanding, the notation used in this and subsequent sections is as follows:

P_i = Reliability of ith component
Q_i = Unreliability of ith component
R = System reliability
Q = System unreliability
X_i = Number of total components connected in parallel at stage-i
n = Number of stages
K_j = Available resource for constraint-j

m = Total number of different types of constraints
$C_{ij}(X_i)$ = Resource-j consumed in stage-i with X_i components connected

7.51 Method I

For a system having n stages in series with X_i redundant components at stage-i, the system reliability is given by

$$R = \prod_{i=1}^{n} 1 - (1-P_i)^{X_i}$$

$$= \prod_{i=1}^{n} 1-Q_i^{X_i} \qquad (7.2)$$

The problem is to maximise R subject to

$$\sum_{i=1}^{m} C_{ij}\ (X_i) \le K_j\ ; \qquad j = 1,2,\ldots\ldots., m \qquad (7.3)$$

The sequential steps involved in solving the problem are as follows:

1. Find the stage which is most reliable.
2. Find the sum of each constraint over all stages.
3. Find the ratios of the maximum values of the constraints to the sum of the constraints.
4. The redundancy for this particular stage can be obtained by finding the minimum of the ratios.
5. Find the stage amongst the remaining stages which is most reliable.
6. Find the sum of all the constraints of a particular type excluding that of the stages for which redundancies have already been calculated.
7. Find the ratios of the maximum value of the constraints which are still available (less the sum of the products of redundancies and constraints of the previous stages) to the sum found in step 6.
8. The redundancy at this stage can be obtained by finding the minimum of the ratios.
9. Continue from step 5 till the redundancies for all the stages have been found out.

Example 7.1

Consider a two stage linear constraint problem for optimum redundancy allocation with the following data :

$n = 2, \qquad K_1 \le 5$

$P_1 = 0.60, \quad C_{11} = 2$

$P_2 = 0.65, \quad C_{21} = 1$

Solution

Stage 2 has the highest reliability. Therefore, we first calculate optimum X_2 such that,

$$(2+1)\ X_2 \le 5 \quad \text{or} \quad X_2 = 1$$

Eliminating stage 2 from further considerations, X_1 is calculated such that,

$$2\ X_1 \le 5 - 1(1) = 4 \quad \text{or} \quad X_1 = 2$$

Therefore, optimum solution is

$$X = [2 \quad 1]$$

$$R = 0.546$$

* * *

Example 7.2

Consider a four stage system for optimum redundancy allocation with two linear constraints. The data are:

$n = 4, \qquad K_1 \le 56, \qquad K_2 \le 120$

$P_1 = 0.80, \quad C_{11} = 1.2, \quad C_{12} = 5$

$P_2 = 0.70, \quad C_{21} = 2.3, \quad C_{22} = 4$

$P_3 = 0.75, \quad C_{31} = 3.4, \quad C_{32} = 8$

$P_4 = 0.85, \quad C_{41} = 4.5, \quad C_{42} = 7$

Solution

Stage 4 has highest reliability. Therefore, optimum value of X_4 is chosen such that,

$$(1.2 + 2.3 + 3.4 + 4.5)\, X_4 \leq 56$$

and $$(5 + 4 + 8 + 7)\, X_4 \leq 120$$

or $$X_4 = 4$$

Eliminating stage 4 and then considering stage 1,

$$(1.2 + 2.3 + 3.4)\, X_1 \leq 56 - 4\,(4.5) = 38$$

$$(5 + 4 + 8)\, X_1 \leq 120 - 4\,(7) = 92$$

or $$X_1 = 5$$

Eliminating stage 1 also and now considering stage 3,

$$(2.3 + 3.4)\, X_3 \leq 38 - 5\,(1.2) = 32$$

$$(4 + 8)\, X_3 \leq 92 - 5\,(5) = 67$$

or $$X_3 = 5$$

Lastly considering stage 2,

$$2.3\ X_2 \leq 32 - 5\,(3.4) = 15$$

$$4\ X_2 \leq 67 - 5\,(8) = 27$$

or $$X_2 = 6$$

Therefore,optimum solution is

$$X = [\,5\ \ 6\ \ 5\ \ 4\,]$$

$$R = 0.99747$$

* * *

7.52 Method II

For a system having n stages in series with X_i redundant components at stage-i, the system unreliability is given by

$$Q = 1 - \prod_{i=1}^{n} 1 - Q_i^{X_i} \tag{7.4}$$

which can be approximated as

$$Q \cong \sum_{i=1}^{n} Q_i^{X_i} \qquad (7.5)$$

The problem is to minimise Q subject to

$$\sum_{i=1}^{n} C_{ij}(X_i) \leq K_j\,; \quad j = 1,2,\ldots\ldots\ldots, m \qquad (7.6)$$

The sequential steps involved in solving the problem by this method are as follows:

1. Assign $X_i = 1$, for $i = 1,2, \ldots\ldots\ldots, n$.
2. Find the stage which is most unreliable. Add one redundant component to that stage.
3. Check the constraints:
   (a) If any constraint is violated, go to step 4.
   (b) If no constraint has been violated, go to step 2.
   (c) If any constraint is exactly satisfied stop. The current X_i's are the optimum values for the system.
4. Remove the redundant component added in step 2. The resulting number is the optimum allocation for that stage. Remove this stage from further consideration .
5. If all the stages have been removed from consideration the current X_i's are the optimum values for the system; otherwise go to step 2.

Example 7.3 (Data same as in example 7.1)

The solution is shown in table 7.2. Optimum solution, therefore is :

Table 7.2
(Solution of Example 7.3)

Stage		Unreliability		
X_1	X_2	I	II	Cost
1	1	0.40*	0.35	3
2	1	0.16	0.35	5+

(+ Constraint exactly satisfied)

X = [2 1]

* * *

Example 7.4 (Data same as in Example 7.2)

The solution is shown in table 7.3 and Optimum solution is :

Table 7.3
(Solution of Example 7.4)

X_1	X_2	X_3	X_4	Stage Unreliability I	II	III	IV	K_1	K_2
1	1	1	1	0.2000	0.3000*	0.2500	0.1500	11.4	24
1	2	1	1	0.2000	0.0900	0.2500*	0.1500	13.7	28
1	2	2	1	0.2000*	0.0900	0.0625	0.1500	17.1	36
2	2	2	1	0.0400	0.0900	0.0625	0.1500*	18.3	41
2	2	2	2	0.0400	0.0900*	0.0625	0.0225	22.8	48
2	3	2	2	0.0400	0.0270	0.0625*	0.0225	25.1	52
2	3	3	2	0.0400*	0.0270	0.0156	0.0225	28.5	62
3	3	3	2	0.0080	0.0270*	0.0156	0.0225	29.7	65
3	4	3	2	0.0080	0.0081	0.0156	0.0225*	32.0	69
3	4	3	3	0.0080	0.0081	0.0156*	0.0034	36.5	76
3	4	4	3	0.0080	0.0081*	0.0039	0.0034	39.9	84
3	5	4	3	0.0080*	0.0024	0.0039	0.0034	42.2	88
4	5	4	3	0.0016	0.0024	0.0039*	0.0034	43.4	93
4	5	5	3	0.0016	0.0024	0.0010	0.0034*	46.8	101
4	5	5	4	0.0016	0.0024*	0.0010	0.0005	51.3	108
4	6	5	4	0.0016*	0.0007	0.0010	0.0005	53.6	112
5	6	5	4	0.0003	0.0007	0.0010	0.0005	54.8	117

(No addition now possible without violating the constraints)

$$X = [5 \quad 6 \quad 5 \quad 4]$$

* * *

7.53 Method III

Two methods presented earlier emphasised adding redundancy to the stage having highest unreliability so far. These algorithms did not depend upon the values of the constraints associated with each stage at any instant. In certain cases, the constraints dictate the addition of only one element to a particular stage (having lowest reliability) ;but these constraints permit the addition of more than one element to another stage (having higher reliability). It may so happen that the net increase in reliability because of later approach may be more than the net increase because of former approach. This situation is particularly common if the stages have components of almost similar reliability, but different in cost (or any other constraint). In any complex practical system invariably there shall be components with almost same reliability but widely differing cost because of different nature of components.

In the following method, a component is added to the stage where its addition has maximum value for a factor defined as " the ratio of decrement in unreliability to the product of increments in constraints". Mathematically, $F_i\ (X_i)$ is expressed as (7.7) and is termed as *stage selection factor*.

$$F_i(X_i) = \frac{P_i\ Q_i^{\ Xi}}{\prod_{j=1}^{m} \Delta\ C_{ij}\ (X_i)} \tag{7.7}$$

It is observed that $F_i(X_i)$ is a function of i, the particular stage; and X_i, the number of elements in that stage. Hence in the process of computation, the value of this factor keeps changing even for a fixed i. In a problem with linear constraints, however, $\Delta C_{ij}(X_i)$ is independent of X_i. Therefore,

$$F_i(X_i + 1) = \frac{P_i\ Q_i^{\ Xi+1}}{\prod_{j=1}^{m} \Delta\ C_{ij}\ (X_i)}$$

$$\text{or, } F_i(X_i + 1) = Q_i \cdot F_i\ (X_i) \tag{7.8}$$

In a linear constraint problem, it is therefore very convenient to evaluate all $F_i(X_i)$ by using recursive relation (7.8), which simply requires successive multiplication by Q_i. The use of this relation makes the proposed method very attractive for practical problems with linear constraints.

The procedure outlined above can be summarised in the form of various steps for an algorithm as:

1. It is a series system; therefore initially let $X_i = 1$ for all i .

2. Calculate $F_i(X_i)$ for all i using (7.7)

3. Mark the stage (*) having highest value of stage selection factor $F_i(X_i)$. A redundant component is proposed to be added to that stage.

4. Check constraints:

   (a) If the solution is still within the permissible region, add the redundant component. Modify the value of X_i and hence $F_i(X_i)$ and go back to step 3.

   (b) If at least one constraint is exactly satisfied, the current value of X is the optimum solution.

   (c) If at least one constraint is violated, cancel the proposal; remove the stage from further consideration and go back to step 3; when all the stages are removed from further consideration the value of X is the optimum solution.

Example 7.5 (Data same as in Example 7.1)

This is a single linear constraint problem. Therefore, stage selection factor $F_i(X_i)$ is:

$$F_i(X_i) = \frac{P_i\, Q_i^{\,Xi}}{C_{i1}}$$

$$\text{So, } F_i(1) = \frac{P_i\, Q_i}{C_{i1}}$$

Hence, $F_1(1) = 0.1200$, $F_2(1) = 0.2275$

The solution is shown in table 7.4. It may be observed that $F_2(2)$ has been obtained by multiplying $F_2(1)$ by $Q_2 = 0.35$. Solution obtained is:

Table 7.4

(Solution of Example 7.5)

X_1	X_2	$F_1(X_i)$	$F_2(X_i)$	$\Sigma X_i C_{i1}$	$F_i(X_i+1)$
1	1	0.1200	0.2275*	3	0.0796
1	2	0.1200+	0.0796*	4	0.0279
1	3	0.1200	0.0279	5	

(+ addition causes violation of constraints)

$X = [1 \quad 3]$
$R = 0.599$

* * *

Example7.6 (Data same as in Example 7.2)

There are two linear constraints in this case. Therefore,

$$F_i(X_i) = \frac{P_i\ Q_i^{\ Xi}}{C_{i1}C_{i2}}$$

$$\text{So, } F_i(1) = \frac{P_i\ Q_i}{C_{i1}C_{i2}}$$

Therefore,

$F_1(1) = 0.02667,$ $F_2(2) = 0.02283$

$F_3(1) = 0.00689,$ $F_4(1) = 0.00404$

For convenience in numerical computations, all those factors can be multiplied by any constant. Multiplying by 100, these values can be taken as 2.667, 2.283, etc. The solution is shown in table 7.5 and is

Table 7.5

(Solution of Example 7.6)

X_1	X_2	X_3	X_4	$F_1(X_i)$	$F_2(X_i)$	$F_3(X_i)$	$F_4(X_i)$	ΣX_iC_{i1}	ΣX_iC_{i2}	$F_i(X_i+1)$
1	1	1	1	2.667*	2.283	0.689	0.404	11.4	24	0.533
2	1	1	1	0.533	2.283*	0.689	0.404	12.6	29	0.685
2	2	1	1	0.533	0.685	0.689*	0.404	14.9	33	0.172
2	2	2	1	0.533	0.685*	0.172	0.404	18.3	41	0.205
2	3	2	1	0.533*	0.205	0.172	0.404	20.6	45	0.107
3	3	2	1	0.107	0.205	0.172	0.404*	21.8	50	0.061
3	3	2	2	0.107	0.205*	0.172	0.061	26.3	57	0.062
3	4	2	2	0.107	0.062	0.172*	0.061	28.6	61	0.043
3	4	3	2	0.107*	0.062	0.043	0.061	32.0	69	0.021
4	4	3	2	0.021	0.062*	0.043	0.061	33.2	72	0.018
4	5	3	2	0.021	0.018	0.043	0.061*	35.5	78	0.009
4	5	3	3	0.021	0.018	0.043*	0.009	40.0	85	0.011
4	5	4	3	0.021*	0.018	0.011	0.009	43.7	93	0.004
5	5	4	3	0.004	0.018*	0.011	0.009	44.6	98	0.005
5	6	4	3	0.004	0.005	0.011*	0.009	46.9	102	0.003
5	6	5	3	0.004	0.005	0.003	0.009*	50.3	110	0.001
5	6	5	4	0.004	0.005	0.003	0.001	54.8	117	

(No addition now possible without voilating the constraints)

X = [5 6 5 4]

* * *

8

MAINTAINABILITY AND AVAILABILITY

8.1 INTRODUCTION

The principal objectives of maintenance can be defined as follows:

1. To extend the useful life of assets. This is particularly important in view of the lack of resources.
2. To ensure the optimum availability of installed equipments for production (or service) and obtain the maximum possible return on investment.
3. To ensure the operational readiness of all equipment required for emergency use, such as standby units, firefighting and rescue equipment, etc.
4. To ensure the safety of personnel using facilities.

From time to time, statistics are generated which emphasize the costliness of maintenance actions. While estimates of actual costs vary, they invariably reflect the immensity of maintenance expenditures. According to one source, approximately 800,000 military and civilian technicians in U.S.A. are directly concerned with maintenance. Another source states that for a sample of four equipments in each of three classes - radar, communication, and navigation the yearly support cost is 0.6, 12 and 6 times, respectively, the cost of the original equipment. Such figures clearly indicate the need for continually improved maintenance techniques.

In addition to these cost considerations, maintainability has a significant effect on other system-effectiveness characteristics. *System effectiveness* is a function of system performance capability, system dependability and system cost. *Performance capability* includes the capacity to meet specified requirements such as range, power output, sensitivity and the like. *Dependability* is a measure of the degree of consistency of performance and is essentially the same as operational availability. *Availability* is, in turn, a function of reliability and maintainability. *System cost* must include the total amount for development, production and service-life support of the equipment.

Maintainability, then is only one part - although a very important part - of the measurement of over- all system worth. The US Department of Defence definition of maintainability is quoted as follows :

Maintainability is a quality of the combined features and characteristics of equipment design which permits or enhances the accomplishment of maintenance by personnel of average skills, under the natural and environmental conditions, in which it will operate.

The definition above is highly qualitative and is not subject to quantification without further specification. The above report states further that

The search for a single definition that encompasses all the attributes of maintainability in a quantitatively measurable term is, for the present, unrewarding. It is first necessary to identify and measure the most relevant factors that make up this end measurement. It is likely that no single final measurement will adequately serve all purposes.

In line with this reasoning, several possible indices were suggested which may be useful in the quantitative description of maintenance activity. Among these are:

1. Ratio of satisfactory operation to total required time.
2. Average down time per unit of calendar time (or any other stated time).
3. Mean time to repair.
4. Man-hour requirements per unit of operating time.
5. Total man-hour requirements per unit of calendar time.
6. Waiting time per unit of time (calendar or other stated time).
7. Material requirements per unit of time.
8. Cost of support per unit of calendar time.

It is probable that any or all of the indices above may be needed in one situation or another, plus, perhaps, other special indices.

For purposes of quantitative prediction, Radio Corporation of America has elected to express maintainability in terms of time required to perform a maintenance action.

It is assumed that this time is a function of such governing factors as equipment design, supply and logistics, test equipment, training, technical orders, operational circumstances, malfunction criticality, and personnel requirements. This concept may be stated symbolically as:

$$M_t = f(X_1, X_2, \ldots, X_n) \quad (8.1)$$

Where,

M_t = maintenance time

$X_1, \ldots, X_n$ = values which quantitatively express the n governing factors described above.

Maintenance is one of the effective ways of increasing the reliability of a system. Repair maintenance is considered to be beneficial when the repair cost in terms of time and money spent is considerably low compared to the cost of the equipment. A low repair time will minimize the ill-effects of the failure. Reliability alone cannot describe the usefulness (or service utility) of such equipments. Factors such as the repair time, the number of failures in a specified period, and the fraction of time the equipment is in operation also refer to the usefulness of an equipment subject to renewals (repairs).

8.2 FORMS OF MAINTENANCE

Maintenance work can either be planned or unplanned. There is only one form of unplanned maintenance and that is emergency maintenance, in which case it is necessary to take up maintenance actions immediatly to avoid serious consequences such as loss of production, extensive damage to assets, or for safety reasons.

Planned maintenance is split up in two main activities, preventive (also referred to as scheduled) and corrective (also referred to as unscheduled).

The major part of preventive maintenance involves inspection on the basis of *look, feel and listen.* It involves lubrication, refuelling, cleaning, adjustment, alignment, etc. at predetermined intervals and the replacement of minor components which are nearing a wear-out condition found as a result of such inspections.

Corrective maintenance involves minor repairs, that may crop up between inspections. This also involves planned overhauls such as yearly or two - yearly overhauls, the extent of which has been planned in detail on a long term basis as a result of prenventive inspection.

A schematic diagram is given (Fig.8.1) to show the relationship between various forms of maintenance .

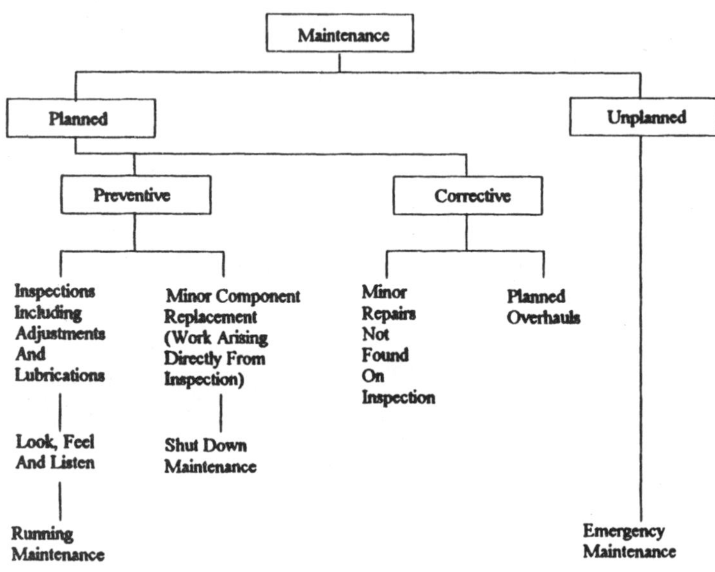

Fig.8.1 Relationship Between Various Forms of Maintenance.

8.21 Cost Analysis

A graphic example of the relationship of the amount of maintenance to the total cost of maintenance is shown in Fig.8.2

A closer study of Fig 8.2 leads to many interesting results. In the first case, as the degree of maintenance increases, the cost of emergency maintenance decreases (shown by a thick line) while the cost for planned maintenance increases with an increase in the degree of maintenance. The total maintenance cost is shown as a dark thick line. By inspection, it is obvious, that there is a point where total maintenance cost is minimum; that is, where the maintenance is economical for a degree of maintenance. The cost figures indicated below the figure show the percentage of cost in three cases. First, before planned maintenance, the major cost involved is of emergency maintenance. In the case of economic maintenance, the interesting point to note is that there is a saving of at least 20 % of the total cost. When the degree of maintenance increases greatly, it becomes uneconomical and the major share is taken by planned maintenance. From this analysis, we may infer that too much maintenance can be as costly as too little maintenance.

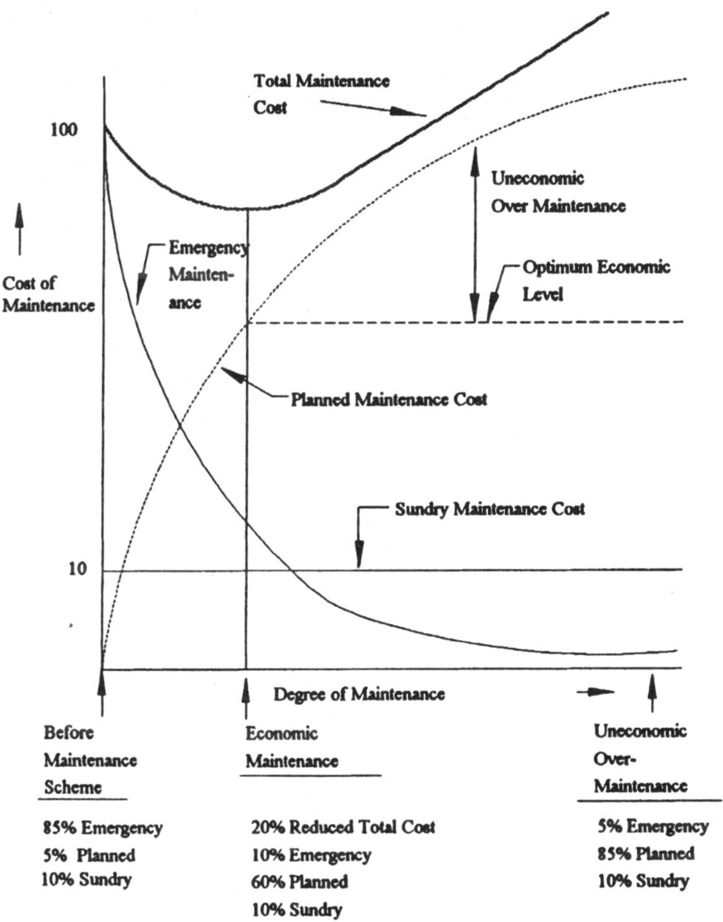

Fig.8.2 Maintenance Cost Relationship.

8.3 MEASURES OF MAINTAINABILITY AND AVAILABILITY

Maintainability is defined as the probability that a failed system is restored to operable condition in a specified down time when maintenance is performed under stated conditions. In the evaluation of any system, the measure of maintainability is quite important; how often the system fails (reliability) and how long it is down (maintainability) are vital considerations in determining its worth. In practice the trade-off between these two concepts is dictated by cost, complexity, weight, operational environment and other requirements.

The first step in measuring maintainability is to define its constituent elements. Down time is the interval during which the system is not in an acceptable operating condition (i.e, the time from initiation of a complaint or most routine maintenance actions to the restoration of the system to satisfactory operating condition). Down time is divided into:

1. Active repair time
2. Logistic time
3. Administrative time

Active repair time is the number of down-time hours during which one or more technicians actually work on a system to restore it to operable condition. *Logistic time* is the number of down-time hours consumed in awaiting parts or units needed to affect a repair. *Administrative time* is that portion of down time not covered by active repair time or logistic time. Based on a 24 hr day, it includes overnight time, weekends,and normal administrative delays.

Active repair time is usually indicative of the complexity of the system, the nature of its design and installation, the adequacy of test facilities, and the skill of maintenance personnel.

Logistic time is generally a function of the supply methods associated with the operational activity, but it can be influenced by the design of the system. For example, if large numbers of non-conventional parts are used in a system, then the supply organization must handle greater quantities of special items; this situation could cause delays in the receipt of spares or replacement parts.

Administrative time is a function of the structure of the operational organization involved; it is influenced by work schedules and the assignment of nontechnical duties to maintenance personnel. Generally, this time can in no way be charged against the manufacturer of the system. In practice, it would be greatly reduced in an emergency or wartime situation.

Repair time can be reduced in most situations by the use of additional manpower. For this reason, records are maintained on the man-hour expended during a given maintenance action. Man-hours is defined as the sum of the times all technicians worked on the system during a given maintenance action. These data can be used effectively to determine the average maintenance support required to maintain a system, often expressed in terms of man-hours per 1000 operation-hours (Maintenance Support Index, or MSI).

We list below the factors which can be provided in the design of a system to achieve optimum maintainability.

1. Design for minimum maintenance skills.
2. Design for minimum tools.
3. Design for minimum adjustment.
4. Use standard interchangeable parts/components.
5. Group subsystems so that they can be easily located and identified.
6. Provide for visual inspection.
7. Provide trouble-shooting indicators such as panel lights.
8. Use colour coding for wires to facilitate tracing faults.
9. Use plug-in rather than solder-in modules.
10. Plan for modular replacement.
11. Orient sockets all in one direction.
12. Use overload indicators, alarms and lighted fuse.
13. Design for safety, use interlocks, safety covers and guarded switches.
14. Make automatic recovery programme for failures (digital computers), wherever possible.
15. Make parts and components easily accessible.

The percentage of time the equipment is under operation is called the *steady-state availability*. It characterizes the mean behaviour of the equipment. The availability function A(t) is defined as the probability that the equipment is operating at time t. Although, this definition appears to be very similar to the reliability function R(t), the two have different meanings. While reliability places emphasis on failure-free operation up to time t, availability is concerned with the status of the equipment at time t. The availability function does not say anything about the number of failures that occur during time t. This means that two equipments A and B can have different number of failures in a given time interval and can still have the same availability. For example, in a period of 100 hr, an equipment of 0.8 availability might have two failures, each causing 10 hr down-time, or three failures, one causing 10 hr down time and the other two 5 hr each.

This brings in another factor known as *frequency of failures*. The frequency of failures is the number of times the equipment enters the failed state per unit time. In many applications, the frequency of failures is of great importance. A failure, irrespective of its length of down-time, can cause serious consequences.

Availability is always associated with the concept of maintainability. The maintainability function M(t) is defined as the probability that the equipment will be restored to operational effectiveness within a specified time when the repair is performed in accordance with the prescribed conditions. It is clearly a function of repair time. Availability therefore depends upon both failure and repair rates.

In general, the availability of a system is a complex function of reliability and maintainability. This can be expressed as

$$A = f(R, M) \tag{8.2}$$

where A = system availability
R = system reliability
M = system maintainability

Equation (8.2) can be viewed as an input and output relation, where R, and M are the inputs and A is the output. Fig.8.3 shows the availability response surface with R and M as inputs.

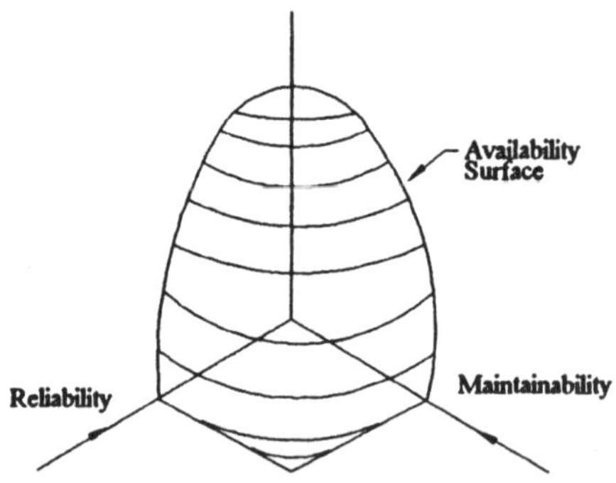

Fig.8.3 Hypothetical Availability Surface: Trade-off between Reliability and Maintainability.

In general, the availability surface is a convex surface from the lowest portion to the highest level of availabilities. Initally, the availability improves rapidly with increase in reliability and maintainability. As these two inputs gradually increase, the rate at which availability increases is slower. If the availability surface is cut by a horizontal plane, we get a constant availability contour generally called an isoavailability curve.

It may also be seen from Fig. 8.3 that along a contour, sccessive incremental increase in reliability (maintainability) require smaller and smaller amounts of maintainability (reliability). This is referred to as *competitive substitution or trade off*.

Generally, the problem is to achieve maximum availability for a given cost or to achieve a required availability at least cost.

Repair can improve the system reliability if the system has redundancy. This is possible because if one equipment fails the other can continue to operate and the system can thus survive. Meanwhile, the failed equipment can be repaired and if it can be brought to operation before the other fails, then the system will continue to operate. Thus, the system can be kept alive continuously if the repair time of the equipment is less than the time between failures.

However, repair has no effect on the reliability of a single equipment (non-redundant) system. This is because when the equipment fails, the system has failed, no matter how soon it is repaired and put back into operation. Also, we know that reliability is defined upto the point of failure-free operation. Nevertheless, the repair will improve the availability of a single equipment system. This applies for a system containing units in series.

In early studies of system maintainability, it was established that any maintenance action can be classified as falling within one of the following categories:

1. Preparation
2. Malfuction verification
3. Fault location
4. Part procurement
5. Repair
6. Final test

The time required to perform each of these tasks varies from zero to several hours, depending on numerous conditions associated with particular maintenance events. Weather, for example, causes great variations in the time required for preparation. Other variables include the skill level of maintenance technicians, their familiarity with the system under repair, and even the manner in which symptoms are reported to them. This variability in preparation time would limit the accuracy of any maintenance-time predictions based on maintenance-category time distributions.

The best time to work towards minimizing maintenance requirements is during the system design and development phases. During this period, any unsatisfactory conditions indicated by a system maintainability analysis can be corrected economically. There is an obvious correlation between the complexity of a system and the time required to maintain it. The bigger and more complex the system, the longer the required maintenance time. Other factors related to hardware, such as accessibility, built-in measuring and metering devices, warning lights, and packaging, may also furnish clues about expected maintenance time. However, the system-hardware approach does not take into account all the factors which influence maintenance time. Human factors, for instance, which are acknowledged to have a considerable effect on maintenance time, would be neglected in the system-hardware approach.

8.4 MAINTAINABILITY FUNCTION

Maintainability is an index associated with an equipment under repair. It is the probability that the failed equipment will be repaired within time t hr. If T is a random variable representing the repair time, then maintainability is defined as

$$M(t) = Pr(T \leq t) \tag{8.3}$$

If the repair time is exponentially distributed with the parameter μ, then the repair-density function is

$$g(t) = \mu \exp(-\mu t) \tag{8.4}$$

and therefore,

$$\begin{aligned} Pr(T \leq t) &= \int_0^t \mu \exp(-\mu t)\, dt \\ &= 1 - \exp(-\mu t) \end{aligned} \tag{8.5}$$

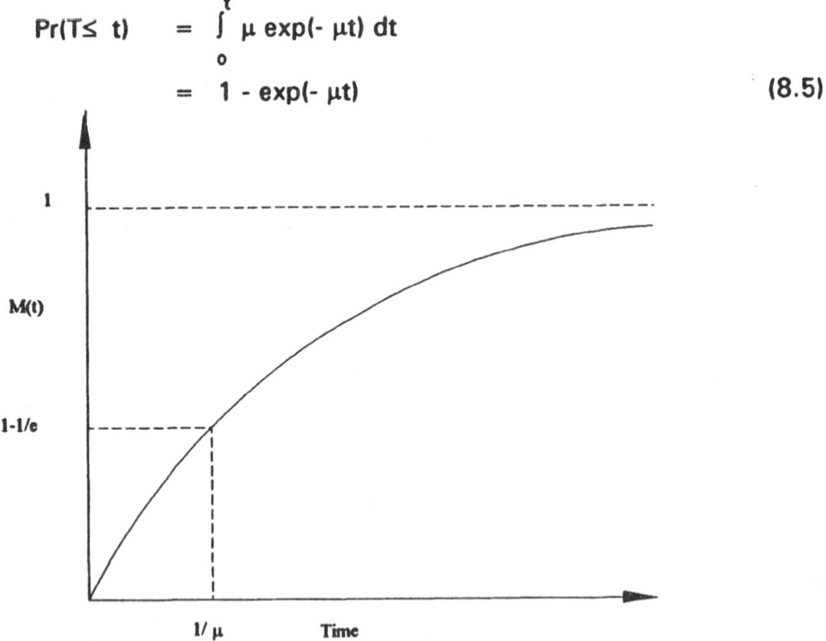

Fig.8.4 Maintainability graph.

Thus the maintainability equation is

$$M(t) = 1 - \exp(-\mu t) \tag{8.6}$$

The graph between M(t) and t is shown in Fig. 8.4.

The expected value of repair-time is called the mean time to repair (MTTR) and is given by

$$\text{MTTR} = \int_0^\infty t\, g(t)\, dt$$

$$= \int_0^\infty \mu\, t \exp(-\mu\, t)\, dt = 1/\mu \tag{8.7}$$

8.5 AVAILABILITY FUNCTION

The availability function can be computed using the familiar Markov model. It is assumed that the failure and repair rates are constant. The Markov graph for the availability of single component with repair is shown in Fig.8.5. The repair starts as soon as the component fails.

λ = failure rate (failures per unit time)
μ = repair rate (repairs per unit time)

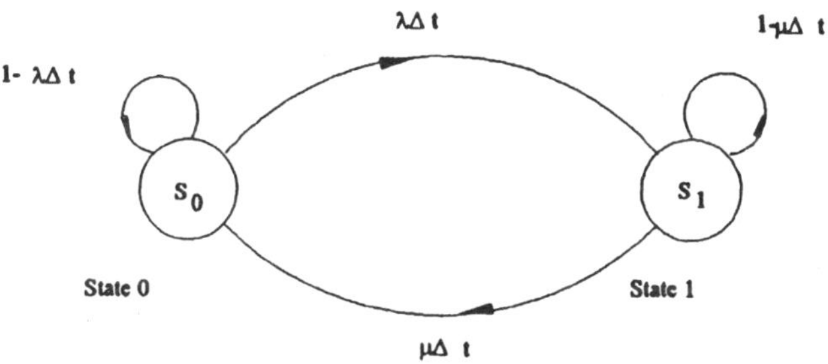

Fig.8.5 Markov graph for availability.

State 0 denotes that no failure has occurred and state 1 denotes that one failure has occurred (i. e. the component is down). If component has not failed at time t, then the probability that the component will fail in the time interval $(t, t+\Delta t)$ is equal to $\lambda\Delta t$. On the other hand, if the component is in state 1 (failed state), then the probability that the compnent will enter into state 0 is equal to $\mu\,\Delta t$.

From the Markov graph, it can be seen that the probability that the component will be in state 0 at time $t+\Delta t$ is

$$P_o(t+\Delta t) = P_o(t)\,(1-\lambda\Delta t) + P_1(t)\,\mu\Delta t \tag{8.8}$$

Similarly, the probability that the component will be in state 1 at time $t+\Delta t$ is

$$P_1(t+\Delta t) = P_1(t)\,(1-\mu\Delta t) + P_0(t)\,\lambda\Delta t \tag{8.9}$$

The above equations can be rewritten as follows:

$$\frac{P_o(t+\Delta t) - P_o(t)}{\Delta t} = -P_o(t)\,\lambda + P_1(t)\,\mu$$

$$\frac{P_1(t+\Delta t) - P_1(t)}{\Delta t} = P_o(t)\,\lambda - P_1(t)\,\mu$$

The resultant differential equations are

$$\frac{dP_o(t)}{dt} = -P_o(t)\,\lambda + P_1(t)\,\mu \tag{8.10a}$$

$$\frac{dP_1(t)}{dt} = P_o(t)\,\lambda - P_1(t)\,\mu \tag{8.10b}$$

At time $t = 0$

$$P_o(0) = 1 \text{ and } P_1(0) = 0$$

The solution of this set of two differential equations yields:

$$P_o(t) = \frac{\mu}{\lambda+\mu} + \frac{\lambda}{\lambda+\mu}\exp[-(\lambda+\mu)t] \tag{8.11a}$$

$$P_1(t) = \frac{\lambda}{\lambda+\mu} - \frac{\lambda}{\lambda+\mu}\exp[-(\lambda+\mu)t] \tag{8.11b}$$

As per the definition of availability,

$$A(t) = P_o(t) = \frac{\mu}{\lambda+\mu} + \frac{\lambda}{\lambda+\mu}\exp[-(\lambda+\mu)t] \tag{8.12}$$

The availibility function is plotted in Fig. 8.6(a).

As time becomes large, the availability function reaches some steady-state value. The steady-state or long term availability of a single component is

$$A(t) = A(\infty) = \mu / (\lambda + \mu) \quad (8.13)$$

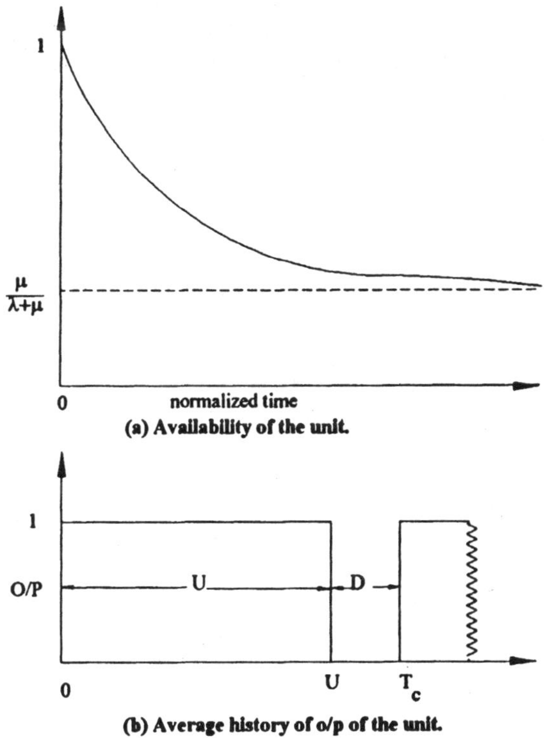

(a) Availability of the unit.

(b) Average history of o/p of the unit.

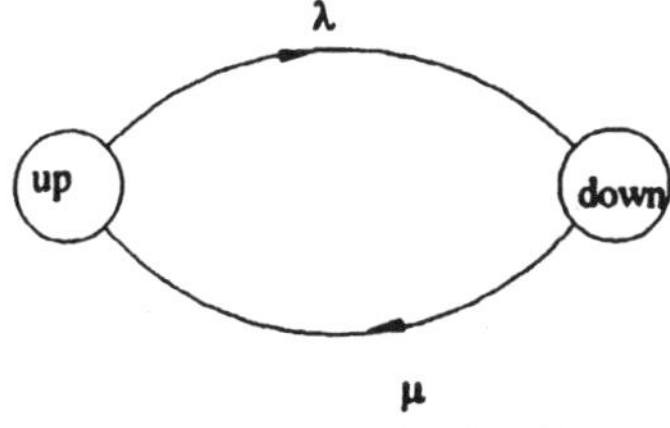

(c) Two state transition diagram.

Fig.8.6 Behaviour of a single repairable unit.

This equation can be modified as

$$A = \frac{1/\lambda}{1/\lambda + 1/\mu} \qquad (8.14)$$

Here, $1/\lambda$ is the mean time between failures (MTBF). It may be noted that this has been defined as the mean time to failure (MTTF) in the case of non-repairable components. $1/\mu$ is the mean repair time or mean time to repair (MTTR). Fig.8.6(b) characterizes the expected or mean behaviour of the component. U represents the mean up-time (MTBF) and D represents the mean down-time (MTTR). T_c is known as cycle time. Here,

$$U = 1/\lambda$$
$$D = 1/\mu$$

The steady-state availability is a number greater than zero and less than one. It is equal to zero when no repair is performed ($\mu=0$) and equal to one when the equipment does not fail ($\lambda=0$). Normally, $1/\mu$ is much smaller than $1/\lambda$ and therefore the availability can be approximated as

$$A = 1/(1 + \lambda/\mu) = 1 - (\lambda/\mu) \qquad (8.15)$$

When λ/μ approaches zero, A approaches unity.

$P_1(t)$ defines the unavailability of the equipment and hence

$$A'(t) = \lambda/(\lambda + \mu)\,[1 - \exp(-(\lambda + \mu)t)]$$

$$A' = A'(\infty) = \lambda/(\lambda + \mu) \qquad (8.16)$$

The number of failures per unit time is called the frequency of failures. This is given by

$$f = 1/T_c = 1/(U + D) \qquad (8.17)$$

The availability, transition rates (λ and μ) and mean cycle time can be related as follows:

$$A = U/(U + D) = fU = f/\lambda \qquad (8.18)$$

$$A' = D/(U + D) = f/\mu \qquad (8.19)$$

$$f = A\lambda = A'\mu \qquad (8.20)$$

Example 8.1

The following data was collected for an automobile:
mean time between failures = 500 hr
mean waiting time for spares = 5 hr
mean time for repairs = 48 hr
mean administrative time = 2 hr
Compute the availability of the automobile.

Solution

Total mean down time = 5 + 48 + 2 = 55 hrs.
Using relation(8.18), we get

$$\text{Availability} = \frac{500}{500+55} = 500/555 = 0.90$$

The automobile would be available 90% of the time.

* * *

Example 8.2

An equipment is to be designed to have a minimum reliability of 0.8 and a minimum availability of 0.98 over a period of 2×10^3 hr. Determine the mean repair time and frequency of failure of the equipment.

Solution

$$R(t) = \exp(-\lambda t)$$

Now, $R(t) = 0.8$ for $t = 2 \times 10^3$ hr

Therefore,

$$\lambda = -0.5 \times 10^{-3} \ln(0.8) = 1.12 \times 10^{-4} \text{ /hr.}$$

Also, steady state availability is given by equation (8.13),

$$0.98 = \frac{\mu}{\mu + \lambda}$$

or, $\mu = 0.98\ \mu + 1.12 \times 10^{-4} \times 0.98$

or, $\mu = 5.49 \times 10^{-3}$ /hr.

Hence, mean repair time is given by

$$MDT = 1/\mu = 10^3 / 5.49 = 182.2 \text{ hrs.}$$

Also, $f = \lambda A = 1.12 \times 10^{-3} \times 0.98 = 1.1 \times 10^{-4}$ /hr.

* * *

8.6 TWO UNIT PARALLEL SYSTEM WITH REPAIR

8.61 System Reliability

The reliability of a parallel system can be influenced by repairs. Consider a simple system having two units in parallel. In such systems when a unit fails it goes to repair and the other unit starts meeting the system demands. The system fails only when the second unit fails before the failed one is restored to operation. A two-unit system can be represented by a three-state Markov model as shown in Fig.8.7. At state 0 both the units are good, at state 1 one unit has failed and at state 2 both units have failed.

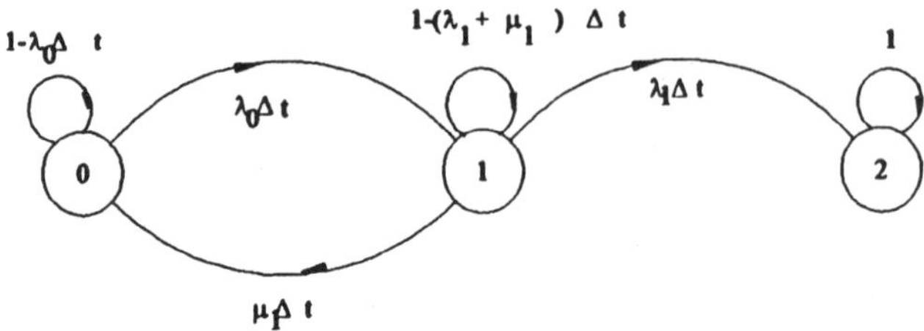

Fig.8.7 Markov reliability model for a two unit parallel system.

The following set of differential equations can be obtained from the state-probability equations,

$$P_o'(t) = -\lambda_o P_o(t) + \mu_1 P_1(t)$$

$$P_1'(t) = -(\lambda_1 + \mu_1) P_1(t) + \lambda_o P_o(t)$$

$$P_2'(t) = \lambda_1 P_1(t)$$

After solving for P's, we find that the system reliability is

$$R(t) = P_o(t) + P_1(t) = 1 - P_2(t) \tag{8.21}$$

$$= \frac{s_1}{s_1 - s_2} \exp(s_2 t) - \frac{s_2}{s_1 - s_2} \exp(s_1 t) \qquad (8.22)$$

Where,

$$s_1, s_2 = 1/2 \, [- (\lambda_o + \lambda_1 + \mu_1) \pm [(\lambda_o + \lambda_1 + \mu_1)^2 - 4 \lambda_o \lambda_1]^{1/2}] \qquad (8.23)$$

The *mean time to first system failure (MTFF)* is another system parameter useful for the analysis of system effectiveness when repairs are performed. This parameter is often referred to as the mean time between failures (MTBF) as the system states alternate between good and bad continuously due to repair.

$$MTFF = \int_0^\infty R(t)\, dt$$

$$= \int_0^\infty \frac{s_1 \exp(s_2 t) - s_2 \exp(s_1 t)}{(s_1 - s_2)}$$

$$= - (s_1 + s_2) / (s_1 s_2) \qquad (8.24)$$

For a two-unit system

$$s_1 + s_2 = - (\lambda_o + \lambda_1 + \mu_1)$$
$$s_1 s_2 = \lambda_o \lambda_1$$
$$MTFF = (\lambda_o + \lambda_1 + \mu_1) / \lambda_o \lambda_1 \qquad (8.25)$$

For the activo - redundant system, this turns out to be,

$$MTFF = (3\lambda + \mu)/2\lambda^2 = 3/(2\lambda) + \mu / (2\lambda^2) \qquad (8.26)$$

For $\mu = 0$, we get $MTFF = 3/(2\lambda)$ which is the mean time to failure of a two-unit non-maintained parallel system. Similarly, for a standby two-unit system

$$MTFF = (2\lambda + \mu)/\lambda^2 = 2/\lambda + \mu/\lambda^2 \qquad (8.27)$$

which reduces to $2/\lambda$ for $\mu = 0$.

8.62 System Availability

The approach to the computation of availability is same as that of reliability computation. However, since availability is concerned with the status of the system at time t, the repair at state 2 is also considered. The Markov-availability model is thus shown in Fig. 8.8.

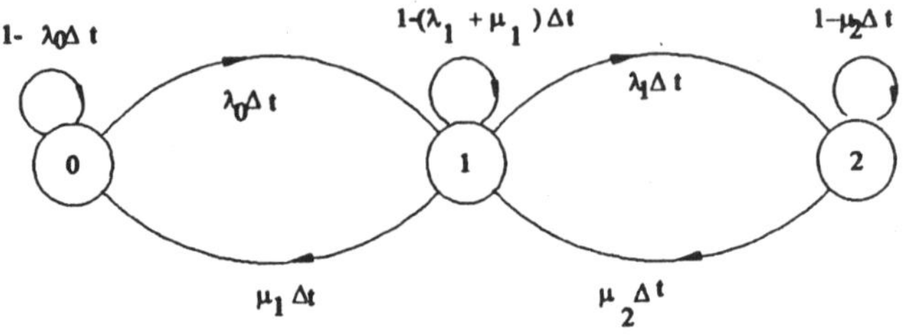

Fig.8.8 Markov availability model for a two unit parallel system.

The steady - state availability of the system is

$$A(\infty) = 1 - (\lambda_o \lambda_1)/(\lambda_o \lambda_1 + \lambda_o \mu_2 + \mu_1 \mu_2) \tag{8.28}$$

For the case of a two-unit active redundant system

$$\lambda_o = 2\lambda, \quad \lambda_1 = \lambda$$

$$\mu_1 = \mu, \quad \mu_2 = 2\mu$$

Therefore,

$$A(\infty) = 1 - \frac{\lambda^2}{\lambda^2 + 2\lambda\mu + \mu^2} = 1 - [\lambda/(\lambda + \mu)]^2 \tag{8.29}$$

For a two-unit series system, the availability becomes

$$A = \mu_1/(\lambda_o + \mu_1) = \mu/(2\lambda + \mu) \tag{8.30}$$

If we have n units in series, then

$$A = \mu/(n\lambda + \mu) \tag{8.31}$$

Example 8.3

Two transmitters are installed at a particular station with each capable of meeting the full requirement. One transmitter has a mean constant failure rate of 9 faults per 10^4 hrs and occurrence of each fault renders it out of service for a fixed time of 50 hours. The other trasmitter has a corresponding failure rate of 15 faults per 10^4 hours and an out of service time per fault of 20 hours. What is the mean availability of the system ?.

Solution

For the first trasmitter,

$$\lambda_1 = 9x10^{-4}/hr$$

$$\mu_1 = 1/50 = 0.02 /hr$$

Hence,

$$A_1 = [\mu_1/(\mu_1 + \lambda_1)] = [0.02/(0.02 + 9x10^{-4})] = 0.9569$$

Similarly, for the second transmitter,

$$\lambda_2 = 15x10^{-4}/hr$$

$$\mu_2 = 1/20 = 0.05 /hr$$

Hence,

$$A_2 = [\mu_2/(\mu_2 + \lambda_2)] = [0.05/(0.05 + 15x10^{-4})] = 0.9800$$

Hence, the system availability for two transmitters in parallel is given by:

$$\begin{aligned} A &= 1 - (1 - A_1)(1 - A_2) \\ &= 1 - (1 - 0.9569)(1 - 0.9800) \\ &= 1 - 0.0431 \times 0.02 = 0.9987 \end{aligned}$$

* * *

8.7 PREVENTIVE MAINTENANCE

Preventive maintenance is sometimes considered as a procedure intended primarily for the improvement of maintenance effectiveness. However, it is more proper to describe preventive maintenance as a particular category of maintenance, designed to optimize the related concepts of reliability and availability.

Preventive maintenance is advantageous for systems and parts whose failure rates increase with time. The cost savings accrue for preventive maintenance (planned replacement) only if the parts under consideration exhibit increasing failure rates. Many types of electron tubes, batteries, lamps, motors, relays and switches fall within this category. Most semiconductor devices and certain types of capacitors exhibit decreasing failure rates, while complex electronic systems generally have constant failure rates. In the latter case, certain classes of parts within the systems display increasing failure rates; consequently, the effectiveness of a preventive maintenance program depends on how well it detects these deteriorating parts.

Reduction of operational failures is the real purpose of scheduled or preventive maintenance. To achieve a balance between reliability and maintenance costs for any equipment, several factors must be weighed simultaneously and a suitable trade-off point selected. The various factors to be considered are:

1. The reliability index and time duration desired.
2. The cost of an in-service failure.
3. The cost of replacement before failure.
4. The most economical point in equipment life to affect this replacement.
5. The predictability of the failure pattern of equipment under consideration.

The ideal procedure would be to replace a unit just prior to failure, and thus realize the maximum of trouble - free life. The relationship used here gives the average hourly cost in terms of two costs, K_1 and K_2 and the failure probability distribution of the particular item.

The model is as follows:

$$A(t) = \frac{K_1 - (K_1 - K_2)G(t)}{\int_0^t G(t)\,dt} \tag{8.32}$$

Where,

$A(t)$ = the average hourly cost
K_1 = the total cost of an in-service failure
K_2 = the total cost of a scheduled replacement
$G(t)$ = the probability that a new unit will last at least t units of time before failure
t = the time to replacement after the last replacement.

Application of this technique enables the optimum replacement interval to be determined if the failure distribution is known. If the replacement interval is too short, considerable loss of useful equipment life would result and the average hourly cost would be high. However, if the replacement interval is too long, then the cost of an in-service failure, in terms of mission aborts and manpower, are quite intolerable. The ratio of K_1 (the cost of in-service failure) to K_2 (the cost of scheduled replacement) is the critical factor in arriving at a decision regarding scheduled replacement policy. As the ratio increases, the lowest average hourly cost is realized by replacing the part after a shorter life, as shown in Fig.8.9.

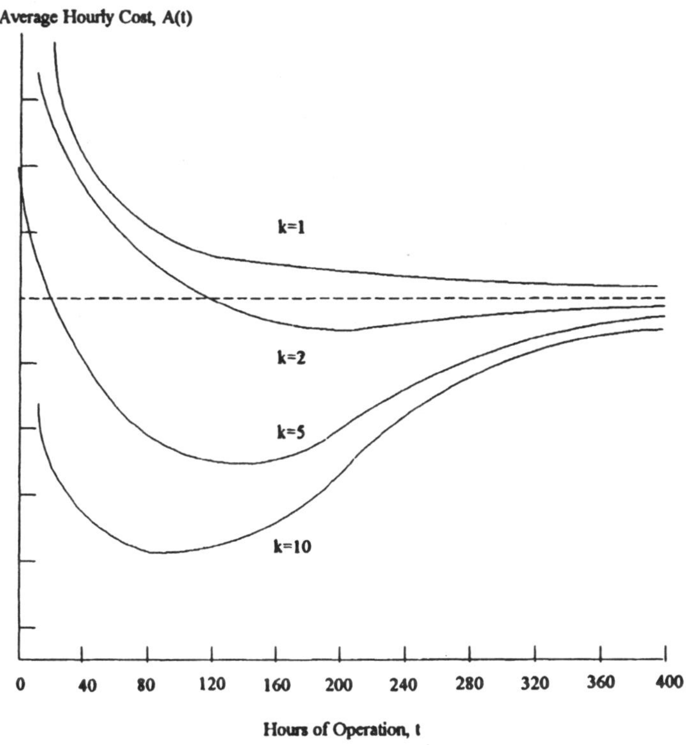

Fig.8.9 Average hourly cost of scheduled replacement.

In the figure, a model for aircraft engine was considered and the family of curves is plotted for various ratios of K_1 to K_2 which is denoted as K. When $K=1$ there is no advantage in scheduled replacement, and the equipment should be allowed to run to failure. When $K>1$, there is an advantage in scheduled replacement. If, for example, the cost of in-service failure was 10 times the cost of a scheduled replacement, then the $K=10$ curve shows that replacement should be scheduled at approximately 80 hr in this case as the cost would be the least at this point.

Preventive maintenance achieves its purpose by inspection, servicing, and minor and major overhauls during which the actions fit into three classes:

1. Regular care of normally operating subsystems, devices, and components which require attention (lubrication, refuelling, cleaning,

adjustment,etc).
2. Checking for, with replacement or repair of, failed redundant components.
3. Replacement or overhaul of components or devices which are nearing wearout.

The frequency of performing class 1 actions to prevent degradation of system reliability depends on the characteristics of the components. The frequency of performing class 3 actions depends on wearout characteristics and the number of components in a system. While these differ for various kinds of devices, an optimum replacement time table can be established in advance. The frequency of performing class 2 actions is a function of probabilities since it depends on failure rates of redundant components in a system and on the reliability required of the system.

An indication of the effect of preventive maintenance on a system is given by the following example.

Example 8.4

Compare the MTBF, reliability for a 10 hr mission, and number of system failures (assume 1000 missions) for

(a) A single unit with an MTBF of 100 hr.
(b) Three such units in parallel with off-schedule maintenance.
(c) Three such units in parallel with *perfect* i.e. periodic maintenance after each 10 hr mission.
(d) A single unit having the MTBF of case (c).

Periodic maintenance is performed every T hr, starting at time zero. Each device and component is checked. Each one which has failed is replaced by a new, statistically indentical component. For useful life, the system is restored to *as good as new* condition after each maintenance operation, since there has been no deterioration of components.

Solution

(i) MTBF

(a) Given as 100 hr.

(b) MTBF $= (1/\lambda + 1/(2\lambda) + 1/(3\lambda)) = (1 + 1/2 + 1/3)(100)$
$= 183.3$ hr.

(c) $R = 1 - Q^3 = 1 - (1 - R_1)^3 = 1 - (1 - \exp(-0.1))^3 = 0.999138$

Expect 0.862 failure/1000 missions of 10 hours each, i.e. 0.862 failure/10,000 hr.

MTBF = 10,000/0.862 = 11,600 hr.

(d) MTBF = 11,600 hr.

(ii) Reliability

(a) $R = \exp(-\lambda t) = 0.90484$.

(b) MTBF = 183.3 hr.

Expect one system failure for every 18.33 missions. 1000/18.33 = 54.56 system failures per 1000 missions, or, R = 0.94544. This is an *average*. When all three units are good, R = 0.999138; when two are good, R = 0.991; when only one is good, R = 0.90484.

(c) R (previously calculated) = 0.999138.

(d) Same as (c).

(iii) Number of System Failures

(a) $N_f = NQ = 1000(1-R) = 1000\,(1-0.90484) = 95.16$. Hence, 96 system failures/1000 missions.

(b) Calculated in determining reliability, 55 system failures /1000 missions.

(c) Calculated in determining MTBF, 1 system failure /1000 missions.

(d) Same as (c).

In a real sense, the effect of successful repairs is equivalent to standby redundancy. The repair is instantaneous, so far as system and mission performance is concerned, if it is accomplished within the maintenance time constraint.

* * *

8.8 PROVISIONING OF SPARES

The number of spares that should always be at hand to maintain a system properly and effectively is of major importance. Therefore, the determination of the number of spares should be made on scientific basis. Too few spares on hand may affect the missions success because of their non-availability in urgent situations. On the other hand ,storing too many spares increases the expenditure and pay loads.

Today, with the knowledge of reliability principles, it is possible to forecast spare part requirements in a more scientific manner. Basically, the simplest method is to divide the expected life or mission requirement by the mean time between failures. However, the average itself is not always adequate, because there is a definite probability that more than the average number of spares may be required for the specific period.

To determine more precisely the number of spares, we use Poisson distribution to derive the formula:

$$S = \lambda T + Z (\lambda T)^{1/2} \tag{8.33}$$

where S = number of spares

T = mission time

λ = failure rate per hour

Z = confidence level (a variable measured from the mean in standard deviations which determines the area under normal curve from $-\infty$ to Z). Values of Z are to be found from standard statistical tables.

Example 8.5

Consider a system consisting of 10 tubes. The failure rate for each tube is $\lambda = 0.01$/hr. How many spares are necessary to satisfy a 99.73% confidence level,that there will be no stock out for a mission time of 1000 hr.

Solution

Here, $\lambda = 0.01$/hr, T = 1000 hr

Z = 3.0 for 99.73 % confidence level (From standard tables)

Using relation (8.33), we get

$$S = \lambda T + Z (\lambda T)^{1/2} = 0.01 \times 1000 + 3(0.01 \times 1000)^{1/2}$$
$$= 19.49 \approx 20 \text{ tubes as spares for each tube.}$$

The total number of tubes required are 20 x 10 = 200 tubes.

* * *

The number of *spares needed* can also be easily determined from the cumulative Poisson distribution.

$$P = \sum_{i=0}^{i=n} [\exp(-\lambda T)(\lambda T)^i] / i! \qquad (8.34)$$

Where,

P = probability of subsystem mission success
λ = equipment failure rate
T = mission time
n = number of spares required

Design for long-term missions cannot depend on high reliability alone, but must be optimized around the availability concept which requires establishing an appropriate balance among system performance, reliability, maintenance complexity, and spares weight/volume.

9
RELIABILITY TESTING

9.1 INTRODUCTION

Reliability tests measure or demonstrate the ability of an equipment to operate satisfactorily for a prescribed period of time under specified operational conditions. The objective of a reliability test program is to gain information concerning failures, i.e., the tendency of systems to fail and the resulting effects of failure. Thus, in a sense, reliability tests are distinguished from most other types of tests, which are generally concerned with normal operation of equipment.

The overall test program for a product can be considered to be the most important single phase of a well-planned and executed reliability program, requiring the largest expenditure of reliability/ quality funds and manpower. It provides the vital inputs on which the designer bases his design and subsequent redesign or design refinement. It is the source of almost all meaningful data from the inception of the project throughout the entire life of the hardware, the springboard for corrective action on design, process, and use, and the only sound basis on which logistics planning can proceed to ensure that the necessary parts and maintenance capability are available to support the equipment in actual use. It provides project management with the most vital information on the technical progress and problems of the project.

The importance of a complete, integrated, planned, documented, and vigorously prosecuted test program cannot be overemphasized, and it is essential that the the most qualified personnel available be assigned to all phases of it.

Although the details differ with the product under consideration, reliability testing at any point in the life cycle is often severely limited by both money and time. Unless the subject of the test is a very inexpensive mass-produced component, it is costly to devote enough units to testing to make the sample size as large as one would like, particularly when the test is likely to cause wear and even destruction of the test units. The time over which the test units must be operated in order to obtain sufficient failure data also may be severely restricted by the date at which the design must be frozen, the manufacture commenced, or the product delivered. Finally, there is a premium attached to having reliability information early in the life cycle when there are few test prototypes available. The later design, manufacture, or operating modifications are made, the more expensive they are likely to be.

9.2 KINDS OF TESTING

Since a comprehensive reliability test program encompasses all tests on the hardware from inception of the project through the final use and disposition of hardware, it follows that the test program includes many kinds of tests. Intelligent planning of an overall test program, then, requires an understanding of the kinds of tests that are available in order that optimum choices can be made. In the following paragraphs we shall subdivide tests into five categories by different factors.

1. *Destructive vs. Nondestructive Testing*

Simply speaking, a destructive test is one that will leave the tested hardware unfit for further use, whereas a non-destructive test is one that will not. In most cases, as with tests of explosives, this simple definition will suffice. However, in some rather rare instances the hardware may still be usable for limited purposes, as with a complete design or production qualification test which leaves the hardware unfit for delivery to a customer but perfectly good for testing to failure to determine failure modes. Hence it is important that the possible or potential *further use* be examined early in deciding on the exact elements of any test program so that a trade-off can be made whenever it is economically feasible.

Other factors being equal, economically it is always desirable to utilize nondestructive testing instead of destructive, provided the net cost to the program is not adversely affected because more nondestructive tests are required to achieve the same purpose as might be achieved with a small number of destructive tests. Furthermore, non-destructive testing leaves the test sample in condition to permit meaningful failure diagnosis, enhancing considerably the potential value of the test.

Most ordinary production testing is nondestructive, but in high reliability programs these tests are backed up with destructive tests performed on samples drawn at regular intervals from the production line. With such a project, it is frequently possible to test only the critical parameters for every unit of a product, leaving for the sample production-assessment program those less critical parameters which can be safely sampled. Thus an economy of testing is achieved by combining in the sample testing both noncritical nondestructive tests and the destructive tests.

2. *Ambient vs. Environmental Testing*

Ambient testing is usually considered to include that testing performed under existing static conditions found in the laboratory or on the factory floor, while environmental testing includes all testing in which the specimen is subjected to some nonambient condition. However, some testing performed under actual-use conditions at existing environmental conditions, particularly when the locale is deliberately chosen to provide extremes in temperature, vibration, humidity, dust, etc., is also considered to be environmental.

Ambient tests are usually used for production testing, largely because of their simplicity and economy. (They may run one tenth to one hundredth the cost of an environmental test.) To be useful in high-reliability production projects, it is essential that they be developed in the R&D phase, in conjunction with environmental tests, to determine their validity for separating out material which will not function in the actual environments that will be encountered by the hardware after delivery.

It should be apparent that the ambient test is only a substitute, dictated by economy, for an environmental test, unless the actual-use conditions for the hardware are approximately equivalent to the factory ambient, as in the case of household appliances or computers destined for use in a protected environment. The comparative economy of ambient tests, however, makes them the most widespread of the two, and when properly correlated with expected performance in use environment, they provide a very high degree of assurance that the hardware will properly function in use. Because of their relative simplicity, they can be and are used at all levels of assembly. Environmental testing is necessary in a high reliability project to determine in absolute terms the performance of the hardware in actual use.

3. *Actual Conditions vs. Simulated (Laboratory)*

Environmental testing can be classified according to the method by which the environmental exposure is created, i.e, naturally in the actual use environment or artificially in a laboratory. Consideration of these subdivisions is worthwhile, since the cost of testing and the usefulness of the data accumulated may vary markedly between them. The decision to choose one over the other is complex and is based on many factors, a few of which follow:

a) Size of Parts
b) Nature of the Parts
c) Frequency of Testing
d) Complexity of Instrumentation
e) Complexity of the Test
f) Accessibility of Natural Environments
g) Relative Costs
h) Relative Time

4. *Levels of Tests*

A fourth convenient way to classify testing is by the level of assembly. Tests can be performed at all levels, but for practical purposes the levels generally chosen are parts, subassemblies, assemblies, subsystems, and systems. Two opposing functions of each class of test operate in check and balance to require some testing at each level, and they are the principal factors dictating the selection of particular attributes to be tested at a specific level. Thus, for example, in production test programs it is desirable to test every attribute as soon as possible after it is created to preclude further investment in nonconfirming hardware. This is the true quality control function. Opposing this function, however, is the necessity for testing attributes at the last possible time before the attribute is covered up to ensure that nothing in the production process has degraded the attribute. This is generally called acceptance testing, and is the customer's assurance that the product being delivered meets the functional requirements.

A facet of the problem of integrating tests, however, needs to be considered with production testing, particularly with electronic or hydraulic functional hardware. Many functional attributes in this kind of hardware drift with time, handling, or functional cycling. If the acceptance limits on these attributes are set identically at successively higher levels of test, there will be a measurable percentage of hardware with attributes just inside the limits at one level of test which will drift outside the limits in the next test and be rejected back to the lower level for rework. To preclude the resulting circulation of hardware in a properly integrated series of successive-level tests, the tolerances of a single attribute are established in a funnel arrangement, with the tightest tolerance at the lowest level of assembly as shown in Fig.9.1.

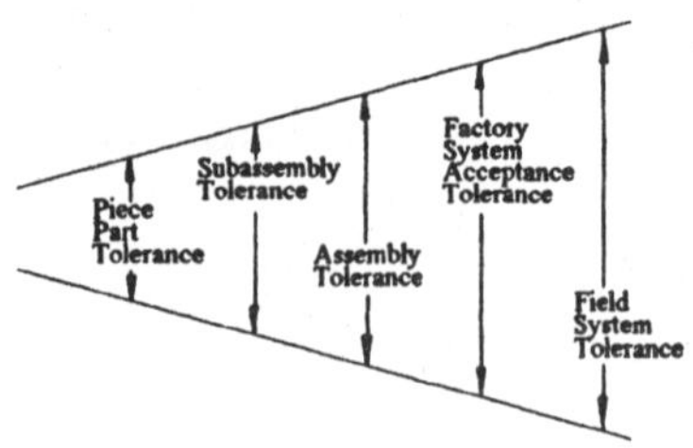

Fig.9.1 Tolerance funneling.

5. *Tests by Purpose*

When one suggests that a test program is needed, the first question is generally *What kind of test?* meaning a test for what purpose. It is natural to think of testing in terms of the intended purpose for which it is being run, since this is the usual departure point for all of the planning, funding, assignment of responsibility, and use of the resulting data. In a comprehensive test program associated with a high reliability project, it is convenient to consider the many purposes for which tests are conducted in groups, named as evaluation; simulated use; quality; reliability; consumer research, and investigations.

9.21 Reliability Tests

Although all testing contributes data for reliability calculations and hence could be considered in a larger sense to be reliability testing, there are specific tests which are performed for no other purpose than to gather these data. These are the tests referred to in this section, and for purposes of this discussion they have been grouped into peripheral testing, life testing, accelerated life testing, service-life evaluation testing, and surveillance testing. The data from reliability testing are used to determine mean time or cycles to and between failure, to calculate or verify attained reliability, to establish storage and operating life limits on critically age-sensitive parts (and from both of these come the depth requirements for spare parts), and to determine modes of failure. Reliability tests are performed at all stages of the project and on all levels of assembly. They are performed both in ambient and environmental conditions, and they include both destructive and nondestructive tests, inspections, and examinations. They may also include some actual-use tests, although they are usually confined to the laboratory to ensure control of input conditions.

1. *Peripheral Testing*

In testing the parts are subjected to environments and input conditions which simulate as nearly as possible the actual range of use conditions, and a successful test is one in which the part functions properly in these conditions. From such testing, however, it is not possible to determine how much margin of safety has been designed and built into the product, since the part has not been stressed to functional yield. It is useful in predicting reliability of a population from data gathered on a limited sample to test the parts to environments and input conditions which are more rigorous than the expected service conditions by a substantial enough margin that failures can clearly be attributed to the peripheral conditions. This margin should be at least 10 to 15 per cent of the spread from ambient to the limit of the service environment.

2. *Life Testing*

Reliability prediction and reliability assessment are vitally concerned with the determination of the mean time (or cycles) to and between failures, since this number is basic in reliability calculations. The number can be computed directly from the data gathered from the life test program, where tests are performed not only on samples of completed assemblies but on spares and piece parts as well. The tests are generally performed in the laboratory on test equipment which, for economy of testing cost, is designed to operate continuously or cycle the hardware automatically. The operation is interrupted at regular intervals, and functional tests or nondestructive inspections are made to find out whether there has been any degradation of the operability of the part with time or cycles of operation. Generally, the most severe expected service environments are chosen and a number of samples are utilized in a statistical design of experiments which permit the interpretation of results.

Life testing is slow and expensive and may take six months to a year to complete. In some situations, where real time is the same as operating time, the test program may take years; typical of these are tests of paint, where the actual service conditions are exposure to outdoor weather, or of submarine cable and equipment, where the actual service condition is exposure to ocean depths. In these situations it is essential that the life-testing program be instituted on the earliest production prototypes, so that field failures of service equipment delivered at a later time can be predicted prior to occurrence or that corrective action on the design or production process can be instituted before production actually begins.

3. *Accelerated Life Testing*

In a tightly compressed schedule, where R&D is hardly finished (or sometimes is not completely finished) before production starts, some assurance must be obtained relatively quickly that the hardware has an adequate life and that no gross weaknesses exist in the design that has been released on high risk basis to production.

Life tests are ordinarily too drawn out to provide such gross information quickly enough to permit design corrections to be made expeditiously. In these projects an accelerated life-test program is generally instituted. We shall discuss Accelerated Life Testing in details in a subsequent section of this chapter.

4. *Service-Life Evaluation Testing*

One problem facing top management of high reliability projects is the determination of the amount of useful life left at any given time in equipments which have been delivered for service use. This knowledge is necessary to permit continuing intelligent evaluation of several aspects of the project and to make important decisions concerning them. Among these, perhaps the most important, particularly in a weapon project, is the decision that the tactical field or fleet stock either has sufficient remaining life that no replacement, refurbishing, or reworking action must be instituted or that it has not. The reasons for instituting such action may be that an intolerable degradation in performance has occurred or that the explosives have reached or are reaching a critical point where further degradation may result in explosive hazard.

SLE testing is generally accelerated life testing, since the object of the testing is to provide management with immediate answers on the expected life remaining in the field population. The samples selected should be the oldest or those with the most use in order that the worst material condition can be detected. Functional hardware should be tested at ambient conditions both before and after being exposed to the accelerated-aging environment or cycling, and the results of these ambient tests should be compared with each other as well as with the original factory test data taken at the time the parts were delivered.

5. *Surveillance Testing*

The last test program in the reliability test group is surveillance testing. These tests, which are performed on samples drawn at regular intervals from the actual field service stocks, consist of ambient tests and examinations performed on the samples at progressive levels of disassembly. The object of the testing is to discover evidence of failure or incipient failures in the hardware, including not only shifts in values of components in functional hardware but chemical deterioration of materials, fatigue cracks, corrosion, whiskers, hardening of rings and seals, and any other unanticipated modes of failure.

The two characteristics differentiating surveillance testing from other kinds of reliability testing are the limitation of testing to ambient examinations and the complete disassembly of the specimens.

9.3 COMPONENT RELIABILITY MEASUREMENTS

For this purpose we must differentiate between the probability of chance failures and the probability of wearout failures. These two probabilities provide us with two different types of information, both of which are useful for the reliable application of components. The first probability tells us how reliable components are in their useful life period, the second tells us how long components can be safely used without jeopardizing the reliability of the equipment in which they are installed. The methods of measurement and their statistical evaluation are different for chance failures and for wearout failures.

As to chance failures, we are interested in a single parameter - the mean time between failures. When this parameter is known for a given stress level of operation, the reliability at that stress level for a given mission time t is then calculated from the exponential formula $R = \exp(-t/m)$, where m, the mean time between failures, is the reciporcal of the failures rate λ. The true value of a probability is theoretically never exactly known, but we can come quite close to it when we perform a large number of experiments. This also applies to the mean time between failures; therefore all we expect to obtain in reliability measurements is a reasonably good estimate.

How good an estimate is depends on the amount of available data from which the estimate is computed. We shall see later that we can set so-called *confidence limits* on both sides of the estimate, an upper and a lower confidence limit, but first let us see how to obtain estimates of the mean time between failures.

We have said that an estimate of the mean time between failures is obtained by measuring the times to failure t_i, of a large number n of specimens, forming the sum t_i, and dividing this by the number of observations, i.e., by the number of times to failure. However, in practice we have to bear in mind that components may fail both because of chance and because of wearout. The main problem which we encounter right from the beginning when planning this test is how much time we can afford to spend. It is obvious that we cannot wait until all components fail so as to have

$$m = (\Sigma t_i) / n \qquad (9.1)$$

Even if we had several years time so that we could compute the mean for all components, the question of how many of them had failed because of chance and how many had failed because of wearout would arise. We can safely assume that the majority would fail because of wearout.

We thus have to limit the duration of the test so as to be reasonably certain that no wearout failures will occur during the test period. This means that we can allow only a small fraction of the components to fail, say r, and after the rth failure we discontinue the test. We then have r measurements of times to failure for all n components; assuming that the r failures were chance failures, we can compute the mean time between (chance) failures for this component population assuming constant failure rate.

The optimum estimate for the mean time between failures is given by:

$$m = [t_1 + t_2 + \dots + t_r + (n - r)t_r] / r$$

$$= (1/r)\left[\sum_{i=1}^{r} t_i + (n - r)t_r\right] \qquad (9.2)$$

To avoid component wearout failures during a test, the test truncation time t_r should be chosen as short as possible compared to the wearout time of the components. On the other hand, because the precision of the estimate m, depends on the number of the times to failure measured during the test and therefore on the number of chance failures, it follows that the largest possible samples of components should be tested.

The choice of the sample size, i.e., of the number of components which we should submit to a test, depends on the available test time t_r and on the precision of or confidence in the test result which we wish to achieve.

When the available test time for a nonreplacement test is t hours and the expected failure rate of the specimens is λ, and m has to be measured with a precision corresponding to r chance failures, the number of specimens n to be submitted to the test is

$$n = r/[1 - \exp(-\lambda t)] = r/Q(t) \qquad (9.3)$$

where Q(t) is the expected unreliability of the components for a test operating time t.

If no provisions are made for an exact measurement of the times of each chance failure during a nonreplacement test and the test is truncated at the time t_r, when the rth failure occurs in an initial lot of n components, the mean time between failures can be estimated from the formula for the probability of failure.

$$r/n = Q(t_r) = 1 - \exp(-t_r/m) \qquad (9.4)$$

Since the time t_r of the test duration is known and r chance failures have been counted during the test, the estimate m is obtained as

$$m = t_r / [\ln(n) - \ln(n-r)] \qquad (9.5)$$

The corresponding estimate of the per-hour failure rate is then

$$\lambda = [\ln(n) - \ln(n-r)] / t_r \qquad (9.6)$$

The ratio of the failing components r to the total initial number of components n can also be expressed in terms of the percentage of failed components during the test, a, so that a = 100r/n. Equation (9.6) then assumes the form:

$$\lambda = [\ln(100) - \ln(100 - a)] / t_r = [4.60517 - \ln(100 - a)]/t_r \qquad (9.7)$$

where a is the per cent of failed components.

To avoid the use of logarithms in quick estimating work, the following thumb-rule approximation of Equation (9.7) is sometimes used:

$$\lambda = a/[(100 - a/2)t] \qquad (9.8)$$

where a is the per cent of components failing in a test of duration of t hours. This formula is not quite exact because it assumes that the r = na/100 failures occurred at an average time of t/2, i.e., halfway through the test, but it can be used for quick estimating work when the per cent of failing components a is small.

9.31 Verification of Exponential Law

In the reliability test's analysis, described above, it is frequently assumed that an item's failure times are exponentially distributed. In order to verify this assumption, various tests are available in the published literature. In this section, we describe one such test known as the Bartlett test. The Barlett test statistic is defined as:

$$S_{bk} = 12k^2 [\ln X - (Y/k)] / (6k + k + 1) \qquad (9.9)$$

where,

$$X = (1/k) \sum_{i=1}^{k} t_i \qquad (9.10)$$

$$Y = \sum_{i=1}^{k} \ln(t_i) \qquad (9.11)$$

where t_i is the ith time to failure and k is the total number of failures in the sample.

A sample of at least 20 failures is necessary for the test to discriminate effectively. If the failure times are exponentially distributed, then S_{bk} is distributed as chi-square with (k-1) degrees of freedom. Thus, a two-tailed chi-square approach (criterion) is utilized.

Example 9.1

A sample of 20 failure times (in days) of an air traffic control system is given in Table 9.1. Determine with the aid of Barlett's test that the data are representative of an exponential distribution.

TABLE 9.1

Failure Times (in days)

7	35	85	142
8	46	86	186
20	45	111	185
19	63	112	266
34	64	141	267

Solution

Substituting the specified data into Equation (9.10) yields

$$X = [1/20](7 + 8 + 20 + 19 + 34 + 35 + 46 + 45 + 63 + 64 + 85 + 86 + 111 + 112 + 141 + 142 + 186 + 185 + 266 + 267)$$

$$= 96.10$$

Similarly, from Equation (9.11) we get Y = 82.8311

With the aid of the above results from Equation (9.9) we get

$$S_{b20} = 12(20)^2 [\ln(96.10) - (82.8311/20)] / [6(20) + 20 + 1]$$
$$= 14.43$$

From Table 9.2 for a two-tailed test with 90 percent confidence level, the corresponding values are:

$\chi^2 [\theta/2, (k-1)] = \chi^2[(0.1/2), (20-1)] = 30.14$

where $\theta = 1 - (\text{confidence level}) = 1 - 0.90 = 0.1$

$\chi^2 [(1-\theta/2), (k-1)] = \chi^2 [(1 - 0.1/2), (20-1)] = 10.12$

TABLE 9.2

Chi- Square Distribution

Degree of Freedom	Probability			
	0.975	0.950	0.05	0.025
1	0.001	0.004	3.840	5.020
2	0.050	0.100	5.990	7.380
3	0.220	0.350	7.820	9.350
4	0.480	0.710	9.490	11.14
5	0.830	1.150	11.07	12.83
6	1.240	1.640	12.59	14.45
7	1.690	2.170	14.07	16.01
8	2.180	2.730	15.51	17.54
9	2.700	3.330	16.92	19.02
10	3.250	3.940	18.31	20.48
11	3.820	4.580	19.68	21.92
12	4.400	5.230	21.92	23.34
13	5.010	5.890	22.36	24.74
14	5.630	6.570	23.69	26.12
15	6.260	7.260	25.00	27.49
16	6.910	7.960	26.30	28.85
17	7.560	8.670	27.59	30.19
18	8.230	9.390	28.87	31.53
19	8.910	10.12	30.14	32.85
20	9.590	10.85	31.41	34.17
21	12.40	13.85	36.42	39.36

The above results exhibit that there is no contradiction to the assumption of exponential distribution.

* * *

9.4 PARAMETRIC METHODS

Plotting the reliability or other quantities versus time as in the Chapter-1 often yields valuable information. In general, however, it is more desirable to fit the reliability data to some particular distribution, such as the exponential, normal, or Weibull. For if this can be accomplished, a great deal more can often be determined about the nature of the failure mechanisms, and the resulting model can be used more readily in the analytical techniques.

In order to obtain parametric models for failure distributions, we must first determine what distribution will adequately represent the data and then determine the parameters. There are a variety of advanced statistical methods for determining the goodness of fit of data to a particular distribution, for estimating the parameters for the distribution, and for calculating confidence levels for each parameter. In what follows, however, we confine our attention to relatively simple graphical methods. Such techniques allow us to evaluate the goodness of fit visually, without using advanced mathematics, and at the same time to estimate the parameters that define the distribution.

In general, the procedure that we follow consists of choosing a distribution and then plotting ungrouped failure data on the appropriate graph paper for this distribution. If the data are described by the distribution, the data points will be clustered along a straight line. The parameters are then estimated from the slope and intercept of the line.

9.41 Exponential Distribution

Often the exponential distribution or constant failure rate model is the first to be used when we attempt to parameterize data. In addition to being the only distribution for which only one parameter must be estimated, it provides a reasonable starting point for considering other two or three parameter distributions. For as will be seen, the distribution of the data may indicate whether the failure rate is increasing or decreasing, and this in turn may provide insight whether another distribution should be considered.

To plot data, we begin by:

$$\ln R = -\lambda t \tag{9.12}$$

or,

$$\ln (1/R) = \lambda t \tag{9.13}$$

It is customary to construct graph paper in terms of Q = 1 -R. Thus we have

$$\ln[1/(1-Q)] = \lambda t \tag{9.14}$$

An exponential distribution probability paper is shown in Fig.9.2. The numerical values labeled on the vertical axis are those of $Q(t_i)$, which may be obtained from

$$Q(t_i) = i / (N + 1) \qquad (9.15)$$

where N is the number of test units. It will be noted that $\lambda t = 1$ when $1 - Q =$

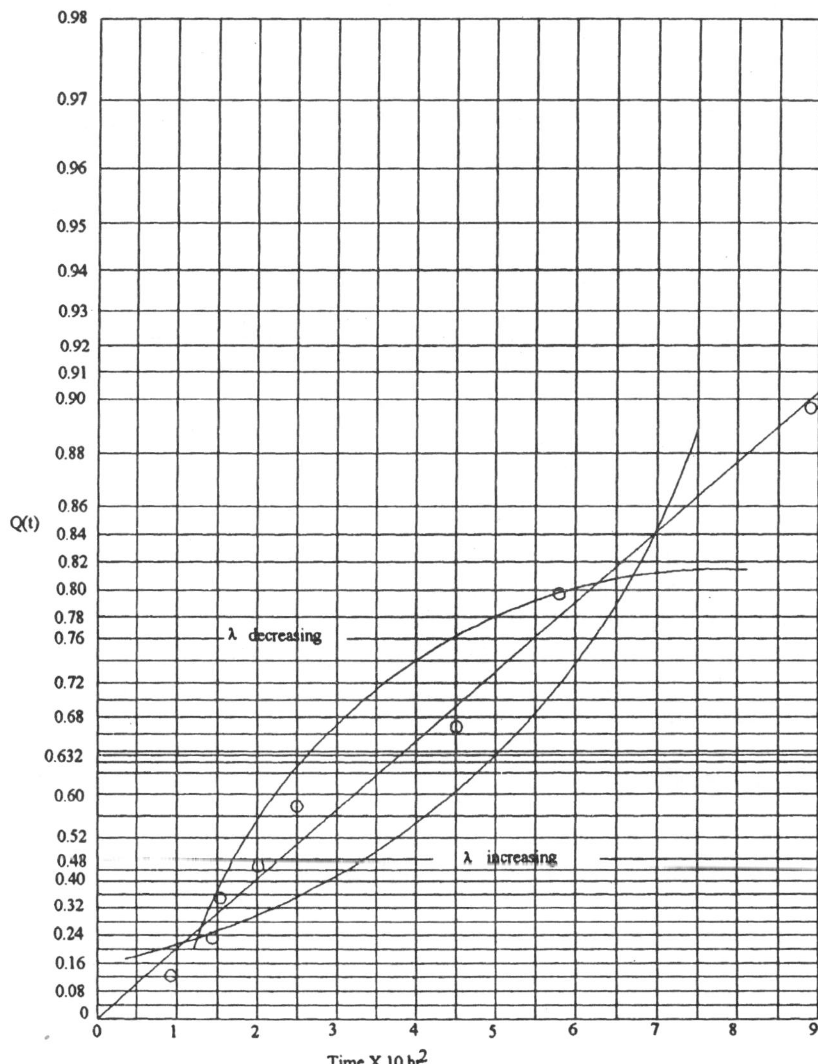

Fig. 9.2: Graphical parameter estimation for the exponential distribution.

e^{-1} or $Q = 0.632$. Thus the value of $1/\lambda$ is equal to the time at which $Q = 0.632$. The data through which the straight line is drawn on Fig.9.2 come from the following example.

Example 9.2

The following are the failure times from eight control circuits in hours: 80, 134, 148, 186, 238, 450, 581, and 890. Estimate the failure rate by making a plot on exponential distribution probability paper.

Solution

The calculations are carried out in Table 9.3. From Fig.9.2 we see that $Q = 0.632$ when $t = 400$ hr. Therefore we estimate $\lambda = 0.0025$/hr.

TABLE 9.3

Exponential Calculations

i	t_i	i/N+1	i	t_i	i/N+1
1	80	0.111	5	238	0.555
2	134	0.222	6	450	0.666
3	148	0.333	7	581	0.777
4	186	0.444	8	890	0.888

* * *

The following is an important feature of plotting failure times on logarithmic paper. If the failure rate is not constant, the curvature of the data may indicate whether the failure rate is increasing or decreasing. The dotted lines on Fig.9.2 indicate the general pattern that the data would follow were the failure rate increasing (concave upward) or decreasing (concave downward) with time.

9.42 Weibull Distribution

The two-parameter Weibull distribution may also be estimated by plotting failure times on specially constructed graph paper. To arrange the Weibull data on a straight line, we first take the logarithm of the Weibull expression for the reliability and obtain;

$$(t/\theta)^m = \ln(1/R) \tag{9.16}$$

Then, taking the logarithm again, we obtain

$$\ln(t) = (1/m)\ln[\ln(1/R)] + \ln(\theta) \tag{9.17}$$

If we rewrite this equation as

$$\ln[\ln(1/R)] = m\ln(t) - m\ln(\theta) \tag{9.18}$$

we see that it has the form $y = mx + b$, where the ordinate is $\ln[\ln(1/R)]$ and the abscissa is $\ln(t)$. Once again it is the convention to number the vertical axis values for $Q = 1 - R$ rather than R. Thus, we plot

$$\ln[\ln\{1/(1-Q)\}] = m \ln(t) - m \ln(\theta) \quad (9.19)$$

The two Weibull parameters are then estimated directly from the straight line. The slope m is obtained by drawing a right triangle with a horizontal side of length one; the length of the vertical side is then the slope. The value of θ is estimated by noting that the ordinate vanished when $Q = 0.632$ yielding $t = \theta$.

9.43 Normal Distribution

Graphical methods may also be used to determine whether a sequence of failure times or other data may be approximated by a normal or a lognormal distribution. We begin with the standardized CDF for the normal distribution given by

$$Q(t) = \Phi\,[(t-\mu)/\sigma] \quad (9.20)$$

where μ is the mean and σ is the standard deviation. Normal paper is based on inverting the above equation to obtain

$$\Phi^{-1}(Q) = (t/\sigma) - (\mu/\sigma) \quad (9.21)$$

Here the inverse of the standardized normal distribution, $\Phi^{-1}(Q)$, is plotted on the vertical axis and time is plotted on the horizontal axis. If the failure data are normally distributed, the line will be straight.

9.44 Bayesian Analysis

Reliability engineering is frequently faced with a paucity of available test data. If only a very few units can be tested to failure because of expense or production deadlines, the foregoing graphical methods are no longer very helpful for estimating parameters.

In such situation engineers may turn to the alternative of using reliability models and parameters from similar equipment. Similarly, the experienced engineers may extrapolate values from previous equipment models with which they are familiar. This indeed is frequently done, particularly if the constant failure rate or some other standard model can be assumed applicable. Nevertheless, it would be unfortunate if test results on the system under consideration- however few- could not be factored into the reliability estimates.

Bayesian analysis makes this possible, for given an estimate of a parameter such as a failure rate, the test results can be used to upgrade that estimate in a systematic way. More specifically, the engineer utilizes handbooks, expert opinion, and previous experiences to formulate a probability distribution expressing the uncertainty in the true value of a parameter. This is referred to as the prior distribution. The best point estimate of the parameter would normally appear as the mean or median of this distribution. With Bayesian analysis the test data are used to modify the distribution, yielding the so-called posterior distribution. Since the posterior distribution represents the new state of knowledge, its mean or median represents an improved point estimate, given the availability of the test results.

Bayesian analysis may be applied to upgrading estimates for a wide variety of reliability problems. We discuss here one typical application.

The Bayesian formula stems from the fact that the intersection of two probabilities can be written in terms of two different conditional probabilities;

$$Pr\{X_i|Y\} = [Pr\{Y|X_i\}\, Pr\{X_i\}] \,/\, Pr\{Y\} \qquad (9.22)$$

We may give the following interpretations to these probabilities: $Pr\{X_i\}$ is our estimate of a probability that **X** has a value of X_i, and Y represents the outcome of an experiment. The probability $Pr\{X_i|Y\}$ is our upgraded estimate, given the outcome of the experiment. To evaluate this result, we must be able to estimate the probability of an experimental outcome Y given that **X** has a value X_i. Finally, $Pr\{Y\}$ is determined as follows:

Suppose that $X_1, X_2, \ldots\ldots, X_n$ are the only possible values that **X** may take on. Since **X** can have only one value, the events X_i are mutually exclusive, and therefore,

$$\sum_{i=1}^{n} Pr\{X_i\} = 1, \qquad (9.23)$$

Also, the Bayes equation, may be written in the form of Total Probability as

$$Pr\{X_i|Y\} = \frac{Pr\{Y|X_i\}Pr\{X_i\}}{\sum_{j=1}^{n} Pr\{Y|X_j\}\, Pr\{X_j\}} \qquad (9.24)$$

The use of the Bayes equation is best understood through a simple example.

Example 9.3

An engineer calls in two experts to estimate the MTTF of a new process computer. Expert 1 estimates 30 months and expert 2 estimates 12 months. Since the engineer gives their opinions equal weight, he estimates the MTTF to be

$$\text{MTTF} = 0.5 \times 30 + 0.5 \times 12 = 21 \text{ Months.}$$

Subsequently, a 6-month test is run, and the prototype for the new computer does not fail. In the light of these test results, (a) how should the experts' opinions be weighed, and (b) how should the estimated MTTF be upgraded?

Solution

Let $\Pr\{X_1\} = \Pr\{X_2\} = 0.5$ be the prior probabilities that the MTTF estimates of experts 1 and 2 are correct. If the experts' opinions are correct, the probability of 6-month operation without failure is

$$\Pr\{Y| X_i\} = \exp(-t / \text{MTTF}_i),$$

assuming that the constant failure rate model is adequate. Thus

$$\Pr\{Y| X_1\} = e^{-6/30} = 0.819,$$
$$\Pr\{Y| X_2\} = e^{-6/12} = 0.607$$

Thus, the revised probabilities that each of the experts are correct are:

$$\Pr\{X_1| Y\} = \frac{0.819 \times 0.5}{0.819 \times 0.5 + 0.607 \times 0.5} = 0.574$$

$$\Pr\{X_2 | Y\} = \frac{0.607 \times 0.5}{0.819 \times 0.5 + 0.607 \times 0.5} = 0.426$$

With these weights the upgraded estimate is

$$\text{MTTF} = 0.574 \times 30 + 0.426 \times 12 = 22.3 \text{ Months}$$

* * *

9.5 CONFIDENCE LIMITS

The estimates of the mean time between failures m, or any other statistical parameter, are so called point estimates to the true unknown parameter. How reliable are such estimates and what confidence can we have in them?

We know that statistical estimates are more likely to be close to the true value as the sample size increases. Thus, there is a close correlation between the accuracy of an estimate and the size of the sample from which it was obtained. Only an infinitely large sample size could give us a 100 per cent confidence or certainty that a measured statistical parameter coincides with the true value. In this context, *confidence* is a mathematical probability relating the mutual positions of the true value of a parameter and its estimate.

When the estimate of a parameter is obtained from a reasonably sized sample, we may logically assume that the true value of that parameter will be somewhere in the neighborhood of the estimate, to the right or to the left. Therefore, it would be more meaningful to express statistical estimates in terms of a range or interval with an associated probability or confidence that the true value lies within such interval than to express them as point estimates. This is exactly what we are doing when we assign confidence limits to point estimates obtained from statistical measurements.

Frequently, this concept is misunderstood to be a much more complicated statistical tool than it actually is. This presentation is deliberately developed here in nonstatistical language so that the casual reader can quickly understand the important implications for testing. In general, a confidence interval is bounded by upper and lower confidence limits. Generally speaking, the broader the limits the higher the confidence that a particular group of events is enclosed. This is illustrated further in the following discussion.

To clarify the principle of confidence limits, consider the illustration of the time that a guest will arrive in town by train on a certain day. If all you know is the date of arrival, you can express 100 per cent confidence that he will arrive sometime between the limits 12:01 A.M. and 12:00 midnight. If you know that no night trains stop in town between the hours of 6:00 P.M. and 6:00 A.M., you can narrow your confidence interval to 100 per cent confidence that the time of arrival will be 12:00 noon ± 6 hours. If you happen to know that all the train arrivals from his direction are in the morning, this again narrows your 100% confidence interval. Your limits are then 6:00 A.M. and 12:00 noon. However, if this were all you knew about the arrival schedule, your confidence would be very low that he would arrive at any specific minute. Even if you knew the train number and the expected time of arrival, your 100 per cent confidence interval would have to be broad enough to allow for any possible exigency that might affect this particular event.

Suppose the train is scheduled to arrive at 11:00 A.M. You might investigate the record of this particular train and find that eight out of ten days, on the average, the 11:00 o'clock train arrives within five minutes of 11:00 A.M. Your confidence would then be 80 per cent that the particular train would arrive 11:00 A.M. ± 5 min. Putting this another way, you would have an 80 per cent confidence that the exact time of arrival would be between 10:55 and 11:05 A.M. The 80 per cent confidence interval would be 10 min long and extend from the upper limit of 11:05 A.M. to the lower limit of 10:55 A.M. This, in statistical language, is described as a two sided confidence interval, meaning that there are both upper and lower limits.

But suppose you want to make sure that the particular train is typical of those which arrive normally within the average confidence interval. You could check at the information window or with the stationmaster sometime before train time to see if this particular train is running on time at earlier stops. Twenty per cent of the trains normally arrive at times outside the 80 per cent confidence interval because of events which make them nontypical. This is the equivalent engineering action of evaluating a test result in terms of ancillary factors to determine mitigating circumstances or system interaction factors.

Suppose also that you are out of town on business and cannot get to the railroad station until a specific time. In that case you might want to know the confidence that the train will arrive some time after you do, so that you will be on hand to greet your guest. If you arrive an hour or more ahead of the normal train time, your confidence will be almost 100 per cent that the train will arrive later than you do. However, as the two times of arrival approach coincidence, the confidence in your arriving first will approach 50 per cent. Under these conditions the variability in the train arrival is a major factor. This example illustrates a statistical approach described as a one sided confidence determination or interval.

Both one sided and two sided confidence intervals are illustrated in the Fig.9.3 and Fig.9.4 respectively.

9.51 Estimation of Confidence Limits

This section is concerned with the estimation of confidence limits on exponential mean life. The chi-square distribution is utilized in establishing the confidence interval limits on mean life.

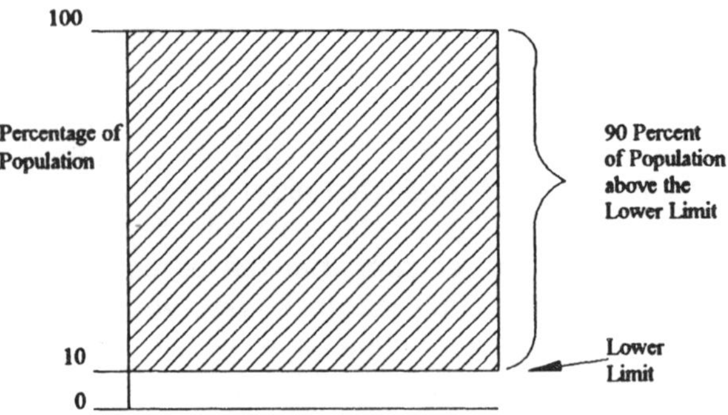

Fig. 9.3: One-sided confidence interval.

Usually sampled data are used when estimating the mean life of a product. If one draws two separate samples from a population for the purpose of estimating the mean life, it will be quite unlikely that both samples will yield the same mean life results. Therefore, the confidence limits on mean life are computed to take into consideration the sampling fluctuations. In this section the confidence limit formulations for the following two types of test procedures are presented.

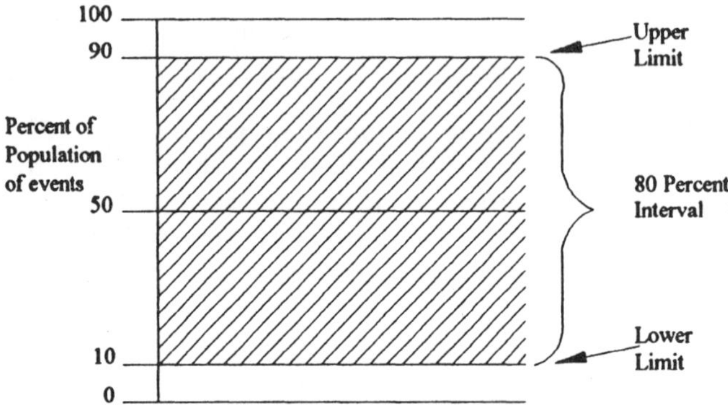

Fig. 9.4: Two-sided confidence interval.

9.511 Test Procedure I

In this situation, the items are tested until the preassigned failures occur. The formulas for one-sided (lower limit) and two sided (upper and lower limits) confidence limits, respectively in this case are as follows:

$$[\ \{2t/\ \chi^2(\theta,2k)\},\ \infty\ ] \tag{9.25}$$

and

$$\left[\frac{2t}{\chi^2(\theta/2,\ 2k)},\ \frac{2t}{\chi^2(1-\theta/2,\ 2k)} \right] \quad (9.26)$$

where k is the total number of failures and θ is the probability that the interval will not contain the true value of mean life [thus $\theta = 1-$ (confidence level)].

The value of t is given by

$$t = xy \text{ (for replacement tests, i.e., failed items replaced or repaired)} \quad (9.27)$$

and

$$t = \sum_{j=1}^{k} y_j + (x-k)y \quad (9.28)$$

(for nonreplacement tests, i.e., failed items are not replaced)

where x is the total items, at time zero, placed on test; y is the time at the conclusion of life test; and y_j is the time of failure j.

Example 9.4

A sample of 25 identical electronic components were tested until the occurrence of the twelfth failure. Each failed component was replaced. The last component failure occurred at 150 hr. At 97.5 percent confidence level, compute the value of the one sided (lower) confidence limit (i.e., the minimum value of mean life.)

Solution

Substituting the given data in equation (9.27) leads to

$$t = (25)(150) = 3{,}750 \text{ hr}$$

The acceptable risk of error is

$$\theta = 1 - (\text{confidence level}) = 1 - 0.975 = 0.025$$

Hence, with the aid of Equation (9.25) we get

$$\left[\frac{2(3750)}{\chi^2[0.025,(2)(12)]}, \infty\right] = [(7500/39.36), \infty] = (190.55, \infty)$$

The minimum value of mean life is 190.55 hr for the 97.5 percent confidence level.

* * *

9.512 Test Procedure II

This is another test procedure in which the testing is terminated at a preassigned number of test hours. The formulas for one sided (lower limit) and two sided (upper and lower limits) confidence limits, respectively, are as follows:

$$[\{2t/\chi^2(\theta,2k+2)\}, \infty] \tag{9.29}$$

and

$$\left[\frac{2t}{\chi^2(\theta/2,\ 2k+2)}, \quad \frac{2t}{\chi^2(1-\theta/2,\ 2k)}\right] \tag{9.30}$$

The symbols k and θ are defined in the previous section.

Example 9.5

A sample of 25 identical components was drawn from a population and put on test at time $t = 0$. The failed components were not replaced and the test was terminated at 120 hr. Six components failed during the test period at 15, 22, 30, 50, 67, and 85 hr. At 97.5 percent confidence level compute the value of one-sided (lower) confidence limit (i.e., the minimum value of mean life).

Solution

By substituting the specified data we get

$$t = (15 + 22 + 30 + 50 + 67 + 85) + (25 - 6)(120) = 2{,}549 \text{ hr}$$

The acceptable risk of error is

$$\theta = 1 - (\text{confidence level}) = 1 - 0.975 = 0.025$$

With the aid of the Equation (9.29) and the above results, we obtain

$$\left[\frac{2(2549)}{\chi^2[0.025, 2(6)+2]}, \infty\right] = [(5098/26.12), \infty] = (195.18, \infty)$$

Thus the minimum value of mean life is 195.18 hr for the 97.5 percent confidence level.

* * *

9.6 ACCELERATED TESTING

If we have enough test data, the conventional testing methods will allow us to fit our choice of a life distribution model and estimate the unknown parameters. However, with today's highly reliable components, we are often unable to obtain a reasonable amount of test data when stresses approximate normal use conditions. Instead, we force components to fail by testing at much higher than the intended application conditions. By this way, we get failure data that can be fitted to life distribution models, with relatively small test sample sizes and practical test times.

The price we have to pay for overcoming the dilemma of not being able to estimate failure rates by testing directly at use conditions (with realistic sample sizes and test times) is the need for additional modeling. How can we go from the failure rate at high stress to what a future user of the product is likely to experience at much lower stresses?

The models used to bridge the stress gap are known as acceleration models. This section develops the general theory of these models and looks in detail at some well known forms of acceleration models, such as the Arrhenius and the Eyring models.

9.61 Accelerated Testing Theory

The basic concept of acceleration is simple. We hypothesize that a component, operating under the right levels of increased stress, will have exactly the same failure mechanisms as seen when used at normal stress. The only difference is *things happen faster*. For example, if corrosion failures occur at typical use temperatures and humidities, then the same type of corrosion happens much quicker in a humid laboratory oven at elevated temperature.

In other words, we can think of time as *being accelerated*, just as if the process of failing were filmed and then played back at a faster speed. Every step in the sequence of chemical or physical events leading to the failure state occurs exactly as at lower stresses; only the time scale measuring event duration has been changed.

When we find a range of stress values over which this assumption holds, we say we have true acceleration.

In theory, any well behaved (order preserving, continuous, etc.) transformation could be a model for true acceleration. However, in terms of practical applicability, we almost always restrict ourselves to simple constant multipliers of the time scale. When every time of failure and every distribution percentile is multiplied by the same constant value to obtain the projected results at another operating stress, we have linear acceleration.

Under a linear acceleration assumption, we have the relationship (time to fail at stress S_1) = AF X (time to fail at stress S_2), where AF is the acceleration constant relating times to fail at the two stresses. *AF* is called the acceleration factor between the stresses.

If we use subscripts to denote stress levels, with U being a typical use set of stresses and S (or S_1, S_2, ...) for higher laboratory stresses, then the key equations in Table 9.4 hold no matter what the underlying life distribution happens to be.

TABLE 9.4

General Linear Acceleration Relationships

1.	Time to fail:	$t_v = AF \times t_s$
2.	Failure probability:	$F_v(t) = F_s(t/AF)$
3.	Density function:	$f_v(t) = (1/AF)\ f_s\ (t/AF)$
4.	Failure rate:	$h_v(t) = (1/AF)\ h_s\ (t/AF)$

In Table 9.4 t_v represent a random time to fail at use conditions, while t_s is the time the same failure would have happened at a higher stress. Similarly, F_u, f_u and h_u are the CDF, PDF, and failure rate at use conditions, while F_s, f_s, and h_s are the corresponding functions at stress S.

Table 9.4 gives the mathematical rules for relating CDFs and failure rates from one stress to another. These rules are completely general, and depend only on the assumption of true acceleration and linear acceleration factors. In the next section, we will see what happens when we apply these rules to exponential distribution as an example.

9.611 Exponential Distribution Acceleration

We add the assumption that $F_s(t) = 1 - \exp(-\lambda_s t)$

By letting $\lambda_u = \lambda_s/AF$, we see that the CDF at use conditions remains exponential, with new parameter λ_s/AF.

This equation demonstrates that an exponential fit at any one stress condition implies an exponential fit at any other stress within the range where true linear acceleration holds. Moreover, when time is multiplied by an acceleration factor AF, the failure rate is reduced by dividing by AF.

Example 9.6

A component, tested at 125°C in a laboratory, has an exponential distribution with MTTF 4500 hr. Normal use temperature for the component is 25°C. Assuming an acceleration factor of 35 between these two temperatures, what will the use failure rate be and what percent of these components will fail before the end of the expected useful life period of 40,000 hr?

Solution

The MTTF is the reciprocal of the failure rate and varies directly with the acceleration factor. Therefore the MTTF at 25°C is 4500 x 35 = 157,500. The use failure rate is 1/157,500 = 0.635%/K. The cumulative percent of failures at 40,000 hr is given by $1-e^{-0.00635 \times 40} = 22.4\%$.

* * *

9.62 Acceleration Models

If we already know the acceleration factor between a laboratory stress test and the field use condition, we can convert the results of our test data analysis to use condition failure rate projections. Indeed, this is often done as an ongoing process monitor for reliability on a lot by lot basis.

But what can be done if an acceleration factor to use conditions is not known, and data can only be obtained in a reasonable amount of time by testing at high stress? The answer is we must use the high stress data to fit an appropriate model that allows us to extrapolate to lower stresses.

There are many models in the literature that have been used successfully to model acceleration for various components and failure mechanisms. These models are generally written in a deterministic form that says that time to fail is an exact function of the operating stresses and several material and process dependent constants.

Since all times to failure are random events that cannot be predicted exactly in advance, and we have seen that acceleration is equivalent to multiplying a distribution scale parameter, we will interpret an acceleration model as an equation that calculates a distribution scale parameter, or percentile, as a function of the operating stress. In the discussion below we use a typical percentile T_{50}, as is the convention for these models.

9.621 The Arrhenius Model

When only thermal stresses are significant, an empirical model, known as the Arrhenius model, has been used with great success. This model takes the form

$$T_{50} = A\, e^{\Delta H/kT} \tag{9.31}$$

where A and ΔH are unknown constants, k is Boltzmann's constant, and T is temperature measured in degrees Kelvin at the location on the component where the failure process is taking place.

Note that we can write the Arrhenius model in terms of T_{50}, or the l/λ parameter (when working with an exponential), or any other percentile of the life distribution we desire. The value of the constant A will change, but this will have no effect on acceleration factors.

We solve for the acceleration factor between temperature T_1 and temperature T_2 by taking the ratio of the times it takes to reach any specified CDF percentile. In other words, the acceleration factor AF between stress 1 and stress 2 is defined to be the ratio of time it takes to reach P% failures at stress 1 divided by the time it takes to reach P% failures at stress 2. The assumption of true acceleration makes this factor the same for all P. Using the Arrhenius model and the fiftieth percentile, we have

$$AF = \frac{T_{50(1)}(\text{at } T_1)}{T_{50(2)}(\text{at } T_2)} = \frac{A \exp(\Delta H/kT_1)}{A \exp(\Delta H/kT_2)} \tag{9.32}$$

from which

$$AF = \exp\left[(\Delta H/k)\, \{(1/T_1) - (1/T_2)\}\right] \tag{9.33}$$

This shows that knowing ΔH alone allows us to calculate the acceleration factor between any two temperatures. Conversely, if we know the acceleration factor, we can calculate ΔH as follows:

$$\Delta H = k[\, \ln(T_{50(1)}/T_{50(2)})\,]\, [(1/T_1) - (1/T_2)]^{-1} \tag{9.34}$$

This last equation shows us how to estimate ΔH from two cells of experimental test data consisting of times to failure of units tested at temperature T_1 and times to failure of units tested at temperature T_2. All we have to do is estimate a percentile, such as T_{50}, in each cell, then take the ratio of the corresponding times and use the preceding equation to estimate ΔH. This procedure is valid for any life distribution.

9.622 The Eyring Model

The Arrhenius model is an empirical equation that justifies its use by the fact that it *works* in many cases. It lacks, however, a theoretical derivation and the ability to model acceleration when stresses other than temperature are involved.

The Eyring model offers a general solution to the problem of additional stresses. It also has the added strength of having a theoretical derivation based on chemical reaction rate theory and quantum mechanics. In this derivation, based on work by Eyring, the parameter ΔH has a physical meaning. It represents the amount of energy needed to move an electron to the state where the processes of chemical reaction or diffusion or migration can take place.

The Eyring model equation, written for temperature and a second stress, takes the form.

$$T_{50} = A\, T^{\alpha} \exp(\Delta H/kT) \exp[(B + C/T)S_1] \qquad (9.35)$$

The first exponential is the temperature term, while the second exponential contains the general form for adding any other type of stress. In other words, if a second nonthermal stress was needed in the model, a third exponential multiplier exactly the same as the second, except for replacing B and C by additional constants D and E, would be added to the equation. The resulting Eyring model for temperature and two other stresses would then be

$$T_{50} = A\, T^{\alpha} \exp(\Delta H/kT) \exp[(B + C/T)S_1] \exp[(D + E/T)S_2] \qquad (9.36)$$

It is interesting to look at how the first term, which models the effect of temperature, compares to the Arrhenius model. Except for the T^{α} factor, this term is the same as the Arrhenius. If α is close to zero, or the range over which the model is applied is small, the term T^{α} has little impact and can be absorbed into the A constant without changing the practical value of the expression. Consequently, the Arrhenius model is successful because it is a useful simplification of the theoretically derived Eyring model.

9.623 Other Acceleration Models

There are many other models, most of which are simplified forms of the Eyring, which have been successful. A model known as the power rule model has been used for paper impregnated capacitors. It has only voltage dependency, and takes the form AV^{-B} for the mean time to fail (or the T_{50} parameter).

Another way to model voltage is to have a term such as Ae^{-BV}. This kind of term is easy to work with after taking logarithms.

Humidity plays a key role for many failure mechanisms, such as those related to corrosion or ionic metal migration. The most successful models including humidity have terms such as $A(RH)^{-B}$ or $Ae^{-B(RH)}$, where RH is relative humidity.

A useful model for electromigration failures uses current density as a key stress parameter.

$$T_{50} = A\ J^{-n}\ e^{\Delta H/kT} \tag{9.37}$$

with J representing current density. This mechanism produces open short failures in metal thin film conductors owing to the movement of ions toward the anode at high temperature and current densities. A typical ΔH value is 0.5 eV, while $n = 2$ is common.

9.63 Accelerated Testing Methods

9.631 Magnified Loading

Use of magnified load does reduce testing time and possibly the number of items required for test. A major problem is that of correlation. For example, if we wish to know the performance of an engine in normal use of 5000 h, we can get much the same performance in 2830 h at full throttle, or in 100 h at 23 percent overload. This correlation is possible, since much information exists. In many situations, however, establishing such correlation is difficult, since we must first know what normal means and then we must have enough overload data to correlate with normal.

As an example, suppose that the MTTF is estimated at the number of different elevated stress levels. Such stress might typically be temperature, voltage, radiation intensity, mechanical stress, or any number of other variables. The MTTF or other reliability parameter is then plotted versus the stress level, as indicated in Fig. 9.5. A curve is fitted to the data, and the

MTTF is estimated at the stress level that the device is expected to experience during normal operation.

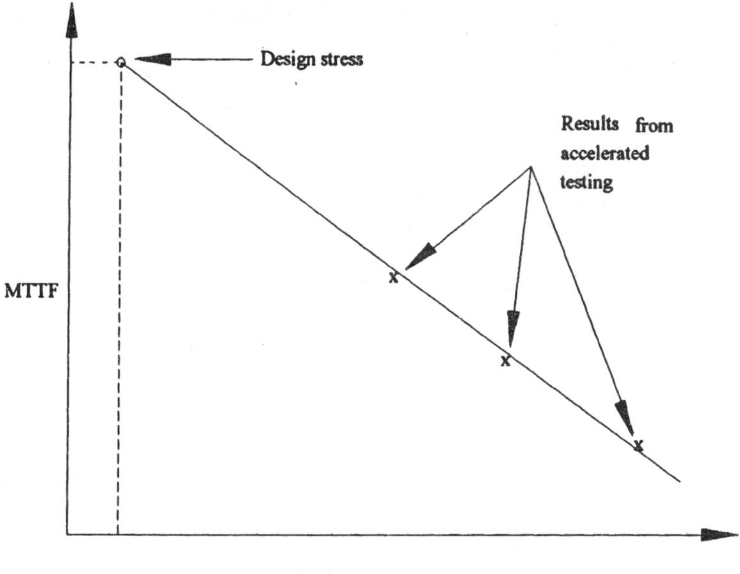

Fig.9.5 Estimate of MTTF from accelerated test data.

Accelerated testing is useful, but it must be carried out with great care to ensure that results are not erroneous. We must know for sure that the phenomena for which the acceleration factor has been calculated are the failure mechanisms. Experience gained with similar products and a careful comparison of the failure mechanisms occurring in accelerated and real time tests will help determine whether we are testing the correct phenomena.

One common type of accelerated test stresses the test sample to the maximum ratings for the part. Acceleration factors are then applied to achieve a probable failure rate which would have been applicable at considerably derated conditions. For example, paper capacitors commonly exhibit a fifth-power acceleration factor with voltage. Most other parts exhibit close to a third power acceleration factor. A standard third power is frequently used for acceptance tests. For example, suppose a test is performed to demonstrate a failure rate of 1.0%/ 1,000 hours while operated at full rated voltage. This could be interpreted as the equivalent of 0.008%/ 1,000 hours at 20 per cent of the full voltage rating. This is calculated as follows:

$$\text{Derated failure rate, } d = \frac{\text{full rating}}{(\text{rated voltage/derated voltage})^3}$$

$$d = \frac{1.0\%/K \text{ hours}}{(V_R/0.2V_R)^3} = (1.0/5^3) = 0.008\%/K \text{ hours}$$

9.632 Step-Stress Tests

A second important form of accelerated test is known as the step-stress test. This type reveals the uniformity and strength of a product but does not normally yield failure-rate data. The step-stress test repeatedly employs increased stresses according to a prearranged test plan. One or more types of stress such as temperature and voltage can be combined in this test with increments of time. After testing at each step or level of stress for the prescribed interval of time, the parameters are measured and the number of rejects or failures is determined. The increased stressing is continued according to the plan until the entire sample has failed.

A typical test plan is illustrated in Fig. 9.6. Any changes or differences in the material, processes, or design are quickly revealed by changes they promote in the step-stress data plot.

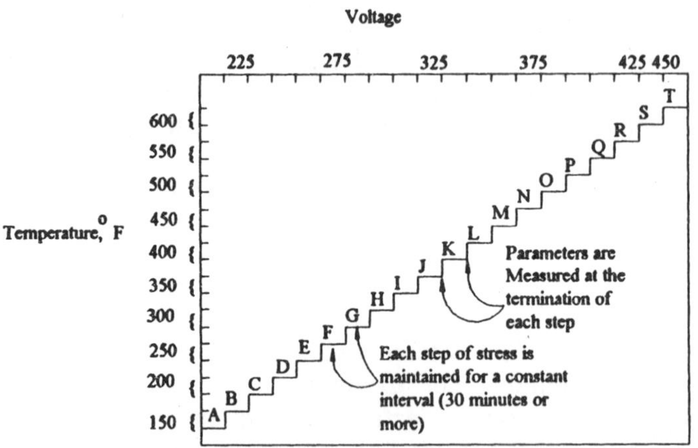

Fig.9.6 Typical step-stress test plan.

The conditions of environment and electrical stress to be imposed at each step are planned to start at near the maximum rating for the item being tested and be increased regularly according to the plan until 100 per cent failure of the sample results. The failure data are then smoothed and plotted as a density distribution to reveal the step-stress fingerprint.

9.7 EQUIPMENT ACCEPTANCE TESTING

The figure of merit usually used for measuring equipment reliability is mean time between failures (MTBF). Reliability acceptance testing for equipment generally consists of operational tests performed under simulated end-use conditions with acceptable MTBF and confidence specified.

A question commonly asked is, How much testing is required to be 90 percent confident that the MTBF is greater than x hours? This question cannot be answered without additional information. For example, if the actual MTBF is less than x hours, then no amount of testing can prove that it is greater than it is. If the actual MTBF is just barely over the value x, then a great deal more testing is required than if the actual MTBF is much greater than the x value.

The test measures the most likely value of MTBF, and the amount of statistical data obtained during the test must be evaluated to determine the confidence which can be placed on the measurement. When this has been done, the following statements can be made: The best estimate of the MTBF is B hours; but, based on the amount of data, we can be 90 percent sure, for example, that it is not more than an upper limit of A hours and 90 percent sure that it is not less than a lower limit of C hours. This defines an 80 percent double sided confidence that the true value lies between the values of A and C.

Usually, for acceptance testing, the single sided description stating the cumulative probability that a measured MTBF is greater than a certain specified minimum value has the greatest usefulness. This brings to mind that it is most frequently desired to plan equipment acceptance tests to prove with a known confidence that the MTBF is greater than a certain specified figure.

9.71 Sequential Acceptance Tests

Sequential testing differs from other test procedures in that the length of test is not established before the test begins but depends upon what happens during the test. The test sample is tested while subjected to a prescribed environment and duty cycle until the preassigned limitations on the risks of making wrong decisions based on the cumulative test evidence have been satisfied. The ratio of quantity of failures to the length of test at any test interval is interpreted according to a sequential analysis test plan. Conspicuously good items are accepted quickly; conspicuously bad items are rejected quickly; and items of intermediate quality require more extensive testing.

The major advantage in using sequential test procedure is that it results in less testing on the average than other testing procedures when the preassigned limitations on the risks of making both kinds of wrong decisions are the same for both tests. The chief disadvantage is that the test time required to reach a decision cannot be determined prior to testing.

Characteristic of this method is that the number of observations - in our case, the number of observed times between failures, or, which is the same, the number of actually observed failures - is not predetermined but depends at any instant on the outcome of the preceding observation. The method supplies an exact rule for making one of three decisions at any instant during the test: (1) accept, (2) reject, (3) continue testing. Which of these three decisions is made depends on the outcome of the observation immediately preceding the time at which a decision is being made. If the rule shows that decision (1) or (2) has to be made, the test is terminated. If it shows that decision (3) must be made, the test continues to obtain more information, such as more failure-free operating time or an additional failure; that is why the test is called *Sequential.*

Assume we are interested in the reliability of a certain item of equipment. We desire a certain reliability R_d so that the item will perform satisfactorily. We are willing, however, to accept a somewhat lower reliability R_m, provided there is a high probability that lots with less than R_m will be rejected. We need four parameters:

R_d = desired or specified reliability
R_m = minimum acceptable reliability
α = producer's risk -probability of rejecting a lot whose reliability is equal to or greater than R_d
β = consumer's risk -probability of accepting a lot whose reliability is equal to or less than R_m

As sampling and testing progress, the number of failed items is plotted against the number of successful items (Fig. 9.7). Testing is continued until the plotted step function crosses one of the two decision lines. The decision lines are obtained as follows:

$$\text{Accept: } F \ln[(1-R_m)/(1-R_d)] + S \ln(R_m/R_d) = \ln[(1-\beta)/\alpha] \quad (9.38)$$
$$\text{Reject: } F \ln[(1-R_m)/(1-R_d)] + S \ln(R_m/R_d) = \ln[\beta/(1-\alpha)] \quad (9.39)$$

where F represents cumulative number of failures and S cumulative number of successes.

It might be thought that a sequential testing plan could lead, on occasion, to an interminable test. It can be shown, however, that the test will eventually terminate. In fact, sequential testing will generally require testing of fewer items, on the average, than single or multiple sampling.

9.8 RELIABILITY GROWTH TESTING

The experience gained from the investigations carried out on several systems are in conformity with general experience of not always being able to get it right first time. Hence it is clear that there needs to be a process of growth

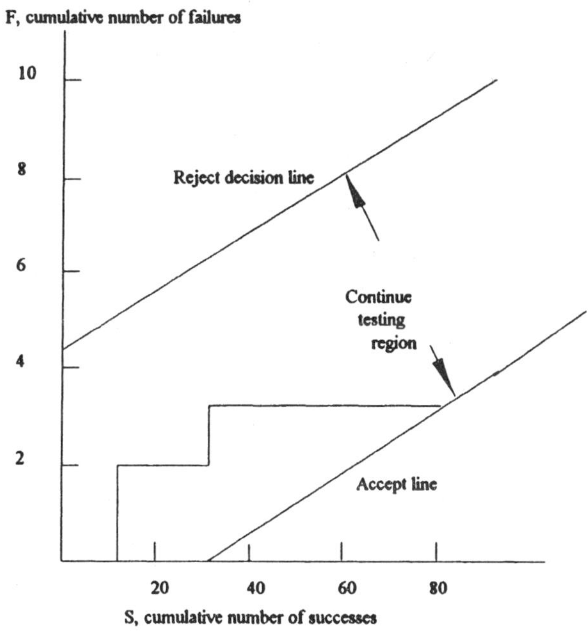

Fig.9.7 Schematic diagram of sequential testing procedure.

and improvement until some target or criterion is met. In newly designed military equipment it was quoted that investigation showed that the mean time between failures (MTBF) achieved was often no more than 8 per cent to 20 per cent of its ultimate potential level. It was also shown that a properly organized *test and fix* programme can be of great assistance in closing this reliability gap. Obviously this argument of the extent of the reliability gap goes on through all the various phases following manufacture and production. However, quite often it is the first time that *the paper-work system* is converted into some form of *material or hardware system* and from an engineering point of view it is required to make the most use of any trials or tests carried out.

One of the features of this transformation from paper to materialistic form is to reveal hidden weaknesses and deficiencies in the equipment for example the systematic type of failure. The reliability growth process may be illustrated as shown in the simplified block diagram of Fig.9.8. Broadly, the failures which are of the systematic type will require careful examination and those which are found significant will lead to some rectification process in conjunction with the designer and other interested parties such as the reliability assessor. The other failures which may be classed as random will require a restoration process so that the equipment may be repaired for further trials. This is a fairly standard procedure which is described particularly for electronic equipment but similar principles hold for other types of equipment.

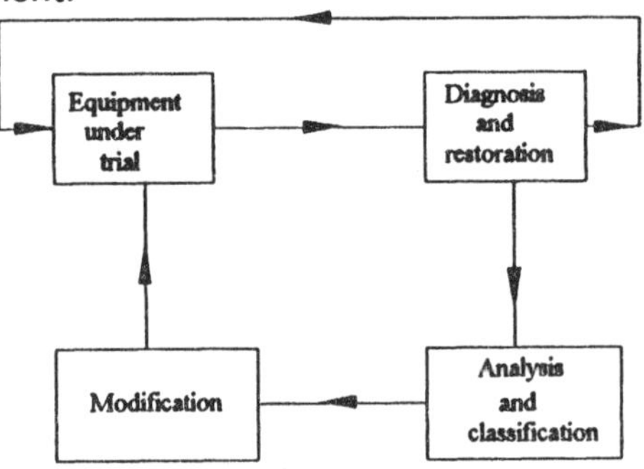

Fig.9.8 Reliability growth process.

These types of trials and tests can be used to obtain initial reliability information but the conditions of the tests require to be carefully studied. The tests themselves may not be under the same environmental conditions, it is often not easy to set up correctly the true conditions. In the case of life testing as already illustrated in the previous section, this may represent accelerated testing particularly where the equipment is of very high reliability and catastrophic failure information is required. Various techniques of analysis exist for estimating the reliability characteristic of interest such as failure rate and some of the techniques have already been illustrated. Typically two periods of testing time may be selected, one at the start of the test and the other at the termination of the test, selecting periods with approximately equal numbers of failures.

Suppose that we define the following:

T = total operation time accumulated on all the prototypes
n(T) = number of failures from the beginning of testing through time T.

If more than one prototype is tested, then T is calculated depending on whether replacement or nonreplacement testing is used. Finally, we assume that as failures occur, the design is modified to eliminate the failure modes.

Duane observed that if n(T) is plotted versus T on log-log paper, the result, as indicated in Fig.9.9 tends to be a straight line, whatever the type of electromechanical equipment under consideration. From such empirical relationships, referred to as Duane plots, we may estimate the growth of MTBF with time and therefore also extrapolate how much reliability is likely to be gained from further development work.

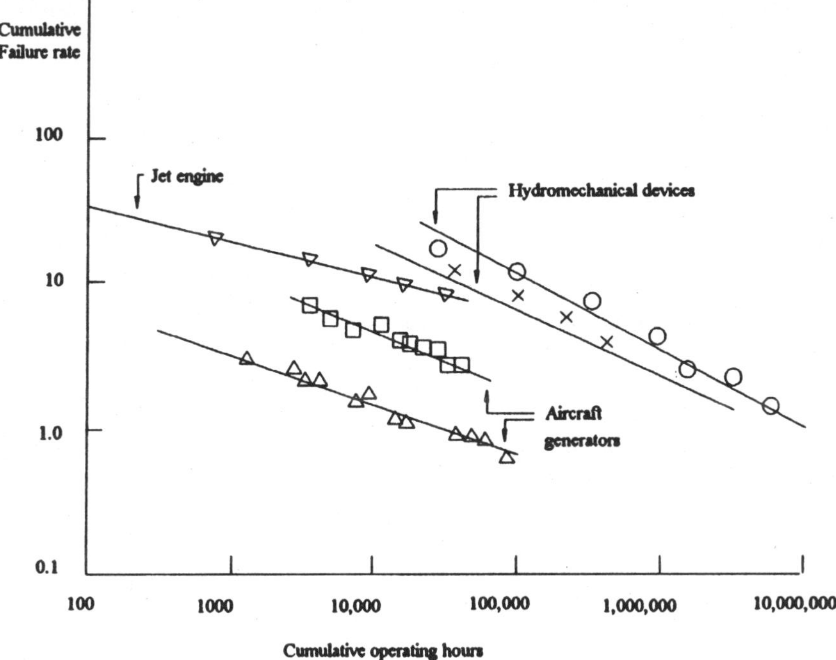

Fig.9.9 Duane's data on a log-log scale.

Thus, according to Duane,

$$\log m_c = \log m_s + \beta (\log t - \log t_s) \qquad (9.40)$$

$$m_c = \frac{\text{total time}}{\text{(total number of product failures)}} = t / k \qquad (9.41)$$

where m_c is the cumulative mean time between failures, m_s is the cumulative mean time between failures at the beginning of the reliability monitoring time period, t_s, and β is the slope parameter (usually it takes values between 0.2 and 0.4)

From Equation (9.40), we get

$$m_c = m_s t^\beta /(t_s^\beta) \quad (9.42)$$

By rewriting Equation (9.41), we get

$$k = t/m_c \quad (9.43)$$

Substituting Eq. (9.42) into Equation (9.43) results in

$$k = tt^\beta/m_s t^\beta = (t^{1-\beta} \; t_s^\beta)/m_s \quad (9.44)$$

Differentiating the above equation with respect to t leads to

$$dk/dt = (1-\beta)[t_s^\beta/(t^\beta \; m_s)] \quad (9.45)$$

With the aid of Eq. (9.42), the above equation reduces to

$$dk/dt = (1-\beta)/m_c \quad (9.46)$$

The left-hand side of Equation (9.46) is the reciprocal of instantaneous mean time between failure, m_{int}, of the population. Thus we let

$$dk/dt = (m_{int})^{-1} \quad (9.47)$$

Substituting Equation (9.47) into Equation (9.46) leads to

$$(1/m_{int}) = (1-\beta)/m_c \quad (9.48)$$

Therefore, $m_{int} = m_c / (1-\beta)$ (9.49)

Example 9.7

A prototypo modcl of an engineering system was initially tested for a 300-hr period during which 5 failures occurred. The specified mean time between failures of the system is 800 hr. Assume that the value of the Duane model slope parameter β is 0.4. Compute the value of additional system test hours.

Solution

Utilizing the specified data in Equation (9.49) yields value for

$$m_c = m_{int}(1-\beta) = 800(1-0.4) = 480 \text{ hr}$$

Substituting the given data into Equation (9.41) yields the estimated value for

$$m_s = 300/5 = 60 \text{ hr}$$

Thus from Equation (9.42) we get

$$480 = (60)\,(t/300)^{0.4} = (6.1278)\,t^{0.4}$$

Therefore, $t = [480/6.1278]^{1/0.4} = 54{,}305.8$ hr

Additional system test hours = (54,305.8) - (300) = 54,005.80 hr

Thus the system has to be tested for another 54,005.80 hr.

* * *

10
SOFTWARE RELIABILITY

10.1 IMPORTANCE

Information processing is probably the most significant industry in the world economy today and in the foreseeable future. It has expanded and continues to expand at a rapid rate. This expansion is, in part, related to the increase in cost-effectiveness of computer hardware. Cost-effectiveness has increased by a factor of about 1000 every decade. As long as this rate of change continues, the range of tasks that can be handled most economically by computing is likely to grow rapidly. Since software is the major part of most computer systems, the field of software engineering is experiencing similar rapid growth.

The major forces affecting the software engineering field include:

*the increasing level and truly international nature of business competition.
*the increasing cost of both information system development and information system failure.
*the increasing pace of change in computing technology,
*the increasing complexity of managing information system development.

Since there is more competition among software producers, software customers are more aware of the products and services available to them. These customers once relatively naive and dependent on their suppliers have become increasingly sophisticated and demanding. Three of the most significant needs are level of quality required, time of delivery, and cost.

At the same time, both the development and operational cost of software have increased substantially. The size, complexity, and degree of distribution of systems are increasing. We have many multicomputer systems linked by networks. A growing proportion of the systems operate in real time. The operational effects of failure are large and often critical. For example, consider the effects of breakdown of airline reservations, banking, automatic flight control, military defense, and nuclear power plant safety control systems, to name just a few. The economic consequences can be large and even catastrophic. Costs of failure include not only direct expenses but also product liability risks and damage to a company's reputation.

With the cost and schedule pressures, it is becoming increasingly impossible to create a software product that is *generous* in the sense of simultaneously providing high quality, rapid delivery, and low cost. The view that such characteristics are simply desired objectives to be achieved is obsolete. For example, it would be nice to have programs that are *correct*, but in this real world we must settle for something less. The need for trade-offs is pressing, and the characteristics of the software must be carefully selected to meet customer needs. This means that measurement and prediction of software product characteristics is essential.

It has been indicated that three of the most important software product characteristics are quality, cost and schedule. Note that these are primarily user-oriented rather than developer-oriented attributes. Quantitative measures exist for the latter two characteristics, but the quantification of quality has been more difficult. It is most important, however, because the absence of a concrete measure for software quality generally means that quality will suffer when it competes for attention against cost and schedule. In fact, this absence may be the principal reason for the well known existence of quality problems in many software products.

Reliability is probably the most important of the characteristics inherent in the concept *software quality*. It is intimately connected with defects, and defects represent the largest cost element in programming. Software reliability concerns itself with how well the software functions to meet the requirements of the customer. We define software reliability simply as the probability that the software will work without failure for a specified period of time. *Failure* means the program in its functioning has not met user requirements in some way. *Not functioning to meet user requirements* is really a very broad definition. Thus reliability subsumes, totally or partially, many properties that are often quoted as aspects of quality. One example is correctness. Another is program behavior that is not user-friendly. Some properties like software safety are actually specialized aspects of software reliability. Two of the few aspects of quality that probably cannot be related to reliability are modifiability and understandability of documentation.

Reliability represents a user-oriented view of software quality. Initial (and many present) approaches to measuring software quality were based on attempting to count the faults or defects found in a program. This approach is developer -oriented. Also, what was usually counted were either failures (the occurrences of malfunction) or repairs (for example, maintenance or correction reports), neither of which are equivalent to faults. Even if faults found are correctly counted, they are not a good status indicator(is a large number good or bad?). Faults remaining may be.

This does not mean that some attention to faults is without value. But the attention should be focused on faults as predictors of reliability and on the nature of faults. A better understanding of faults and the causative human error processes should lead to strategies to avoid, detect and remove, or compensate for them.

10.2 SOFTWARE RELIABILITY AND HARDWARE RELIABILITY

The field of hardware reliability has been established for some time. Hence, one might ask how software reliability relates to it. In reality, the division between hardware and software reliability is somewhat artificial. Both may be defined in the same way. Therefore, one may combine hardware and software component reliabilities to get system reliability. Both depend on the environment. The source of failures in software is design faults, while the principal source in hardware has generally been physical deterioration. However, the concepts and theories developed for software reliability could really be applied to any design activity,including hardware design. Once a software (design) defect is properly fixed, it is in general fixed for all time. Failure usually occurs only when a program (design) is exposed to an environment that it was not developed or tested for. Although manufacturing can affect the quality of physical components, the replication process for software (design) is trivial and can be performed to very high standards of quality. Since introduction and removal of design faults occurs during software development, software reliability may be expected to vary during this period.

The *design reliability* concept has not been applied to hardware to that extent. The probability of failure due to wear and other physical causes has usually been much greater than that due to an unrecognized design problem. It was possible to keep hardware design failures low because hardware was generally less complex logically than software. Hardware design failures had to be kept low because retrofitting of manufactured items in the field was very expensive. Awareness of the work that is going on in software reliability, plus a growing realization of the importance of design faults, may now be having an effect on hardware reliability too. This growing awareness is strengthened by the parallels that people are starting to draw between software engineering and chip design.

A final characteristic of software reliability is that it tends to change continually during test periods. This happens either as new problems are introduced when new code is written or when repair action removes problems that exist in the code. Hardware reliability may change during certain periods, such as initial burn-in or the end of useful life. However, it has a much greater tendency than software toward a constant value.

Despite the foregoing differences, we can develop software reliability theory in a way that is compatible with hardware reliability theory. Thus system reliability figures may be computed using standard hardware combinatorial techniques. Hardware and software reliability share many similarities and some differences. One must not err on the side of assuming that software always presents unique problems, but one must also be careful not to carry analogies too far.

10.3 FAILURES AND FAULTS

What do we mean by the term software failure? It is the departure of the external results of program operation from requirements. So our *failure* is something dynamic. The program has to be executing for a failure to occur. The term failure relates to the behavior of the program. This very general definition of failure is deliberate. It can include such things as deficiency in performance attributes and excessive response time.

A *fault* is the defect in the program that, when executed under particular conditions, causes a failure. There can be different sets of conditions that cause failures, or the conditions can be repeated. Hence a fault can be the source of more than one failure. A fault is a property of the program rather than a property of its execution or behavior. It is what we are really referring to in general when we use the term *bug*. A fault is created when a programmer makes an error. It's very important to make the failure-fault distinction!

Reliability quantities have usually been defined with respect to time, although it would be possible to define them with respect to other variables. We are concerned with three kinds of time. The *execution time* for a program is the time that is actually spent by a processor in executing the instructions of that program. The second kind of time is *calendar time.* It is the familiar *time* that we normally experience. Execution time is important, because it is now generally accepted that models based on execution time are superior. However, quantities must ultimately be related back to calendar time to be meaningful to engineers or managers. Sometimes the term *clock time* is used for a program. It represents the elapsed time from start to end of program execution on a running computer. It includes wait time and the execution time of other programs. Periods during which the computer is shut down are not counted. If computer utilization by the program, which is the fraction of time the processor is executing the program, is constant, clock time will be proportional to execution time.

There are four general ways of characterizing failure occurrences in time:

1. time of failure,
2. time interval between failures,
3. cumulative failures experienced up to a given time,
4. failures experiences in a time interval.

These are illustrated in Tables 10.1 and 10.2.

TABLE 10.1

Time based failure specification

Failure number	Failure time (sec)	Failure interval (sec)
1	8	8
2	18	10
3	25	7
4	36	11
5	45	9
6	57	12
7	71	14
8	86	15
9	104	18
10	124	20
11	143	19
12	169	26
13	197	28
14	222	25
15	250	28

Note that all the foregoing four quantities are random variables. By *random*, we mean that the values of the variables are not known with certainty. There are many possible values, each associated with a probability of occurrence. For example, we don't really know when the next failure will occur. If we did, we would try to prevent or avoid it. We only know a set of possible times of failure.

TABLE 10.2

Failure based failure specification

Time(sec)	Cumulative failures	Failures in interval (sec)
30	3	3
60	6	3
90	8	2
120	9	1
150	11	2
180	12	1
210	13	1
240	14	1

There are at least two principal reasons for this randomness. First, the commission of errors by programmers, and hence the introduction of faults, is a very complex, unpredictable process. Hence the locations of faults within the program are unknown. Second, the conditions of execution of a program are generally unpredictable. For example, with a telephone switching system, how do you know what type of call will be made next? In addition, the relationship between program function requested and code path executed, although theoretically determinable, may not be so in practice because it is so complex. Since failures are dependent on the presence of a fault in the code and its execution in the context of certain machine states, a third complicating element is introduced that argues for the randomness of the failure process.

Table 10.3 illustrates a typical probability distribution of failures that occur within a time period of execution. Each possible value of the random variable of number of failures is given along with its associated probability. The probabilities, of course, add to 1. Note that here the random variable is discrete, as the number of failures must be an integer. Note that the most probable number of failures is 2 for $t=1$ hr. The mean or average number of failures can be computed. You multiply each possible value by the probability it can occur and add all the products. The mean is 3.04 failures for $t=1$ hour.

We will look at the time variation from two different viewpoints, the mean value function and the failure intensity function. The mean value function represents the average cumulative failures associated with each time point. The failure intensity function is the rate of change of the mean value function or the number of failures per unit time. For example, you might say 0.01 failure/hr or 1 failure/100 hr. Strictly speaking, the failure intensity is the derivative of the mean value function with respect to time, and is an instantaneous value.

A random process whose probability distribution varies with time is called nonhomogeneous. Most failure processes during test fit this situation. Fig.10.1 illustrates the mean value and the related failure intensity functions at time t_A and t_B. Note that the mean failures experienced increases from 3.04 to 7.77 between these two points, while the failure intensity decreases.

TABLE 10.3

Probability distribution at times t_A and t_B

Value of random variable (failures in time period)	Probability	
	Elapsed time $t_A = 1$hr	Elapsed time $t_B = 5$hr
0	0.10	0.01
1	0.18	0.02
2	0.22	0.03
3	0.16	0.04
4	0.11	0.05
5	0.08	0.07
6	0.05	0.09
7	0.04	0.12
8	0.03	0.16
9	0.02	0.13
10	0.01	0.10
11	0	0.07
12	0	0.05
13	0	0.03
14	0	0.02
15	0	0.01
Mean failures	3.04	7.77

Failure behavior is affected by two principal factors:

1. the number of faults in the software being executed,
2. the execution environment or the operational profile of execution.

The number of faults in the software is the difference between the number introduced and the number removed.

Faults are introduced when the code is being developed by programmers. They may introduce the faults during original design or when they are adding new features, making design changes, or repairing faults that have been identified. The term developed code, defined as instructions that have been created or modified, is used deliberately. In general, only code that is new or modified results in faults introduction. Code that is inherited from another application does not usually introduce any appreciable number of faults, except possibly in the interfaces. It generally has been thoroughly debugged in the previous application. Note that the process of faults removal introduces some new faults because it involves modification or writing of new code.

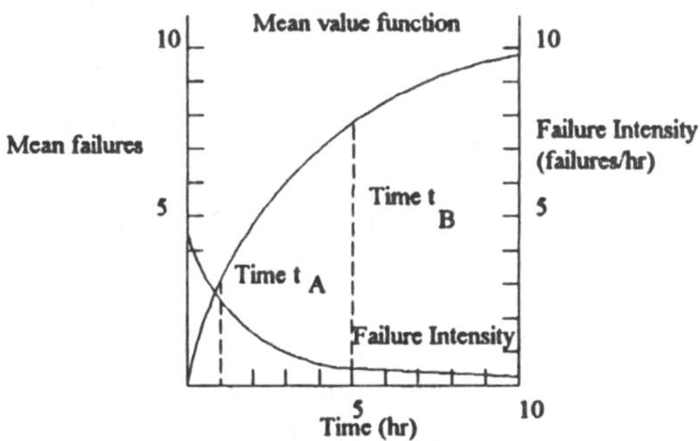

Fig.10.1 Mean value & failure intensity functions

Fault removal obviously can't occur unless you have some means of detecting the fault in the first place. Thus fault removal resulting from execution depends on the occurrence of the associated failure. Occurrence depends both on the length of time for which the software has been executing and on the execution environment or operational profile. When different functions are executed, different faults are encountered and the failures that are exhibited tend to be different; thus the environmental influence. We can often find faults without execution. They may be found through inspection, compiler diagnostics, design or code reviews, or code reading.

10.31 Environment

Let us scrutinize the term environment. The environment is described by the operational profile. We need to build up to the concept of the operational profile through several steps. It is possible to view the execution of a program as a single entity. The execution can last for months or even years for a real time system. However, it is more convenient to divide the execution into runs. The definition of run is somewhat arbitrary, but it is generally associated with some function that the program performs. Thus, it can conveniently describe the functional environment of the program. Runs that are identical repetitions of each other are said to form a run type. The proportion of runs of various types may vary, depending on the functional environment. Examples of a run type might be:

1. a particular transaction in an airline reservation system or a business data processing system,

2. a specific cycle in a closed loop control system (for example, in a chemical process industry), or

3. a particular service performed by an operating system for a user.

During test, the term test case is sometimes used instead of run type.

We next need to understand the concept of the input variable. This is a variable that exists external to the program and is used by the program in executing its function. For an airline reservation system, *destination* might be an input variable. One generally has a large quantity of input variables associated with the program, and each set of values of these variables characterize an input state. In effect, the input state identifies the particular run type that you're making. Therefore, runs can always be classified by their input states. Again, taking the case of the airline reservation system, the input state might be characterized by particular values of origin, destination, airline, day and flight number. The set of all possible input states is known as the input space.

Similarly, an output variable is a variable that exists external to a program and is set by it. An output state is a set of values of all output variables associated with a run of a program. In the airline reservation system, an output state might be the set of values of variables printed on the ticket and on different reports used in operating the airline. It can now be seen that a failure involves a departure of the output state from what it is expected to be.

The run types required of the program by the environment can be viewed as being selected randomly. Thus, we define the operational profile as the set of runtypes that the program can execute along with probabilities with which they will occur. In Fig.10.2, we show two of many possible input states. A and B, with their probabilities of occurrence. The part of the operational profile for just those two states is shown in Fig.10.3. In reality, the number of possible input states is generally quite large. A realistic operational profile is illustrated in Fig.10.4. Note that the input states have been located on the horizontal axis in order of the probabilities of their occurrence. This can be done without loss of generality. They have been placed close together so that the operational profile would appear to be a continuous curve.

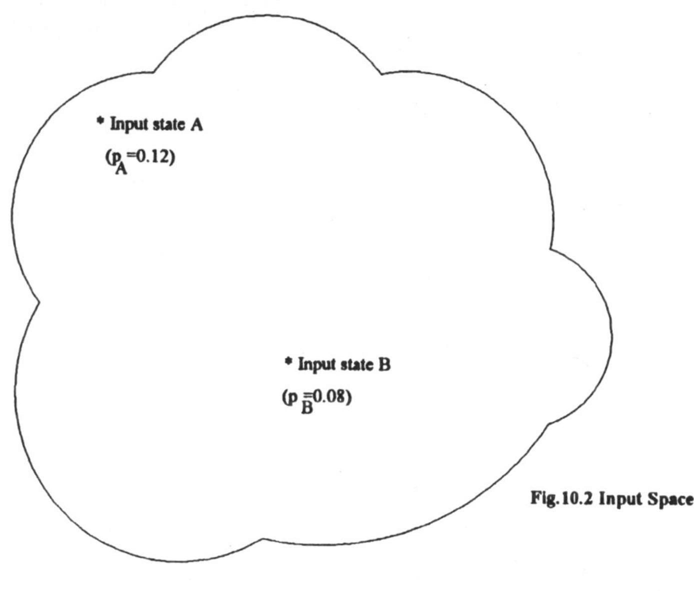

Fig.10.2 Input Space

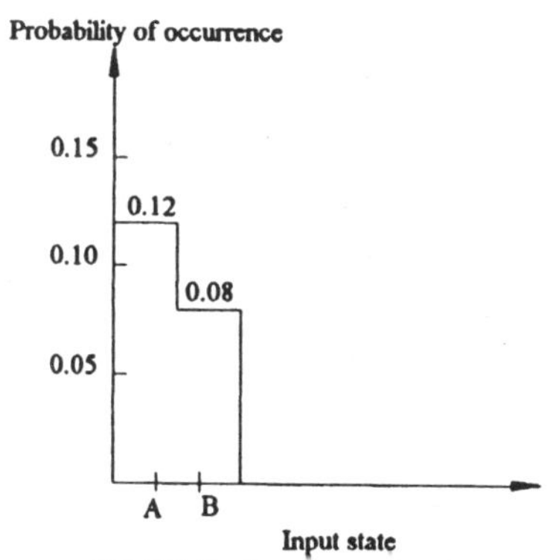

Fig.10.3 Portion of operational profile

10.4 SOFTWARE RELIABILITY

The definition that we will present here for software reliability is one that is widely accepted throughout the field. It is the probability of failure-free operation of a computer program for a specified time in a specified environment. For example, a time-sharing system may have a reliability of 0.95 for 10 hr when employed by the *average user*. This system, when executed for 10 hr, would operate without failure for 95 of these periods out of 100. As a result of the general way in which we defined failure, note that the concept of software reliability incorporates the notion of performance being satisfactory. For example, excessive response time at a given load level may be considered unsatisfactory, so that a routine must be recoded in more efficient form.

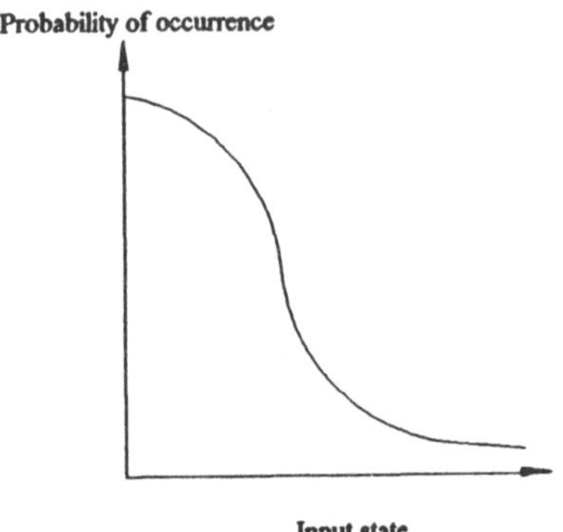

Fig.10.4 Operational profile

Failure intensity is an alternative way of expressing reliability. We just gave the example of the reliability of a particular system being 0.95 for 10 hr of time. An equivalent statement is that the failure intensity is 0.05 failure/hr. Each specification has its advantages. The failure intensity statement is more economical, as you only have to give one number. However, the reliability statement is better suited to the combination of reliabilities of components to get system reliability. If the risk of failure at any point in time is of paramount concern, failure intensity may be the more appropriate measure. Such would be the case for a nuclear power plant. When proper operation of a system to accomplish some function with a time duration is required reliability is often best. An example would be a space flight to the moon. Fig.10.5 shows how failure intensity and reliability typically vary during a test period, as faults are removed. Note that we define failure intensity, just like we do reliability, with respect to a specified environment.

10.41 Uses of Reliability Studies

Pressures have been increasing for achieving a more finely tuned balance among product and process characteristics, including reliability. Trade-offs among product components with respect to reliability are also becoming increasingly important. Thus an important use of software reliability measurement is in system engineering. However, there are at least four other ways in which software reliability measures can be of great value to the software engineer, manager, or user.

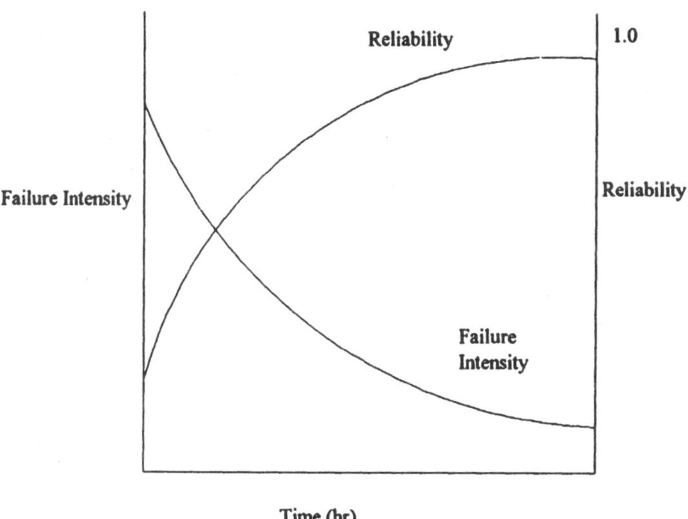

Fig.10.5 Reliability & Failure Intensity

First, you can use software reliability measures to evaluate software engineering technology quantitatively. New techniques are continually being proposed for improving the process of developing software, but unfortunately they have been exposed to little quantitative evaluation. The inability to distinguish between good and bad, new technology has often led to a general resistance to change that is counterproductive. Software reliability measures offer the promise of establishing at least one criterion for evaluating the new technology. For example, you might run experiments to determine the decrease in failure intensity (failures per unit time) at the start of system test resulting from design reviews. A quantitative evaluation such as this makes the benefits of good software engineering technology highly visible.

Second, a software reliability measure offers you the possibility of evaluating development status during the test phases of a project. Methods such as intuition of designers or test team, percent of tests completed, and successful execution of critical functional tests have been used to evaluate testing progress. None of these have been really satisfactory and some have been quite unsatisfactory. An objective reliability measure(such as failure intensity) established from test data provides a sound means of determining status. Reliability generally increases with the amount of testing. Thus, reliability can be closely linked with project schedules. Furthermore, the cost of testing is highly correlated with failure intensity improvement. Since two of the key process attributes that a manager must control are schedule and cost, reliability can be intimately tied in with project management.

Third, one can use a software reliability measure to monitor the operational performance of software and to control new features added and design changes made to the software. The reliability of software usually decreases as a result of such changes. A reliability objective can be used to determine when, and perhaps how large, a change will be allowed. The objective would be based on user and other requirements. For example, a freeze on all changes not related to debugging can be imposed when the failure intensity rises above the performance objective.

Finally, a quantitative understanding of software quality and the various factors influencing it and affected by it enriches insight into the software product and the software development process. One is then much more capable of making informed decisions.

10.5 SOFTWARE RELIABILITY MODELS

To model software reliability one must first consider the principal factors that affect it: fault introduction, fault removal, and the environment. Fault introduction depends primarily on the characteristics of the developed code (code created or modified for the application) and development process characteristics include software engineering technologies and tools used and level of experience of personnel. Note that code can be developed to add features or remove faults. Fault removal depends upon time, operational profile, and the quality of repair activity. The environment directly depends on the operational profile. Since some of the foregoing factors are probabilistic in nature and operate over time, software reliability models are generally formulated in terms of the random processes. The models are distinguished from each other in general terms by the nature of the variation of the random process with time.

A software reliability model specifies the general form of the dependence of the failure process of the factors mentioned. We have assumed that it is, by definition, time based (this is not to say that non-time-based models may not provide useful insights). The possibilities for different mathematical forms to describe the failure process are almost limitless. We have restricted ourselves to considering well-developed models that have been applied fairly broadly with real data and have given reasonable results. The specific forms can be determined from the general form by establishing the values of the parameters of the model through either:

1. estimation-statistical inference procedures are applied to failure data taken for the program, or

2. prediction- determination from properties of the software product and the development process (this can be done before any execution of the program).

10.51 Comparison of Software Reliability Models

It is recommended that software reliability models be compared by the criteria discussed below. It is expected that comparisons will cause some models to be rejected because they meet few of the criteria discussed here. On the other hand, there may or may not be a clear choice between the more acceptable models. The relative weight to be placed on the different criteria may depend on the context in which the model is being applied. When comparing two models, we should consider all criteria simultaneously. We should not eliminate models by one criterion before considering other criteria, except if predictive validity is grossly unsatisfactory. It is not expected that a model must satisfy all criteria to be useful.

The proposed criteria include predictive validity, capability, quality of assumptions, applicability, and simplicity. We will discuss each of the criteria in more detail in the following sections.

10.511 Predictive Validity

Predictive validity is the capability of the model to predict future failure behavior from present and past failure behavior (that is, data). This capability is significant only when failure behavior is changing. Hence, it is usually considered for a test phase, but it can be applied to the operational phase when repairs are being regularly made.

There are at least two general ways of viewing predictive validity. These are based on the two equivalent approaches to characterizing the failure random process, namely;

1. the number of failures approach and
2. the failure time approach.

We may apply various detailed methods, some representing approximations for predictive validity. It has not been determined if one is superior at the present time.

The number of failures approach may yield a method that is more practical to use than the failure time approach. In the former approach, we describe the failure random process by $[M(t), t \geq 0]$, representing failures experienced by time t. Such a counting process is characterized by specifying the distribution of M(t), including the mean value function $\mu(t)$.

Assume that we have observed q failures by the end of test time t_q. We use the failure data up to time $t_e(\leq t_q)$ to estimate the parameters of $\mu(t)$. Substituting the estimates of the parameters in the mean value function yields the estimate of the number of failures by the time t_q. The estimate is compared with the actually observed number q. This procedure is repeated for various values of t_e.

We can visually check the predictive validity by plotting the relative error against the normalized test time. The error will approach 0 as t_e approaches t_q. If the points are positive (negative), the model tends to overestimate (underestimate). Numbers closer to 0 imply more accurate prediction and hence a better model.

10.512 Capability

Capability refers to the ability of the model to estimate with satisfactory accuracy quantities needed by software managers, engineers, and users in planning and managing software development projects or running operational software systems. We must gauge the degree of capability by looking at the relative importance of the quantities as well as their number. The quantities, in approximate order of importance, are:

1. present reliability, mean time to failure (MTTF), or failure intensity,
2. expected date of reaching a specified reliability, MTTF, or failure intensity objective, and
3. human and computer resource and cost requirements related to the achievement of the objective.

Any capability of a model for prediction of software reliability in the system design and early development phases is extremely valuable because of the resultant value for system engineering and planning purposes. We must make these predictions through measurable characteristics of the software (size, complexity, structure, etc.), the software development environment, and the operational environment.

10.513 Quality of Assumptions

The following considerations of quality should be applied to each assumption in turn. If it is possible to test an assumption, the degree to which it is supported by data is an important consideration. This is especially true of assumptions that may be common to an entire group of models. If it is not possible to test the assumption,we should evaluate its plausibility from the view point of logical consistency and software engineering experience. For example,does it relate rationally to other information about software and software development? Finally, we should judge the clarity and explicitness of an assumption. These characteristics are often necessary to determine whether a model applies to particular software system or project circumstances.

10.514 Applicability

Another important characteristic of a model is its applicability. We should judge a model on its degree of applicability across software products that vary in size, structure, and function. It is also desirable that it be usable in different development environments, different operational environments, and different life cycle phases. However, if a particular model gives outstanding results for just a narrow range of products or development environments, we should not necessarily eliminate the model.

There are at least four special situations that are encountered commonly in practice. A model should either be capable of dealing with them directly or should be compatible with procedures that can deal with them. These are:

1. program evolution,
2. classification of severity of failures into different categories,
3. ability to handle incomplete failure data or data with measurement uncertainties (although not without loss of predictive validity),
4. operation of the same program on computers of different performance.

Finally, it is desirable that a model be robust with respect to departures from its assumptions, errors in the data or parameters it employs, and unusual conditions.

10.515 Simplicity

A model should be simple in three aspects. The most important consideration is that it must be simple and inexpensive to collect the data required to particularize the model. If this is not the case, we will not use the model. Second, the model should be simple in concept. Software engineers without extensive mathematical background should be able to understand the model and its assumptions. They can then determine when it is applicable and the extent to which the model may diverge from reality in an application. Parameters should have readily understood interpretations. This property makes it more feasible for software engineers to estimate the values of the parameters when data are not available. The number of parameters in the model is also an important consideration for simplicity. It should be pointed out that we need to compare the number of parameters on a common basis (for example, don't include calendar time component parameters for one model and not another).

Finally, a model must be readily implementable as a program that is a practical management and engineering tool. This means that the program must run rapidly and inexpensively with no manual intervention required (does not rule out possibility of intervention) other than the initial input.

On the basis of the above characteristics of a good software reliability model we select two models for presentation and application. Two models were chosen because each has certain advantages not possessed by the other. However, the effort required to learn the application of a model makes presenting more than two a question of sharply diminishing returns. The models are the basic execution time model and the logarithmic Possion execution time model. Both the models have two components, named the execution time component and the calander time component. Each component will be described with respect to both models.

10.6 EXECUTION TIME COMPONENT

The execution time component for both models assumes that failures occur as a random process, to be specific, a nonhomogeneous Poisson process. *Poisson* simply refers to the probability distribution of the value of the process at each point in time. The term *nonhomogeneous* indicates that the characteristics of the probability distributions that make up the random process vary with time. This is exhibited in a variation of failure intensity with time. You would expect that,since faults are both being introduced and removed as time passes.

The two models have failure intensity functions that differ as functions of execution time. However, the difference between them is best described in terms of slope or decrement per failure experienced (Fig.10.6). The decrement in the failure intensity function remains constant for the basic execution time model whether it is the first failure that is being fixed or the last. By contrast, for the logarithmic Poisson execution time model, the decrement per failure becomes smaller with failures experienced. In fact, it decreases exponentially. The first failure initiates a repair process that yields a substantial decrement in failure intensity, while later failures result in much smaller decrements.

The failure intensity for the basic model as a function of failures experienced is

$$\lambda(\mu) = \lambda_0[1-\mu/ v_0] \tag{10.1}$$

The quantity λ_0 is the initial failure intensity at the start of execution. Note that μ is the average or expected number of failures experienced at a given point in time. The quantity v_0 is the total number of failures that would occur in infinite time.

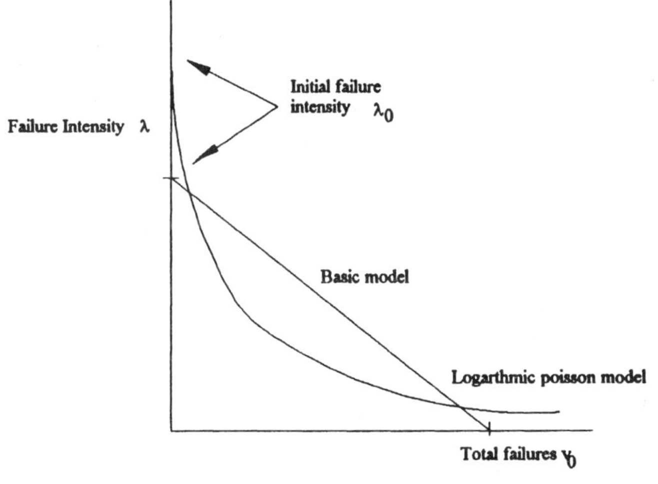

Fig.10.6 Failure intensity functions

Example 10.1

Assume that a program will experience 100 failures in infinite time. It has now experienced 50. The initial failure intensity was 10 failures/CPU hr. Determine the value of the current failure intensity.

Solution

$\lambda(\mu) = \lambda_0[1-\mu/ v_0] = 10\ [1-50/100] = 5$ failures/CPU hr.

* * *

The failure intensity for the logarithmic Poisson model is

$$\lambda(\mu) = \lambda_0 \exp(-\theta\mu) \qquad (10.2)$$

The quantity θ is called the failure intensity decay parameter. Suppose we plot the natural logarithm of failure intensity against mean failures experienced. Then we can see by transforming (10.2) that the failure intensity decay parameter θ is the magnitude of the slope of the line we have plotted. It represents the relative change of failure intensity per failure experienced.

Example10.2

Assume that the initial failure intensity is again 10 failures/ CPU hr. The failure intensity decay parameter is 0.02/failure. We assume that 50 failures have been experienced. The current failure intensity is to be determined.

Solution

$\lambda(u) = \lambda_0 exp(-\theta\mu) = 10 \exp[-(0.02)(50)] = 3.68$ failures/CPU hr.

* * *

The slope of failure intensity, $d\lambda/d\mu$, is given by

$$d\lambda /d\mu = - \lambda_0 / v_0 \tag{10.3}$$

for the basic model.

Example 10.3

In Example 10.1, the decrement of failure intensity per failure is to be determined.

Solution

$d\lambda /d\mu = - \lambda_0 / v_0 = -10/100 = -0.1$/CPU hr.

* * *

The corresponding slope is

$$d\lambda /d\mu = - \lambda_0 \theta \exp(-\theta\mu) = -\theta\lambda \tag{10.4}$$

for the logarithmic Poisson model.

Example 10.4

In Example 10.2, the decrement of failure intensity per failure is to be determined.

Solution

$$\begin{aligned} d\lambda /d\mu &= - \lambda_0 \theta \exp(-\theta\mu) \\ &= -10(0.02) \exp(-0.02\mu) = -0.2 \exp(0.02\mu)\text{/CPU hr.} \end{aligned}$$

* * *

When no failures have been experienced, we have a decrement of -0.2/CPU hr. greater than that for the basic model. After 50 failures have been experienced, the decrement is -0.0736/CPU hr. Note the decrease to an amount smaller than the corresponding amount for the basic model. The relative change in failure intensity per failure experienced is constant at 0.02. In other words, the failure intensity at a given number of failures experienced is 0.98 of that at the preceding failure.

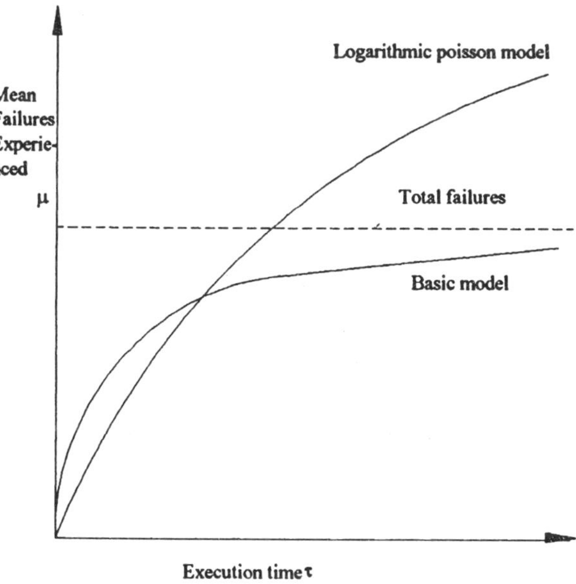

Fig.10.7 Mean failures experienced versus execution time

We can obtain some interesting relationships with some straight forward derivations. The expected number of failures experienced as a function of execution time is illustrated for both models in Fig.10.7. Whether the curve for the logarithmic Poisson model crosses that for the basic model depends on its parameter values. Note that the expected number of failures for the logarithmic Poisson model is always infinite at infinite time. This number can be and usually is finite for the basic model during test, although it is usually infinite during the operational phase. The curve for the former model is logarithmic, hence the name. The curve for the latter is *negative* exponential, approaching a limit. Infinite failures can occur for the logarithmic Poisson model.

Let execution time be denoted by τ. We can then write, for the basic model.

$$\mu(\tau) = v_0[1-\exp(-\lambda_0 t/ v_0)] \qquad (10.5)$$

Example 10.5

Let's again consider a program with an initial failure intensity of 10 failures/CPU hr and 100 total failures. Calculate the failures experienced after 10 and 100 CPU hr of execution.

Solution

For 10 CPU hr,

$$\mu(\tau) = v_0[1\text{-exp}(-\lambda_0 t\ /\ v_0)]$$
$$= 100[1\text{-exp}(-10{*}10/100)] = 63 \text{ failures}$$

For 100 CPU hr, we have:

$$= 100[1\text{-exp}(-10)]$$
$$= 100(1\text{-}0.0000454) = 100 \text{ failures(almost)}.$$

* * *

For the logarithmic Poisson model, we have the corresponding relation for the number of failures as given by:

$$\mu(\tau) = [\ln(\lambda_0\, \theta\tau + 1)]/\ \theta \qquad (10.6)$$

Example 10.6

Use the same parameters as Example 10.2. Let's find the number of failures experienced for the logarithmic Poisson model at 10 and 100 CPU hr of execution.

Solution

$$\mu(\tau) = [\ln(\lambda_0\theta\tau + 1)]/\theta$$
$$= \ln[(10))(0.02)(10) + 1]/0.02$$
$$= 50 \ln(2 + 1) = 55 \text{ failures.}$$

This is smaller than the number of failures experienced by the basic model at 10 CPU hr. At 100 CPU hr we have:

$$\mu(\tau) = \ln[(10)(0.02)(100) + 1]/0.02$$
$$= 50 \ln 21 = 152 \text{ failures.}$$

* * *

The failure intensity as a function of execution time for both models is shown in Fig.10.8. The relationship is useful for determining the present failure intensity at any given value of execution time. For the same set of data, the failure intensity of the logarithmic Poisson model drops more rapidly than that of the basic model at first. Later, it drops more slowly. At large values of execution time, the logarithmic Poisson model will have larger values of failure intensity than the basic model.

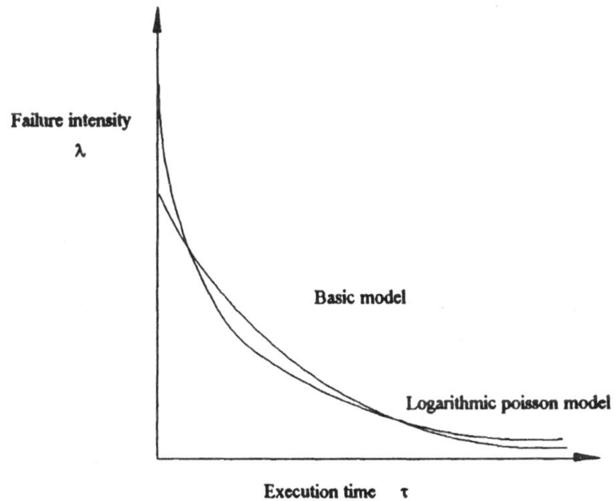

Fig.10.8 Failure intensity versus execution time

We have, for the basic model,

$$\lambda(\tau) = \lambda_0 \exp(-\lambda_0 \tau / v_0) \tag{10.7}$$

Example 10.7

Calculate the failure intensities at 10 and 100 CPU hr, using the parameters of the example 10.1.

Solution

We have, at 10 CPU hr:

$$\lambda(\tau) = \lambda_0 \exp(-\lambda_0 \tau / v_0) = 10 \exp(10 x 10/100)$$
$$= 3.68 \text{ failures/CPU hr.}$$

At 100 CPU hr we have:

$$\lambda(\tau) = 10 \exp(-10 \text{ x } 100/100)$$
$$= 10 \exp(-10) = 0.000454 \text{ failures/CPU hr.}$$

* * *

We can write, for the logarithm Poisson model, the expression for failure intensity as:

$$\lambda(\tau) = \lambda_0/(\lambda_0 \theta\tau + 1) \tag{10.8}$$

Example 10.8

Calculate the failure intensities for the logarithmic Poisson model at 10 CPU hr and 100 CPU hr, using the parameters of Example 10.2.

Solution

We have for 10 CPU hr,

$$\begin{aligned} \lambda(\tau) &= \lambda_0/(\lambda_0 \theta\tau + 1) = 10/[10(0.02)(10) + 1] \\ &= 3.33 \text{ failures/CPU hr.} \end{aligned}$$

This is slightly lower than the corresponding failure intensity for the basic model. At 100 CPU hr we have:

$$\begin{aligned} \lambda(\tau) &= 10/[10(0.02)(100) + 1] \\ &= 0.476 \text{ failure/ CPU hr.} \end{aligned}$$

The failure intensity at the higher execution time is larger for the logarithmic Poisson model.

* * *

10.61 Derived Quantities

Assume that you have chosen a failure intensity objective for the software product being developed. Suppose some portion of the failures are being removed through correction of their associated faults. Then one can use the objective and the present value of failure intensity to determine the additional expected number of failures that must be experienced to reach that objective. The process is illustrated graphically in Fig.10.9. Equations describing the relationship in closed form may be derived for both models so that manual calculations can be performed. They are

$$\Delta_\mu = (v_0/\lambda_0)(\lambda_p - \lambda_f) \tag{10.9}$$

for the basic model and

$$\Delta_\mu = (1/\theta)\ln(\lambda_p/\lambda_f) \tag{10.10}$$

for the logarithmic Poisson model. The quantity Δ_{μ} is the expected number of failures to reach the failure intensity objective, λ_p is the present failure intensity, and λ_f is the failure intensity objective.

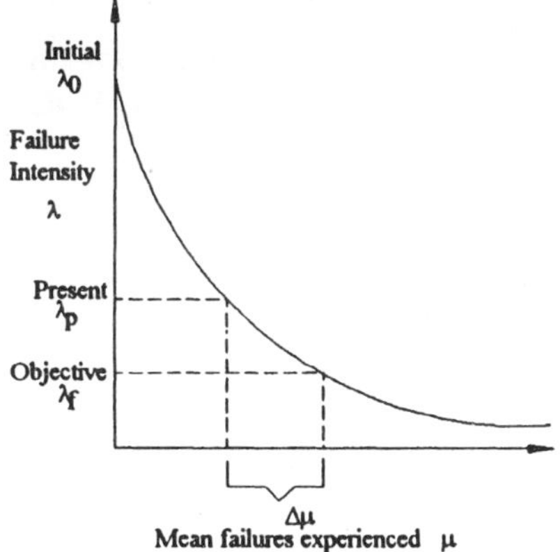

Fig. 10.9 Additional failures to failure intensity objective

Example 10.9

For the basic model, we determine the expected number of failures that will be experienced between a present failure intensity of 3.68 failures/CPU hr and an objective of 0.000454 failure/CPU hr. We will use the same parameter values as in Example 10.1.

Solution

$$\begin{aligned}\Delta_{\mu} &= (v_0/\lambda_0)(\lambda_p - \lambda_f)\\ &= (100/10)(3.68 - 0.000454)\\ &= 10(3.68) = 37 \text{ failures}\end{aligned}$$

* * *

Example 10.10

We will find, for the logarithmic Poisson model, the expected number of failures experienced between a present failure intensity of 3.33 failures/CPU hr and an objective of 0.476 failure/CPU hr. The parameter values will be the same as in Example 10.2.

Solution

$$\begin{aligned}\Delta_{\mu} &= (1/\theta)\ln(\lambda_p/\lambda_f)\\ &= (1/0.02)\ln(3.33/0.476)\end{aligned}$$

$= 50 \ln 6.996 = 97$ failures.

* * *

Similarly, you can determine the additional execution time $\Delta\tau$ required to reach the failure intensity objective for either model. This is

$$\Delta_\tau = (v_0/\lambda_0) \ln(\lambda_p/\lambda_f) \quad (10.11)$$

for the basic model and

$$\Delta_\tau = (1/\theta) [(1/\lambda_f)-(1/\lambda_p)] \quad (10.12)$$

for the logarithmic Poisson model. This is illustrated in the Fig.10.10.

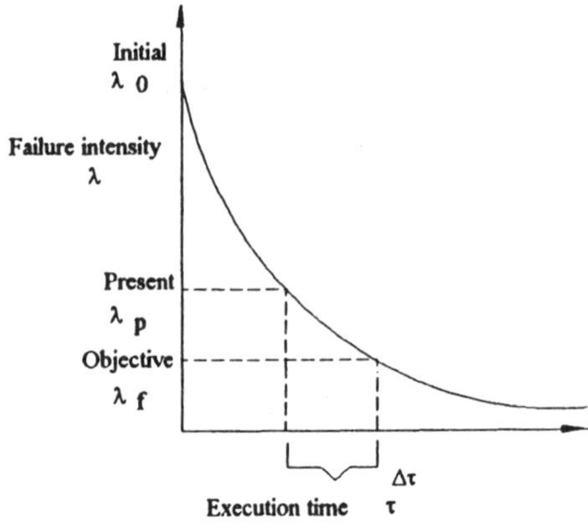

Fig.10.10 Additional execution time to failure intensity objective

Example 10.11

For the basic model, with the same parameter values used in Example 10.1 we will determine the execution time between a present failure intensity of 3.68 failures/CPU hr and an objective of 0.000454 failure/CPU hr.

Solution

$$\begin{aligned}\Delta_\tau &= (v_0/\lambda_0) \ln(\lambda_p/\lambda_f)\\ &= (100/10) \ln(3.68/0.000454)\\ &= 10 \ln 8106 = 90 \text{ CPU hr}\end{aligned}$$

* * *

This result checks with the results of Example 10.7.

Example 10.12

For the logarithmic Poisson model, with the same parameter values used in Example 10.2, we will find the execution time between a present failure intensity of 3.33 failures/CPU hr and an objective of 0.476 failure/CPU hr.

Solution

Δ_τ $= (1/\theta)\ [(1/\lambda_f)-(1/\lambda_p)]$

$= (1/0.02)\ [(1/0.476)-(1/3.33)]$

$= 90$ CPU hr.

* * *

The foregoing quantities are of interest in themselves. The additional expected number of failures required to reach the failure intensity objective gives some idea of the failure correction workload. The additional execution time indicates the remaining amount of test required. However, even more importantly, they are both used in making estimates of the additional calendar time required to reach the failure intensity objective.

10.7 CALENDAR TIME COMPONENT

The calendar time component relates execution time and calendar time by determining the calendar time to execution time ratio at any given point in time. The ratio is based on the constraints that are involved in applying resources to a project. To obtain calendar time, one integrates this ratio with respect to execution time. The calendar time component is of greatest significance during phases where the software is being tested and repaired. During this period one can predict the dates at which various failure intensity objectives will be met. The calendar time component *exists* during periods in which repair is not occurring and failure intensity is constant. However, it reduces in that case to a constant ratio between calendar time and execution time.

In test, the rate of testing at any time is constrained by the failure identification or test team personnel, the failure correction or debugging personnel, or the computer time available. The quantities of these resources available to a project are usually more or less established in its early stages. Increases are generally not feasible during the system test phase because of the long lead times required for training and computer procurement. At any given value of execution time, one of these resources will be limiting. The limiting resource will determine the rate at which execution time can be spent per unit calendar time. A test phase may consist of from one to three periods, each characterized by a different limiting resource.

The following is a common scenario. At the start of testing one identifies a large number of failures separated by short time intervals. Testing must be stopped from time to time to let the people who are fixing the faults keep up with the load. As testing progresses, the intervals between failures become longer and longer. The time of the failure correction personnel is no longer completely filled with failure correction work. The test team becomes the bottleneck. The effort required to run tests and analyze the results is occupying all their time. That paces the amount of testing done each day. Finally, at even longer intervals, the capacity of the computing facilities becomes limiting. This resource then determines how much testing is accomplished.

The calendar time component is based on a *debugging process* model. This model takes into account:

1. resources used in operating the program for a given execution time and processing an associated quantity of failures.
2. resource quantities available, and
3. the degree to which a resource can be utilized(due to bottlenecks) during the period in which it is limiting.

Table 10.4 on the following page will help in visualizing these different aspects of the resources, and the parameters that result.

TABLE 10.4

Calendar time component resources and parameters

	Usage parameters requirements per		Planned parameters	
Resources	CPU hr	Failure	Quantities available	Utilization
Failure identification Personnel	θ_i	μ_i	P_i	1
Failure correction Personnel	0	μ_f	P_f	ρ_f
Computer time	θ_c	μ_c	P_c	ρ_c

10.71 Resource Usage

Resource usage is linearly proportional to execution time and mean failures experienced. Let x_r be the usage of resource r. Then

$$x_r = \theta_r \tau + \mu_r \mu \qquad (10.13)$$

Note that θ_r is the resource usage per CPU hr. It is nonzero for failure identification personnel(θ_i) and computer time (θ_c). The quantity μ_r is the resource usage per failure. Be careful not to confuse it with mean failures experienced μ. It was deliberately chosen to be similar to suggest the connection between resource usage and failures experienced. It is nonzero for failure identification personnel (μ_i), failure correction personnel (μ_f), and computer time (μ_c).

Example 10.13

Suppose the test team runs test cases for 8 CPU hr and identifies 20 failures. The effort required per hr of execution time is 6 person hr. Each failure requires 2 hr on the average to verify and determine its nature. Calculate the total failure identification effort required.

Solution

Using Equation 10.13,

$$x_r = \theta_r \tau + \mu_r \mu = 6(8) + 2(20) = 48 + 40 = 88 \text{ person hr.}$$

* * *

For failure correction (unlike identification), resources required are dependent only on the mean failures experienced. However, computer time is used in both identification and correction of failures. Hence, computer time used will usually depend on both the amount of execution time and the number of failures.

Note that since *failures experienced* is a function of execution time, resource usage is actually a function of execution time only. The intermediate step of thinking in terms of failures experienced and execution time is useful in gaining physical insight into what is happening.

Computer time required per unit execution time will normally be greater than 1. In addition to the execution time for the program under test, additional time will be required for the execution of such support programs as test drivers,recording routines, and data reduction packages.

Consider the change in resource usage per unit of execution time. It can be obtained by differentiating Equation (10.13) with respect to execution time.

We obtain

$$dx_r/ d\tau = \theta_r + \mu_r \lambda \tag{10.14}$$

Since the failure intensity decreases with testing, the effort used per hour of execution time tends to decrease with testing. It approaches the execution time coefficient of resource usage asymptotically as execution time increases.

10.72 Calendar Time to Execution Time Relationship

Resource quantities and utilizations are assumed to be constant for the period over which the model is being applied. This is a reasonable assumption, as increases are usually not feasible.

The instantaneous ratio of calendar time to execution time can be obtained by dividing the resource usage rate of the limiting resource by the constant quantity of resources available that can be utilized. Let t be calendar time. Then

$$dt / d\tau = (1/ P_r \rho_r)\, dx_r/d\tau \tag{10.15}$$

The quantity P_r represents resources available. Note that ρ_r is the utilization. The above ratio must be computed separately for each resource-limited period. Since x_r is a function of τ, we now have a relationship between t and τ in each resource limited period.

The form of the instantaneous calendar time to execution time ratio for any given limiting resource and either model is shown in Fig.10.11. It is readily obtained from Equations (10.14) and (10.15) as

$$dt / d\tau = (\theta_r + \mu_r \lambda) / P_r \rho_r \tag{10.16}$$

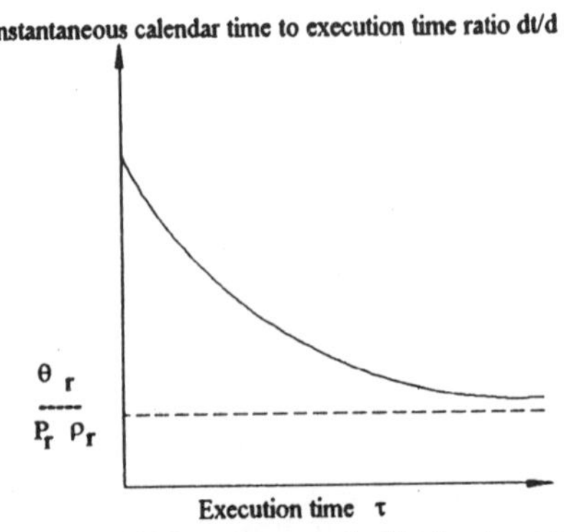

Fig.10.11 Instantaneous calendar time to execution time ratio

The shape of this curve will parallel that of the failure intensity. The curve approaches an asymptote of $\theta_r/P_r\ \rho_r$. Note that the asymptote is 0 for the failure correction personnel resource. At any given time, the maximum of the ratios for the three limiting resources actually determines the rate at which calendar time is expended; this is illustrated in Fig.10.12. The maximum is plotted as a solid curve. When the curve for a resource is not maximum (not limiting), it is plotted thin. Note the transition points FI and IC. Here, the calendar time to execution time ratios of two resources are equal and the limiting resource changes. The point FC is a potential but not true transition point. Neither resource F nor resource C is limiting near this point.

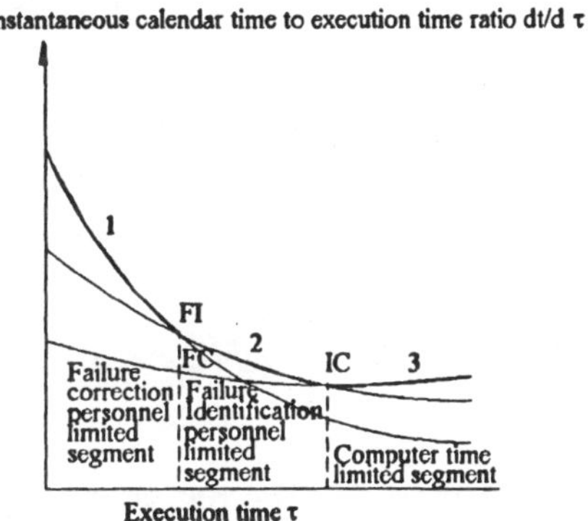

Fig. 10.12 Calendar time to execution time ratio for different limiting resources.

The calendar time component allows you to estimate the calendar time in days required to meet the failure intensity objective. The value of this interval is particularly useful to software managers and engineers. One may determine it from the additional execution time and additional number of failures needed to meet the objective that we found for the execution time component. Second, one now determines the date on which the failure intensity objective will be achieved. This is a simple variant of the first quantity that takes account of things like weekends and holidays. However, it is useful quantity because it speaks in terms managers and engineers understand.

11

RELIABILITY ANALYSIS OF SPECIAL SYSTEMS

11.1 COMPUTER COMMUNICATION NETWORKS

The reliability analysis of a computer communication network (CCN) using graph theoretic approach is based on modeling the network by a linear graph in which the nodes (vertices) correspond to computer centres (Hosts and Interface Message Processors) in the network, and edges correspond to the communication links. The terminal reliability, a commonly used measure of connectivity, is the probability of obtaining service between a pair of operative centres, called source and sink, in terms of reliability for each communication link/node in the network. This calculation obviously does not take into account the communication between any other nodes but for the source and sink. Here, we find the probability of obtaining a situation in which each node in the network communicates with all other remaining communication centres(nodes). In the event that this probability, now onwards called *Network Reliability* of a CCN, is to be calculated using the concepts of terminal reliability only, one can proceed by finding all possible paths between each of the n(n-1)/2 node pairs. Since this is impractical for graphs with a large number of nodes, an alternative procedure is suggested using the concept of spanning trees.

A tree T_i is said to be a *spanning tree* of graph G if T_i is a connected subgraph of G and contains all nodes of G. An edge in T_i is called a branch of T_i, while an edge of G that is not in T_i is called a chord. For a connected graph of n nodes and b edges, and spanning tree has (n-1) branches and (b-n+1) chords.

11.11 Reliability Analysis

From the definition of spanning tree, any T_i will link all n nodes of G with (n-1) branches and hence represents the minimum interconnections required for providing a communication between all computer centres which are represented by nodes. Thus, the problem of studying the network reliability between any of the centres in the CCN is a problem of:

1. Enumerating all T_i's in the reliability graph corresponding to the network.

2. Interpreting Boolean algebraic statement of step #1 as probability expression.

For step#1, a simple approach is to use Cartesian products of (n-1) vertex cutsets C_i whose elements are the branches connected to any of the (n-1) nodes of G. Thus

$$C = C_1 \times C_2 \times \dots \times C_{n-1}$$

$$= \mathop{X}_{i=1}^{n-1} C_i \qquad (11.1)$$

where C is a set of subgraphs of G with (n-1) branches. It has been proved that any circuit of G with (n-1) branches will have an even number of identical appearances in C. If these terms are recognized, then deleted from C, the normalised Cartesian product C^* contains only those subgraphs which do not repeat an even number of times and are of cardinality (n-1). From the concept of spanning tree, C^* is, thus, the set of all T_i's of a connected graph G.

Example 11.1

Enumerate the spanning trees for a bridge network shown in Figure 11.1.

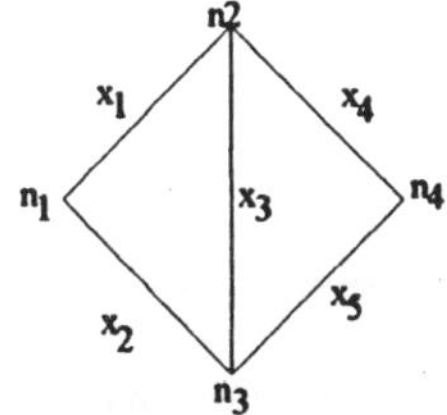

Fig.11.1 A bridge networ

Solution

The three vertex cutsets are:

$C_1 = (X_1, X_2);\ C_2 = (X_4, X_5);\ C_3 = (X_1, X_3, X_4)$

Using (11.1),

$$C = (X_1X_2) \times (X_4, X_5) \times (X_1, X_3, X_4)$$

$$= (X_1X_4, X_1X_5, X_2X_4, X_2X_5) \times (X_1, X_3, X_4)$$

$$= (X_1X_3X_4,\ X_1X_3X_5,\ X_1X_4X_5,\ X_1X_2X_4,\ X_2X_3X_4,\ X_1X_2X_5,\ X_2X_3X_5,\ X_2X_4X_5) \tag{11.2}$$

Since no term in (11.2) has an even number of identical appearances, C* is the same as C. The 8 elements of set C* thus represent 8 different spanning trees.

* * *

In step #2, a Boolean algebraic expression has a one-to-one correspondence with the probability expression if the Boolean terms are modified until they represent a disjoint grouping. We present below an algorithm for finding the probability expression and hence the network reliability of CCN starting from a set of T_i's.

1. For the purpose of network reliability, let system success S, be defined as the event of having at least one spanning tree with all its branches operative.

$$S = T_0 \cup T_i \cup \ldots \cup T_{N-1} \tag{11.3}$$

2. Define F_i for each term T_i such that

$$F_0 = T_0$$

$$F_i = T_0 \cup T_1 \cup \ldots \cup T_{i-1} \Big|_{\text{Each literal of } T_i \to 1,\quad \text{for} \quad 1 \le i \le (N-1)}$$

The literals of T_i are assigned a value 1(Boolean) which is substituted in any predecessor term in which they occur. F_i can be simplified by using elementary Boolean theorems.

3. Use Exclusive Operator ξ to get

$$S(\text{disjoint}) = T_o \bigcup_{i=1}^{N-1} T_i \, \xi \, (F_i) \tag{11.4}$$

Since, all terms in (11.4) are mutually exclusive, the network reliability expression R_s, is obtained from (11.4) by changing X_i to p_i, and X'_i to q_i, viz.,

$$R_s = S(\text{disjoint}) \Big|_{X_i\,(X'_i) \to p_i\,(q_i)} \tag{11.5}$$

Example 11.2

Derive the network reliability expression for a simple bridge network as given in figure 11.1.

Solution

$$S = X_1X_3X_5 \cup X_1X_3X_4 \cup X_1X_4X_5 \cup X_2X_3X_5 \cup X_1X_2X_5 \cup X_2X_3X_4 \cup X_1X_2X_4 \cup X_2X_4X_5.$$

The F'_is and $\xi(F_i)$'s for i = 1, ..., 7 are obtained as shown in Table 11.1.

TABLE 11.1

F_i	$\xi(F_i)$	F_i	$\xi(F_i)$
$F_1 = X_5$	X_5'	$F_5 = X_1 + X_5$	$X_1'X_5'$
$F_2 = X_3$	X_3'	$F_6 = X_3 + X_5$	$X_3'X_5'$
$F_3 = X_1$	X_1'	$F_7 = X_1 + X_3$	$X_1'X_3'$
$F_4 = X_3 + X_4$	$X_3'X_4'$		

From (11.5), the network reliability expression is

$$R_s = p_1p_3p_5 + p_1p_3p_4q_5 + p_1p_4p_5q_3 + p_2p_3p_5q_1 + p_1p_2p_5q_3q_4 + p_2p_3p_4q_1q_5 + p_1p_2p_4q_3q_5 + p_2p_4p_5q_1q_3 \tag{11.6}$$

For the CCN having equal probabilities of survival p for each communication link(11.6) simplifies to

$$R_s = 8p^3 - 11p^4 + 4p^5 \tag{11.7}$$

* * *

In deriving(11.6) we have assumed perfect nodes. As computer outages account for as much as 90% of failures in most CCNs, we have to consider the reliability of nodes as less than 1 in such situations. In such a case. (11.6) is to be multiplied by a factor ($p_{n1}\ p_{n2}\ p_{n3}\ p_{n4}$) where p_{ni} represents the reliability of node n_i.

11.2 PHASED MISSION SYSTEMS

At various times during its life time, the structure of the system may not remain constant throughout the mission but may have a time varying structure due to reconfiguration of the system or changes in the requirements placed on the system. Such systems are called phased mission systems. These systems perform several different tasks during their operational life.

Depending on the varying configuration with time of the system, its mission can be broken down into many phases; each phase corresponding to one configuration. Such a mission is known as phased mission. A phased mission is a task, to be performed by a system, during the execution of which the system is altered such that the logic model changes at a specified time.

Some of the complex and automated systems encountered in nuclear, aerospace, chemical, electronic and other industries perform several different tasks during their operational life. For example, highly integrated modern avionics must perform different functions, or set of functions, during different phases of a mission. Such a system has many subsystems e.g. radar, navigation and communications. A mission relating to an avionics system might be a two-hour bombing mission. Such a mission would be divided into phases such as take off, climb, cruise, attack, descend and land. During each of the mission phases, different subsystems are required to perform the mission.

Reliability evaluation techniques for phased-mission systems are different from reliability evaluation techniques for single mission systems. The reliability of a phased-mission system is the probability that the mission successfully achieves (all) the required objectives in each phase. Reliability analysis of such systems having reconfiguration capabilities is difficult because of the effects of the history of the systems structure, use environment, and user success criteria. The solution of phased-mission systems is equivalent to solving a sequence of uni-phase systems with appropriate initial conditions. If a component C_1 is required only in phase 1, then for system reliability calculation, failure or survival of component C_1 after phase 1, does not affect reliability. If a component C_1 is required to work in phase 4 and it is given that no repair facility exists, then the component should not fail in any of the previous phases.

Phased mission techniques are required for proper analysis of problems when switching procedures are carried out or equipment is reassembled into new system at predetermined times or system performs several different tasks during its operational life. For a given mission to be successful, the system must be available at the start of a mission and the system must complete its mission within the maximum allowable time that this given mission specifies and without failure during this period. During each phase, the system structure must stay the same. The effects of environment and operator can be reflected in the mission duration. In other words, unfavourable environmental conditions and poor operator performance can be taken into account by making the actual mission duration longer than the mission duration under ideal conditions, i.e. adverse effects of the environment and the operator tend to reduce the probability of mission success.

The components can, but need not, be repairable, with specified repair times. Often a system undergoing a phased mission will contain both repairable and non-repairable components. In a mission such as that of an intercontinental ballistic missile, all of the components are non-repairable. During a manned space flight, however, an astronaut might be able to replace or atleast repair a malfunctioning item.

11.21 Reliability Analysis

For reliability calculations, we assume an s-coherent binary system. A binary system is s-coherent if : (1) a component failure cannot cause the system to transit from failed to good, and (2) at least one component is relevant to the state of the system.

The event that the system functions throughout the mission is

$$[\ \phi_1\ (X(t_1)\ ] = 1,\ \ldots,\ \phi_L\ (X\ (t_L)) = 1\ ] \tag{11.8}$$

The exact reliability can be found by transforming the phased-mission problem into an equivalent single-phase system. Following steps are followed in the transformation of block diagram:

(1) Mission cut-set cancellation: A minimal cut-set in a phase is cancelled, i.e., omitted from the list of minimal cut-sets for that phase, if it contains a minimal cut-set of a later phase.

(2) Basic Event Transformation: In the configruation for phase j, basic event C_k is replaced by a series logic in which the basic events C_{k1}, ..., C_{kj} perform s-independently with the probability of failure frtc (k,j).

(3) The transformed phase configurations are considered to be subsystems operating in series logic in a new system involved in a 1-phase mission.

(4) Minimal cut-sets are obtained for this new logic model.

(5) Usual quantitative evaluation techniques are used to obtain system unreliability from these final minimal cutsets.

The method is illustrated with the help of an example. Let us consider the block diagram for a simple three-phased mission as shown in fig.11.2. Cutsets for this example system are given as

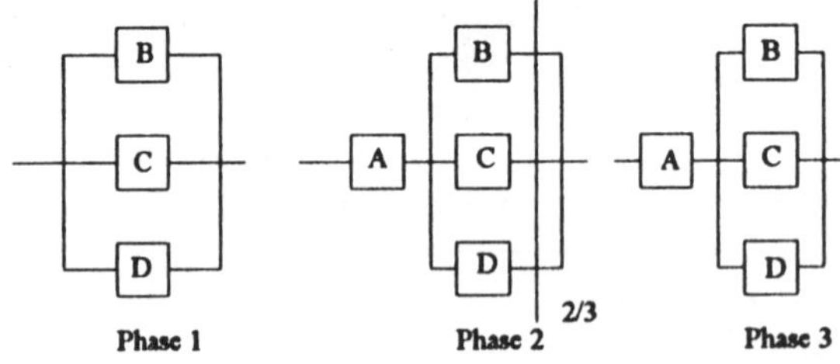

Fig.11.2 Block diagrams for a simple three phased mission.

Phase 1 BCD
Phase 2 A, BC, BD, CD
Phase 3 A, BCD

The solution is obtained in following steps:

(1) Mission Cut-Set Cancellation: The cutset A can be eliminated from phase 2 of Fig. 11.2 because it contains the cutset A from phase 3. In some cases, entire phases can be eliminated by this procedure. For example, phase 1 of Fig. 11.2 can be eliminated because its cutset contains only one cutset which is also a cutset of phase 3.

After cutset cancellation, we obtain

Phase 1
Phase 2 BC, BD, CD
Phase 3 A, BCD

(2) Basic Event Transformation: By applying this step, block diagram as shown in Fig. 11.3 is obtained.

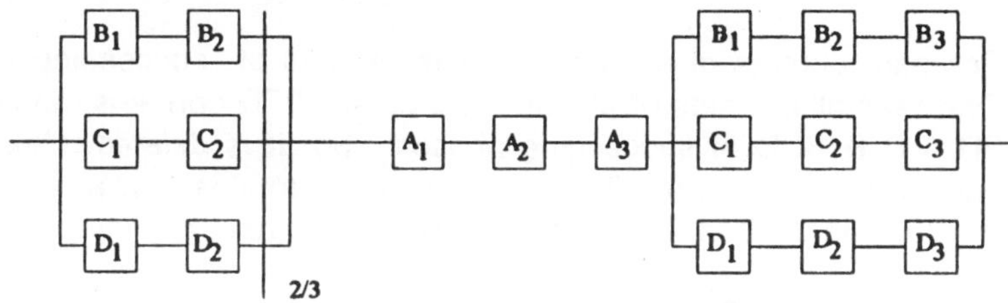

Fig.11.3 Block diagram for the equivalent one-phase system.

(3) Minimal cutsets for this new logic model are

B_1C_1, B_2C_2, B_1D_1, B_2D_2, C_1D_1, C_2D_2,

A_1, A_2, A_3, $B_1C_1D_1$, $B_2C_2D_2$, $B_3C_3D_3$.

(4) The above minimal cutsets are used to obtain total system unreliability.

Example 11.3

For the phased-mission system shown in Fig.11.2, calculate probability of mission success. It is given that each phase lasts for 40,60 and 100 hours respectively i.e.

d(1) = 40 hours
d(2) = 60 hours
d(3) = 100 hours

Cutsets in each phase are given as

Phase 1 BCD
Phase 2 A, BC, BD, CD
Phase 3 A, BCD

Failure rate/hour of each component in each phase is given as

frtc(i,j) =

	Phase1	Phase2	Phase3
Component 1	.001	.001	.003
Component 2	.001	.005	.002
Component 3	.002	.010	.010
Component 4	.010	.030	.020

Solution

Step1: Compare cutsets of each phase with cutsets of succeeding phases. First take all cut sets of phase 1. Its cutset BCD contains cutsets BC, BD, CD & BCD of succeeding phases. So, it is deleted. Next take one by one cutsets of phase 2. Cutset A contains cutset A of phase 3. So, it is deleted. Cutsets after mission cutset cancellation are

Phase 1 0 0 0 0
Phase 2 0 1 1 0, 0 1 0 1 and 0 0 1 1
Phase 3 1 0 0 0 and 0 1 1 1

where 0 indicates absence of element
and 1 indicates presence of element.

First, second, third and fourth positions correspond to elements A, B, C and D respectively.

Step 2: At this step, in any phase j basic event C_k is replaced by a series logic in which the basic events $C_{k1},\ldots,C_{kj}$ perform s-independently. So, after this step, the system can contain upto (n) x (L) unique components. For the example considered, following cutsets are obtained.

	A_1	A_2	A_3	B_1	B_2	B_3	C_1	C_2	C_3	D_1	D_2	D_3
Phase 1												
Phase 2	0	0	0	1	0	0	1	0	0	0	0	0
	0	0	0	0	1	0	0	1	0	0	0	0
	0	0	0	1	0	0	0	0	0	1	0	0
	0	0	0	0	1	0	0	0	0	0	1	0
	0	0	0	0	0	0	1	0	0	1	0	0
	0	0	0	0	0	0	0	1	0	0	1	0
Phase 3	1	0	0	0	0	0	0	0	0	0	0	0
	0	1	0	0	0	0	0	0	0	0	0	0
	0	0	1	0	0	0	0	0	0	0	0	0
	0	0	0	1	0	0	1	0	0	1	0	0
	0	0	0	0	1	0	0	1	0	0	1	0
	0	0	0	0	1	1	0	1	1	0	1	1

Step 3: Disjoint terms are calculated as:

1	0	0	0	0	0	0	0	0	0	0	0
-1	1	0	0	0	0	0	0	0	0	0	0
-1	-1	1	0	0	0	0	0	0	0	0	0

-1	-1	-1	0	1	0	0	0	0	0	1	0
-1	-1	-1	0	-1	0	1	0	0	1	0	0
-1	-1	-1	0	1	0	1	0	0	1	-1	0
-1	-1	-1	0	-1	0	-1	1	0	0	1	0
-1	-1	-1	0	-1	0	1	1	0	-1	1	0
-1	-1	-1	1	-1	0	1	-1	0	-1	0	0
-1	-1	-1	1	-1	0	1	1	0	-1	-1	0
-1	-1	-1	1	1	0	1	0	0	-1	-1	0
-1	-1	-1	0	1	0	-1	1	0	0	-1	0
-1	-1	-1	-1	1	0	1	1	0	-1	-1	0
-1	-1	-1	1	-1	0	-1	-1	0	1	0	0
-1	-1	-1	1	-1	0	-1	1	0	1	-1	0
-1	-1	-1	1	1	0	-1	-1	0	1	-1	0
-1	-1	-1	-1	-1	1	-1	-1	1	0	0	1
-1	-1	-1	1	-1	1	-1	-1	1	-1	0	1
-1	-1	-1	-1	-1	1	-1	1	1	0	-1	1
-1	-1	-1	1	-1	1	-1	1	1	-1	-1	1
-1	-1	-1	-1	-1	1	1	-1	1	-1	0	1
-1	-1	-1	-1	-1	1	1	1	1	-1	-1	1
-1	-1	-1	-1	1	1	-1	-1	1	0	-1	1
-1	-1	-1	1	1	1	-1	-1	1	-1	-1	1
-1	-1	-1	-1	1	1	1	-1	1	-1	-1	1

Step 4: For every component in each phase, calculate reliability and unreliability. For any component in phase j, duration of phase j is considered while calculating reliability.

Component	Reliability		Unreliability
1.	$e^{-(.001)40}$	= .96	.04
2.	$e^{-(.001)60}$	= .94	.06
3.	$e^{-(.003)100}$	= .74	.26
4.	$e^{-(.001)40}$	= .96	.04
5.	$e^{-(.005)60}$	= .74	.26
6.	$e^{-(.002)100}$	= .81	.19
7.	$e^{-(.002)40}$	= .92	.08
8.	$e^{-(.01)60}$	= .54	.46

9.	$e^{-(.01)100}$	=	.36	.64
10.	$e^{-(.01)40}$	=	.67	.33
11.	$e^{-(.03)60}$	=	.17	.83
12.	$e^{-(.02)100}$	=	.14	.86

Step 5: Calculate unreliability for each term. Calculation of unreliability for a sample term is explained below. Let the sample term be

-1 -1 -1 0 -1 0 1 0 0 1 0 0

Unreliability = $P_1 \quad P_2 \quad P_3 \quad P_5 \quad Q_7 \quad Q_{10}$

$= (e^{-frtc(1,1)\,d(1)})\ (e^{-frtc(1,2)\,d(2)})\ (e^{-frtc\,(1,3)\,d(3)})\ (e^{-frtc\ (2,2)\,d(2)})$
$(1-e^{\,-frtc(3,1)\,d(1)})\ (1-e^{\,-frtc(4,1)\,d(1)})$

$= [e^{-(.001)40}]\ [e^{-(.001)60}]\ [e^{-(.003)100}]\ [e^{-(.005)60}]\ [1-e^{-(.002)40}]$
$[1-e^{-(.01)40}]$

$= (.96)\ (.94)\ (.74)\ (.74)\ (.077)\ (.33) = 0.013$

Probability of mission failure

$= Q_1 + P_1Q_2 + P_1\ P_2\ Q_3 + P_1\ P_2\ P_3\ Q_5\ Q_{11} + P_1\ P_2\ P_3\ P_5$
$Q_7\ Q_{10} + \ldots + P_1\ P_2\ P_3\ P_4\ Q_5\ Q_6\ Q_7\ P_8\ Q_9\ P_{10}\ P_{11}\ Q_{12}$

$= .04 + .0576 + .235 + .144 + .013 + \ldots + 9.9 \times 10^{-5}$

$= .72$

* * *

11.3 COMMON CAUSE FAILURES

Computing system reliability is relatively straightforward when components fail independently of each other. Such a system is said to have s-independent components. As an example, let us consider two pumps connected in parallel, then in order to obtain the reliability it is considered that either of them is functioning. Here it is assumed that the event causing the failure of one of the pumps does not affect the failure probability of the other pump i.e, the components fail independently of each other . However, it is not true in all the cases, as the failure of a component might result from the conditions of neighbouring components. As in the above case both the pumps may fail if contaminated fluid flows through them. Also, if external stresses such as earthquakes, floods, thunderstorms, fires, etc. are too excessive, a so called *Common Cause Failure*, which is the failure of several components together almost at the sametime due to the same cause, often results.

A common cause failure is taken to be any design susceptibilty to the occurence of single event which can lead to coexisting failure of multiple channels or interdependent sub-systems such that the system is disabled. Generally, Common Cause Failures represent those wherein failures of redundant systems or components results from a single causative factor or event.

Common Cause Failures can generally be categorized by their cause into following groups :

1. External Normal Environment: Causative factors such as dust, dirt, humidity, temperature, etc. which are normal extremes of the operating environment.

2. Equipment Design Deficiency: Considers design and installation features which give rise to either electrical or mechanical inter- dependence between components.

3. Operation and Maintenance Errors: Includes carelessness, improper adjustment or caliberation, improper maintenance, inadvertant human factors.

4. External Phenomena: Events such as tornado, fire, flood, earthquake, etc.

5. Functional Deficiency: Covers those possibilities where the design may be inadequate either because of erroneous predictions about the behaviour or usefulness of variables monitored or erroneous predictions of the effectiveness of protection action to be taken.

The possibility of common cause failures must be considered whenever a system is provided with redundancy. Identification of common cause failure processes is an important step towards common cause analysis methodology. Each failure cause is examined for its potential to cause multiple components failures.

A common cause failure can have more complex direct consequences than the simple failure of a number of components. In particular, the failure of a component might protect another from the common event's effects. Thus, Commom Cause Analysis cannot proceed in a general manner by substituting specific component failures for component event.

11.31 Reliability Analysis

The method below is very general & is applicable for calculating the reliability of a system composed of non-identical components and depicted by a non-series-parallel reliability block diagram in presence of common-cause failures. However, the calculation for the reliability of a system with identical components in presence of common-cause failures is discussed first.

11.311 Reliability Calculation for a System composed of Identical Components

Expression for reliability of a specified system configuration is derived in three steps :

(1) Find an expression for reliability of a specified system component.

(2) Find the probability that a specified group of m components out of the n components system are all good.

(3) Construct an expression for reliability using results from above two steps and the reliability expression of the system under s-independent assumption.

Reliability of a specified component :

A specific component can fail due to the occurence of following different failure processes.

(1) 1-component process Z_1 for s-independent failure of the specified component.

(2) 2-componont processes that include the specified component. There are a total of ${}^nC_2$ i.i.d. Z_2 failure processes but only ${}^{n-1}C_1$ of these processes include specified component.

(3) In general, there are ${}^nC_r$ i.i.d. Z_r failure processes with parameter λ_r governing the simultaneous failure of r components. Out of these ${}^nC_r$ failure processes, ${}^{n-1}C_{r-1}$ include the specified component.

The $P_n^{(1)}(t)$; the probability that the specific component is operating at time t is :

$$P_n^{(1)}(t) = \prod_{r=1}^{n} ({}^{n-1}C_{r-1} \exp(-\lambda_r t))$$

$$= \exp\left(-\sum_{r=1}^{n} {}^{n-1}C_{r-1}\, \lambda_r t\right) \qquad (11.9)$$

Probability that a group of m components are all good

$$P_n^{(m)}(t) = \Pr(S_1 \cap S_2 \cap S_3 \ldots S_m; t) \qquad (11.10)$$

Probability that both components S_1 and S_2 are good at time t is

$$\Pr(S_1 S_2 \,;\, t) = \Pr(S_1, t)\Pr(S_2 | S_1; t)$$

$\Pr(S_2 | S_1; t)$ = Probability that component 2 is good at time t given no event of any common-cause failure processes associated with the failure of component 1 has occured.

= Probability that component 2 is good at time t for n-1 component system, which is the original system with component 1 excluded.

$= P_{n-1}^{(1)}(t)$

Since the components are i.i.d. ;

$$P_n^{(m)}(t) = P_n^{(1)}(t)\, P_{n-1}^{(1)}(t) \ldots P_{n-m+1}^{(1)}(t)$$

$$= \prod_{k=n-m+1}^{n} P_k^{(1)}(t) \qquad (11.11)$$

Example 11.4

Calculate the reliability for a 1-out-of-3:G system.

Solution

For identically distributed components with s-independent failure processes, the $R_{IIC}(t)$, reliability at time t of a system configuration with i.i.d components is given as :

$$R_{IIC}(t) = 3\,P(t) - 3\,P^2(t) + P^3(t)$$

When the identically distributed components have Common Cause failures, then R_{ICC}, reliability at time t of a system configuration with identically distributed components having common-cause failures is obtained from $R_{IIC}(t)$ by substitution of $P_n^{(m)}(t)$ for $P^{(m)}(t)$.

$$R_{ICC}(t) = 3\,P_3^{(1)}(t) - 3\,P_3^{(2)}(t) + P_3^{(3)}(t)$$

$$= 3\,P_3^{(1)}(t) - 3\,P_2^{(1)}(t)\,P_3^{(1)}(t) + P_1^{(1)}(t)\,P_2^{(1)}(t)\,P_3^{(1)}(t)$$

Substituting the value of $P_n^{(m)}(t)$ from equation (11.11),

$$= 3\exp[-(\lambda_1 + 2\lambda_2 + \lambda_3)t] - 3\exp[-(2\lambda_1 + 3\lambda_2 + \lambda_3)t + \exp[-(3\lambda_1 + 3\lambda_2 + \lambda_3)t]$$

If $\lambda_1 = 0.002$, $\lambda_2 = 0.001$, $\lambda_3 = 0.0005$, we find

$R_{ICC}(10) = 0.99413$, or , $Q_{CC} = 0.00587$

If we consider i.i.d. components having all possible causes of failure, then

$P(t) = \exp -(\lambda_1 + 2\,\lambda_2 + \lambda_3)t$

i.e. $P(10) = 0.955997$

Now,

$R_{IIC}(10) = 3P - 3P_2 + P_3 = 0.9999148$, or , $Q_{II} = 0.0000852$

So, system reliability is considerably poorer if we use CC Methodology.

* * *

Example 11.5

For the system given in the Fig.11.4 below, calculate the system reliability.

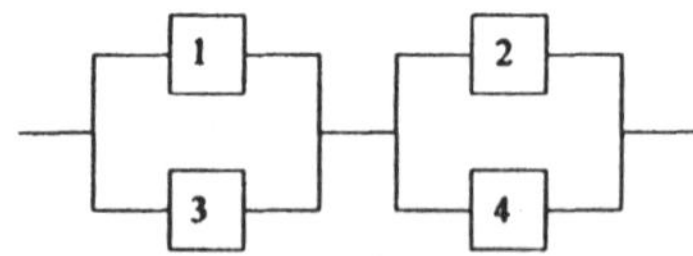

Fig.11.4 Block diagram for Example 11.5.

Solution

$$R_{IIC}(t) = [1 - (1 - P(t))^2]^2$$
$$= 4\,P^2(t) - 4\,P^3(t) + P^4(t)$$

The reliability of a single component in a 4-component system is :

$P_4^{(1)}(t) = \exp(-(\lambda_1 + 3\,\lambda_2 + 3\,\lambda_3 + \lambda_4)t)$

Thus

$$P_4^{(1)}(10) = 0.93473$$

Hence,

$$R_{IIC}(10) = 0.99150, \text{ or, } Q_{II} = 0.0085$$

For the common cause relaibility ;

$$\begin{aligned} R_{ICC}(t) &= 4P_4^{(2)}(t) - 4\,P_4^{(3)}(t) + P_4^{(4)}(t) \\ &= 4\,P_3^{(1)}(t) - 4\,P_2^{(1)}(t)\,P_3^{(1)}(t)\,P_4^{(1)}(t) \\ &\quad + P_1^{(1)}(t)\,P_2^{(1)}(t)\,P_3^{(1)}(t)\,P_4^{(1)}(t) \end{aligned}$$

$$R_{ICC}\,(10) = 0.95566, \text{ or, } Q_{CC} = 0.04434$$

Once again, CC-reliability is much poorer.

* * *

Example 11.6

Calculate the reliability for 1-out-of-3 : F system.

Solution

$$R_{IIC}(t) = P^3(t)$$

Now $P(t) = \exp\{-(\lambda_1 + 2\,\lambda_2 + \lambda_3)\,t\}$

$$P(10) = 0.955997$$

Hence,

$$R_{IIC}(10) = 0.87372, \text{ or, } Q_{II} = 0.12628$$

Now,

$$\begin{aligned} R_{ICC}(t) &= P_3^{(1)}(t) \\ &= P_1^{(1)}(t)\,P_2^{(1)}(t)\,P_3^{(1)}(t) \\ &= \exp\{-(3\,\lambda_1 + 3\,\lambda_2 + \lambda_3)t\} \end{aligned}$$

Thus,

$$R_{ICC}\,(10) = 0.90937, \text{ or, } Q_{CC} = 0.09063$$

In this example, the system reliability assuming s-independence is appreciably lower than that considering common cause failures.

This is because the 2-component and 3-component processes are not applied independently in calculating CC reliability. For example, the 3-

component failure process either causes all components to fail simultaneously or it does not occur, and application of the 3-component process individually to each component unnecessarily degrades the system reliability.

* * *

11.312 Reliability Evaluation Technique for a NSP System consisting of Non-Identical Components

The previous method is now extended for calculating the source-terminal reliability of Non Series Parallel network subjected to common-cause failures. Each failure process is represented by failure combinations and associated failure rate.

Algorithm

Following steps are followed in deriving the reliability of a system subjected to common-cause failures.

(1) If there are any parallel branches in the network, combine them into one i.e. every set of parallel branches is to be replaced by a single branch.

(2) Write the matrix graph for the network. If b is the number of edges in the network, then matrix graph is a b x 4 matrix. There is one to one correspondence between each edge and each row. First column gives the branch number, second column gives the starting node of the branch, third column gives the converging node of the branch and fourth column gives the direction code.

(3) Obtain simple minimal paths from matrix graph .

(4) System reliability is given by the probability of union of all minimal paths. Calculate disjoint terms corresponding to all minimal paths thus replacing *Union* by *Summation.*

(5) Each disjoint term may contain complemented and/or uncomplemented variables corresponding to edges. Simplify the expression by using relation X' = 1 - X. Let the modified expression (containing only uncomplemented variables corresponding to edges) be
$S = t_1 + t_2 + \ldots + t_i + \ldots + t_N$, where N is the number of disjoint terms.

(6) Reliability of the system is
$R(t) = Pr(S) = Pr(t_1) + Pr(t_2) + \ldots + Pr(t_i) + \ldots + Pr(t_N)$

To calculate reliability of any term t_i, it can be taken as the expression for a series subsystem consisting of, say, m components. Then, reliability of term t_i is the probability that all m components of term t_i are good. Let the various failure processes be $Z_1, Z_2, \ldots, Z_r, \ldots, Z_m$ with failure rates $\lambda_1, \lambda_2, \ldots, \lambda_r, \ldots, \lambda_m$.

Calculation of reliability of term t_i

Term t_i can be considered corresponding to a series subsystem consisting of m components. For success of term t_i, all components have to be good. We know that in series subsystem, failure rates of all components in series are added to calculate failure rate of subsystem. Calculate failure rate for the term t_i by adding failure rates of all those failure processes which include one or more components of term t_i (because series subsystem corresponding to t_i will fail even if one component fails). Reliability of term t_i = exp [-(failure rate for term t_i) * time]

(7) Substitute reliabilities of all terms in expression of step (6) to calculate total reliability .

Example 11.7

For the network shown in Fig.11.5, calculate s-t reliability at time 10, 20, ... 100 hours. Source node number is given as 1 and sink node number is given as 4. Components can fail individually as well as under common-cause. Components can fail individually with failure rates .001, .002, .003, .004, and .005/hour respectively. Three common-cause events can occur :

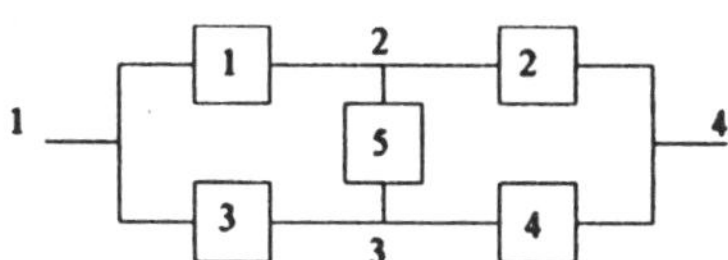

Fig.11.5 Block diagram for Example 11.7.

(1) Component 1 and component 2 can fail under common-cause with failure rate .001 .

(2) Component number 1,3 & 4 can fail under common-cause with failure rate .002.

(3) Component number 1,3,4 & 5 can fail under common-cause with failure rate .0001.

Calculate the reliability of the system.

Solution

Step (1) Matrix Graph

$$\begin{bmatrix} 1 & 1 & 2 & 0 \\ 2 & 2 & 4 & 0 \\ 3 & 1 & 3 & 0 \\ 4 & 3 & 4 & 0 \\ 5 & 2 & 3 & 0 \end{bmatrix}_{bx4}$$

where b = 5 = number of branches in the network.

Step 2 (i) Connection matrix

	1	2	3	4
1	0	1	3	0
2	1	0	5	2
3	3	5	0	4
4	0	2	4	0

node x node

Connection matrix shows that between
nodes 1 & 1 no branch is connected,
nodes 1 & 2 branch number 1 is connected,
nodes 1 & 3 branch number 3 is connected etc. etc.,

(ii) From connection matrix, obtain another matrix of order (node x node) which shows to which all nodes, each node is directly connected. Matrix of order (node x node) is obtained as

	1	2	3	4
1	0	2	3	0
2	1	0	3	4
3	1	2	0	4
4	0	2	3	0

which shows that node number 1 is directly connected to node number 2 & 3, Node number 2 is directly connected to node number 1,3 & 4 etc.

(iii) Obtain minimal paths from above matrix. The process consists of two steps : (a) & (b). In step (a) Minimal paths in node form are obtained and in (b) minimal paths in edge form are obtained.

(a) Start path tracing from node number 1 i.e source node. Node number 1 is directly connected to node number 2. Go to row corresponding to node number 2, which is directly connected to 1,3 & 4. As in the path tracing, node number 1 has already been taken, so we take path from node 2 to 3. Now go to row corresponding to node number 3, which is directly connected to 1,2, & 4. Node number 1 and 2 have already been taken so path from node 3 to 4 is chosen. As the sink node number is reached, stop the process. So the first minimal path obtained is 1234.

Now start moving backwards. Node number 3 was also connected to node number 1 & 2 but they have already been taken into account. Now go to row corresponding to node number 2. Node number 2 is directly connected to 1, 3 & 4. Node number 1 & 3 have directly been taken care of. From node 2, we can reach directly node number 4. As the sink node number is reached, stop process. Thus second minimal path is 124.

Now go to node number 1. It is directly connected to node number 3 also. So path 13 is taken. Node number 3 is directly connected to 2 & 4. Let us take path 132 first. Node number 2 is directly connected to 1, 3 & 4. Node number 1 & 3 have already been taken. So the third path obtained is 1324. Now choose another alternative path from node number 3. So the fourth path obtained is 134. Thus the minimal paths in node form are

$$\begin{matrix} 1 & 2 & 3 & 4 \\ 1 & 2 & 4 & 0 \\ 1 & 3 & 2 & 4 \\ 1 & 3 & 4 & 0 \end{matrix}$$

(b) Minimal paths obtained in node form are converted to minimal paths in edge form by using connection matrix of step (2) part (i). For example, to obtain first minimal path in edge form from minimal path in node form 1234, the steps are as follows :

Between node 1 & 2, branch number 1 is connected.
Between node 2 & 3, branch number 5 is connected.
Between node 3 & 4, branch number 4 is connected.

Therefore, minimal path in edge form is 154. Similarly all other paths are obtained and arranged in order of increasing number of elements present.

Minimal paths are 1 2
3 4
2 3 5
1 4 5

Step (3)

S = Union of all paths
= 12 U 34 U 235 U 145

S(disjoint) is obtained in the following way :

= 12 U 1'34 U 134 U 235 U 145

= 12 U 1'34 U 12'34 U <u>1234</u> U 235 U 145 *Underscored terms are deleted

= 12 U 1'34 U 12'34 U 1'235 U <u>1235</u> U 145

= 12 U 1'34 U 12'34 U 1'234'5 U <u>1'2345</u> U 145

= 12 U 1'34 U 12'34 U 1'234'5 U 12'45 U <u>1245</u>

= 12 U 1'34 U 12'34 U 1'234'5 U 12'3'45 U <u>12'345</u>

= 12 U 1'34 U 12'34 U 1'234'5 U 12'3'45

= 12 + 1'34 + 12'34 + 1'234'5 + 12'3'45

Step(4) Expand the terms which have complemented variables. For each complemented variable in a term, two terms in uncomplemented variables are obtained, e.g.,

(i) 1'34 is expanded into two terms

(a) In first term, eliminate complemented variables
(b) In second term, substitute uncomplemented variable in place of complemented variable and attach opposite sign to it as compared to initial term

i.e. 1'34 = 34 -134

(ii) 1'234'5 is expanded initially into two terms.
1'234'5 = 234'5 - 1234'5

These two terms are further expanded as under

234'5 = 235 - 2345
-1234'5 = -1235 + 12345

After expanding all terms in a similar manner, 13 total terms are obtained, which are listed below :

t_1 = 12
t_2 = 34
t_3 = -134
t_4 = 134
t_5 = -1234
t_6 = 235
t_7 = -2345
t_8 = -1235
t_9 = 12345
t_{10} = 145
t_{11} = -1345
t_{12} = -1245
t_{13} = 12345

Step (5)

R(t) = Pr(12) + Pr(34) - Pr(134) + Pr(134) - Pr(1234) + Pr(235) - Pr(2345) - Pr(1235) + Pr(12345) + Pr(145) - Pr(1345) - Pr(1245) + Pr(12345)

Calculate failure rate for each term t_i.

Calculation of failure rate of t_1

Compare term t_1 with all eight failure processes.

(i) Failure process Z_1 can cause s-independent failure of component number 1.
(ii) Failure process Z_2 can cause s-independent failure of component number 2.
(iii) Failure processes Z_3 to Z_5 have no effect on either component number 1 or 2.
(iv) Failure process Z_6 is a common-cause event causing failure of 1 & 2 simultaneously.
(v) Failure process Z_7 and Z_8 are common-cause events causing failure of component number 1 also in multiple component failures.

So total failure rate for term t_1 = Sum of failure rates of failure processes Z_1, Z_2, Z_6, Z_7, Z_8.

$= 0.001 + 0.002 + 0.001 + 0.002 + 0.0001 = 6.1 \times 10^{-3}$/hr

Similarly failure rates of all terms are calculated. At any time, say 10 hours, reliability for term $t_1 = \exp[-(6.1\times10^{-3})10] = 0.9408232$. Reliability of all other terms can be calculated in a similar manner.

Step (6)

$$R(10) = 0.94 + 0.91 - 0.89 + 0.89 - 0.88 + 0.88 - 0.84 - 0.87 + 0.83 + 0.80 - 0.85 - 0.86 + 0.83 = 0.97$$

* * *

11.4 RELIABILITY AND CAPACITY INTEGRATION

The most common quantitative index in reliability analysis is *s-t reliability*, defined as the probability of successful communication between the source and the terminal node. It is assumed that the required amount of information can always be transmitted from s to t whenever a path is available; thus implying that every link is capable of the required flow- an implication which is neither valid nor economically justifiable in the design of telecommunication networks. The most-used index in capacity analysis finds the maximum possible flow capability of the network and carries out the capacity assignment. The failure probability of links is implicitly neglected in this analysis; which is again an assumption that is neither valid nor feasible.

These two performance measures are thus used independently while neither is a true measure of the performance of the telecommunication network.

In this section, these two important measures are integrated by a *weighted reliability index*. In computing the conventional s-t reliability of a network, the sucess states of the network (considering path availability only) are identified and the probabilities of all these states are added. Multiplication of each probability term by a normalized weight is now suggested before the summation. The normalized weight is defined to be 1 if the particular network state is capable of transmitting full required traffic between the terminal nodes. For a state which permits less than the required traffic capability, the weight is correspondingly reduced to a fractional value.

11.41 Evaluation of Performance Index

The set of all system states can be partitioned as

$$T = S \cup F \qquad (11.12)$$

S: subset corresponding to those system states where at least one path is available from s to t.

F: subset corresponding to those states where no such path is available.

The states of S only are further analyzed because no state in F can contribute to reliability by permitting any transmission, howsoever small.

For each state S_i ($S_i \in S$) define α_i and β_i as:

α_i {i | branch i is *Up* }

β_i {i| branch i is *Down* }

The probability of system state S_i is:

$$P_{si} = Pr\{S_i\} = \prod_{i \in \alpha i} p_i \prod_{i \in \beta i} q_i \qquad (11.13)$$

In the conventional sense, s -t reliability is:

$$R_{st} = \sum_{Si \in S} P_{si} \qquad (11.14)$$

Let the capacity of the subnetwork formed by the up branches in state S_i be C_i. Then define the normalized weight w_i as

$$w_i \cong C_i / C_{max} \qquad (11.15)$$

Then the weighted reliability measure, viz, performance index, is defined as:

$$PI = \sum_{Si \in S} w_i P_{si} \qquad (11.16)$$

Example 11.8

A network with 5 branches is given in Fig. 11.6 where the capacity of each link is also shown. Compute the performance Index.

Solution

The 16 success states are listed in the table 11.2 considering path availability only. The capacity of subnetwork for each success states is also given in the table, $C_{max} = 7$. The performance index, PI is now determined as

$$\begin{aligned} PI &= p_a p_b q_c p_d p_e + p_a p_b p_c p_d q_e + p_a p_b p_c p_d p_e + (4/7)\,(q_a q_b p_c p_d q_e \\ &+ q_a q_b p_c p_d p_e + q_a p_b p_c p_d p_e + q_a p_b p_c p_d q_e + p_a q_b p_c p_d p_e \end{aligned}$$

$$+ \; p_a q_b p_c p_d q_e + p_a q_b q_c p_d p_e) + 3/7(q_a p_b p_c q_d p_e + p_a p_b q_c q_d q_e$$
$$+ \; p_a p_b q_c q_d p_e + p_a p_b q_c p_d q_e + p_a p_b p_c q_d p_e + p_a p_b p_c q_d q_e) \quad (11.17)$$

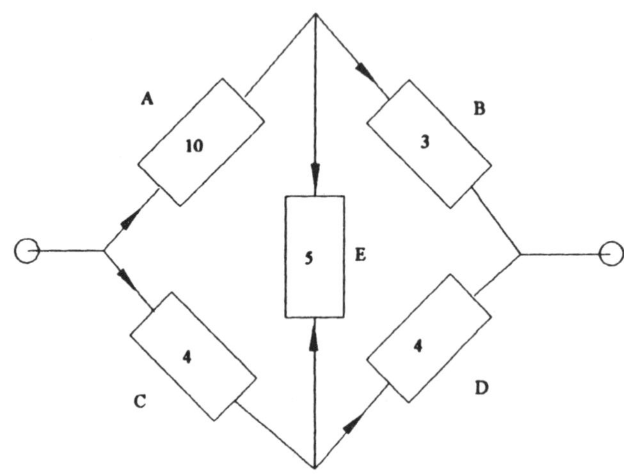

Fig.11.6 A non series parallel network.

If each branch has a reliability, p and unreliability, q ($q \cong 1 - p$); then

TABLE 11.2
System Success States

Element States A	B	C	D	E	Capacity (C_i)	Probability Term(p_i)
0	0	1	1	0	4	$q_a q_b p_c p_d q_e$
0	0	1	1	1	4	$q_a q_b p_c p_d p_e$
0	1	1	0	1	3	$q_a p_b p_c q_d p_e$
0	1	1	1	1	4	$q_a p_b p_c p_d p_e$
0	1	1	1	0	4	$q_a p_b p_c p_d q_e$
1	1	0	0	0	3	$p_a p_b q_c q_d q_e$
1	1	0	0	1	3	$p_a p_b q_c q_d p_e$
1	1	0	1	1	7	$p_a p_b q_c p_d p_e$
1	1	0	1	0	3	$p_a p_b q_c p_d q_e$
1	1	1	1	0	7	$p_a p_b p_c p_d q_e$
1	1	1	1	1	7	$p_a p_b p_c p_d p_e$
1	1	1	0	1	3	$p_a p_b p_c q_d p_e$
1	1	1	0	0	3	$p_a p_b p_c q_d q_e$
1	0	1	1	1	4	$p_a q_b p_c p_d p_e$
1	0	1	1	0	4	$p_a q_b p_c p_d q_e$
1	0	0	1	1	4	$p_a q_b q_c p_d p_e$

$$PI = (2p^4q + p^5) + 4/7(p^2q^3 + 4p^3q^2 + 2p^4q) + (3/7)(4p^3q^2 + p^2q^3 + p^4q) \quad (11.18)$$

For p = 0.9, PI = 0.85478

* * *

12
ECONOMICS OF RELIABILITY ENGINEERING

12.1 INTRODUCTION

Any manufacturing industry is basically a profit making organization and no organization can survive for long without minimum financial returns for its investments. There is no doubt that the expense connected with reliability procedures increases the initial cost of every device, equipment or system. However, when a manufacturer can lose important customers because his products are not reliable enough, there is no choice other than to incur this expense. How much reliability cost is worth in a particular case depends on the cost of the system and on the importance of the system's failure free operation. If a component or equipment failure can cause the loss of a multimillion dollars' system or of human lives, the worth of reliability and the corresponding incurred cost must be weighed against these factors. For the producer, it is a matter of remaining in the business. However, his business volume and profit will be substantially increased once his reliability reputation is established. Therefore, from manufacturer's point of view, two important economic issues are involved:

(i) Financial profit
(ii) Customers' satisfaction

If a manufacturer intends to stay in his business, he has not only to optimize his own costs and profits but to maximize customers' satisfaction as well.

12.2 RELIABILITY COSTS

Reliability costs can be divided into five categories as shown in fig. 12.1.

Components of each classification are described below:

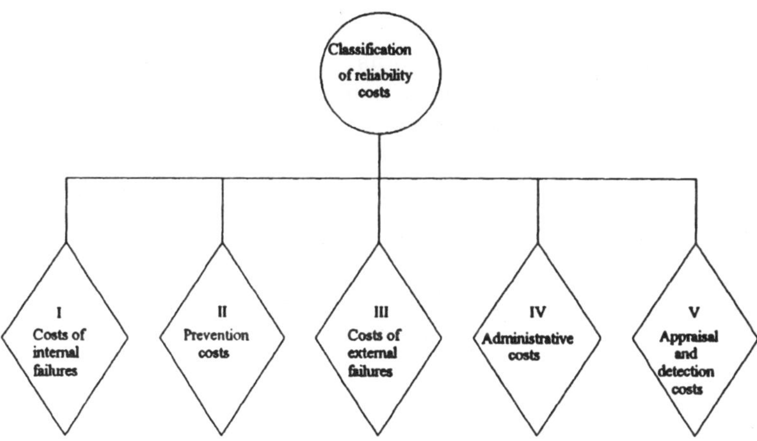

Fig.12.1 Classifications of reliability costs.

Classification I

This classification includes all those costs associated with internal failures, in other words, the costs associated with materials, components, and products and other items which do not satisfy quality requirements.

Furthermore, these are those costs which occur before the delivery of the product to the buyer. These costs are associated with things such as the following :

1. Scrap
2. Failure analysis studies
3. Testing
4. In-house components and materials failures
5. Corrective measures

Classification II

This classification is concerned with prevention costs. These costs are associated with actions taken to prevent defective components, materials, and products. Prevention costs are associated with items such as the following:

1. Evaluating suppliers
2. Calibrating and certifying inspection and test devices and instruments.

3. Receiving inspection
4. Reviewing designs
5. Training personnel
6. Collecting quality-related data
7. Coordinating plans and programs
8. Implementing and maintaining sampling plans
9. Preparing reliability demonstration plans

Classification III

Under this classification are costs associated with external failures - in other words, costs due to defective products shipped to the buyers. These costs are associated with item such as the following :

1. Investigation of customer complaints
2. Liability
3. Repair
4. Failure analysis
5. Warranty charges
6. Replacement of defective items

Classification IV

This category includes all the administrative-oriented costs- for example, costs associated with the following :

1. Reviewing contracts
2. Preparing proposals
3. Performing data analysis
4. Preparing budgets
5. Forecasting
6. Management
7. Clerical

Classification V

This category includes costs associated with detection and appraisal. The principal components of such costs are as follows:

1. Cost of testing
2. Cost of inspection (i.e.,in-process, source, receiving, shipping and so on)
3. Cost of auditing

12.3 EFFECT OF RELIABILITY ON COST

Any effort on the part of manufacturer to increase the reliability of his products will increase reliability design costs and internal failure costs. However, after some time internal failure costs will start decreasing. The external costs like transportation do not depend on reliability but installation and commissioning and maintenance costs will show decline with an increase in reliability.

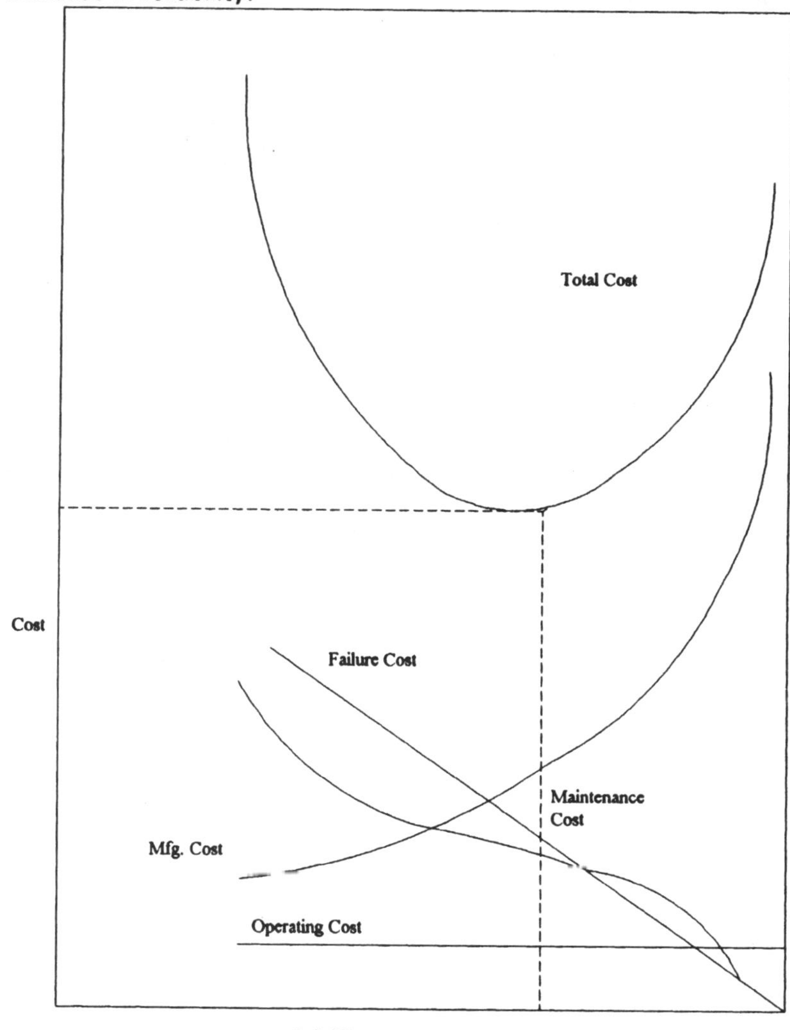

Fig.12.2 Cost curves of a product.

In general, it is not profitable to aim for complete perfection by eliminating all failures (even if it is possible). This is clear from the reliability cost curves given in Figure 12.2 for various categories of costs for an equipment. Upto certain point, it is worth to make appropriate investments for reliability and further investments will be advisable only where the reliability has an over-riding importance.

The subsequent sections describe some reliabililty cost models which show how the equipment life-cost is affected by reliability achievement, utility, depreciation and availability.

12.4 RELIABILITY ACHIEVEMENT COST MODELS

The reliability and cost relationship for any equipment can be described mathematically by suitably choosing a cost-reliability relationship function. A suitable cost-reliability function $C(r_1,r_2)$ must satisfy the following properties:

1. $C(r_1,r_2) \geq 0; \; r_2 \geq r_1$ (12.1)
   where C is the cost required to increase the reliability of the equipment from r_1 to r_2. This means that improvement in reliability always costs some amount of money on the part of the manufacturer.

2. $C(r_1,r_3) = C(r_1,r_2) + C(r_2,r_3); \; r_3 \geq r_2 \geq r_1$ (12.2)
   where r_1, r_2 and r_3 are reliability levels of the equipment. It implies that the cost required to raise reliability from current level by a given amount is the same, irrespective of whether this is achieved directly or by a series of separate improvements, however small.

3. C(r) is differentiable i.e. cost-reliability relationship function should be defined so carefully that cost investment should increase reliability of the equipment.

4. $d^2 C / dr^2 \geq 0$ (12.3)

   meaning thereby that the cost investment becomes progressively higher as reability r increases.

5. $C(r_1,r_2) \rightarrow \infty$ as the reliability level $r_2 \rightarrow 1$ for a fixed reliability level r_1; i.e., perfect reliability is unattainable.

There exist several cost-reliability functions in the literature. Some of them are as follows:

1. *Misra et al Function:*

   $$C(r) = a \exp[b/(1-r)]; \; a, b > 0 \tag{12.4}$$

   where a and b are positive constants and r is the reliability of the equipment.

2. *Tillman et al Function:*

$C(r) = k\ r^a;\ k > 0 \text{ and } 0 < a < 1$ (12.5)

where k and a are positive constants and r is the reliability of the equipment.

3. *Aggarwal et al Function:*

$C(r) = k\ [\tan(\pi\ r/2)]^{h(r)}$ (12.6)

where k is a constant and h(r) is given by:

$h(r) = 1 + r^a;\ 0 \le a \le 1,$ or (12.7)

$h(r) = m;\ \ 1 \le m \le 2$

and r is equipment reliability.

4. *Fratta et al's Function:*

$$C(r_1,r_2) = \begin{cases} a \ln [(1-r_1)/(1-r_2)]; & 0 \le r_1 \le r_2 \le 1, \quad a > 0 \\ 0, & r_1 \ge r_2 \end{cases} \quad (12.8)$$

where a is a positive constant; and r_1 and r_2 are the reliability values of the equipment.

5. *Majumdar et al's Function:*

$C(r) = k\ [r/(1-r)]^a;\ \ k, a > 0$ (12.9)

where k and a are constants and r is the reliability of the equipment.

6. *Llyod and Lipow's Function:*

$$C(r_1,r_2) = \begin{cases} a \ln [(r_2+b)/r_1+b)]; & 0 \le r_1 \le r_2 \le 1, \quad a, b > 0 \\ 0; & r_1 \ge r_2 \end{cases} \quad (12.10)$$

where a and b are constants; and r_1 and r_2 are reliability values of the equipment.

We just illustrate the use of Misra et al's function in detail for understanding the behaviour of cost-reliability function. This function shows that cost increases exponentially with reliability and becomes prohibitively high at higher reliabilities. Figure 12.3 illustrates the nature of cost variation with reliability. There is always a minimum cost involved in developing and manufacturing the equipment even if it is to have a zero reliability. The level of operational reliability is to be decided by cost constraints and operational requirements. However, in this relationship, the constants a and b are equipment dependent and can be estimated as follows:

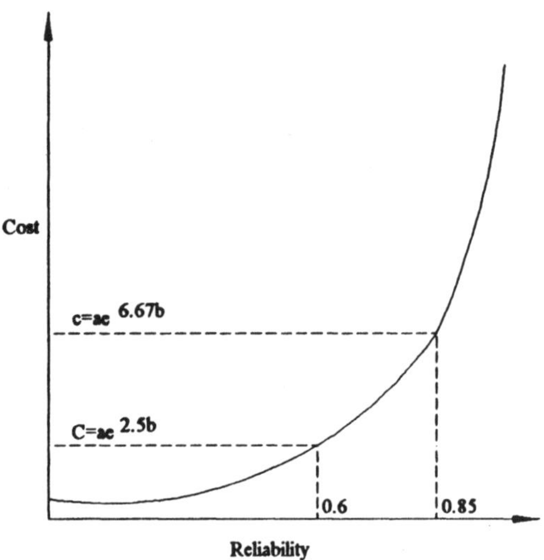

Fig.12.3 Product reliability and cost.

Let us assume that the cost of equipment is known at some reliability, say r_o, then

$$C_o = C(r_o) = a \exp[b/(1-r_o)] \tag{12.11}$$

If the cost of manufacturing the equipment with very low reliability is C_1 then

$$C_1 \approx ae^b \tag{12.12}$$

Therefore,

$$a = C_1 e^{-b} \tag{12.13}$$

and

$$b = [(1-r_o)/r_o] \ln (C_o/C_1) \tag{12.14}$$

Thus, if the equipment cost is known at some value of reliability and the manufacturer intends to improve the reliability of the equipment, the corresponding cost to be incurred can be evaluated by obtaining the constants a and b with the help of the above equations and then by using these values in the equation:

$$C(r) = a \exp[b/(1-r)]; \quad a, b > 0 \tag{12.15}$$

Reliability can also be increased by operating more than one equipment in parallel. In such a case, the cost of the system is

$$C_s = C\,m = a\,e^{b/(1-r)}\,[\ln(1-R)]/[\ln(1-r)] \qquad (12.16)$$

where $C = a\,e^{b/(1-r)}$ (12.17)

is cost of each equipment, and

$$m = [\ln(1-R)] \,/\, [\ln(1-r)] \qquad (12.18)$$

is the number of equipments to be operated in parallel.

An optimum value of equipment reliability can be found by solving the equation

$$dC_s \,/\, dr = 0 \qquad (12.19)$$

i.e. $b \ln(1-r) + (1-r) = 0$ (12.20)

Example 12.1

Consider an equipment with reliability 0.6. The desired reliability is approximately 0.85. The desired reliability level can be obtained either by improving the equipment reliability from 0.6 to 0.85 or by operating two identical equipments of reliability 0.6 each in parallel. Compare the cost incurred in both the cases.

Solution

Case 1 - When the equipment reliability is improved from 0.6 to 0.85.

Let C_1 and C_2 be the respective costs of the equipment when its reliability values are 0.6 and 0.85.

$$C_1 = a\,\{\exp[1/(1-0.6)]\}^b = a\,\exp(2.5b) = a\,12^b$$

and $C_2 = a\,\exp(6.67b) = a\,788^b$

Thus, $C_2/C_1 = (788/12)^b$

and when $b = 0.5$, C_2 is 8 times the C_1.

Case 2- When two equipments of reliability 0.6 are put in parallel.

In this case C_2 is just twice the C_1.

Thus it is clear from the above calculations that sometimes it is not desirable to improve the equipment reliability. Instead, one should think of some other alternative of getting the desired reliability level. Though, in the second case, the total cost is just twice the original cost, but this can not be done as and when we desire. While doing so, several other constraints like, volume, size and weight etc. are also to be taken into account as already discussed in an earlier chapter.

* * *

12.5 RELIABILITY UTILITY COST MODELS

Suppose customer invests money for a product. The costs and benefits accruing from the investment will continue for a number of years. The similar products may have different costs and returns depending upon the manufacturer. A cost utility analysis is required for making comparisons of product values. The customer's investment includes the following categories of costs:

(i) Initial Costs (C_i):
  - (a) Purchase cost
  - (b) Installation cost

(ii) Operating Costs (C_o)

(iii) Cost of failures (C_f):
  - (a) Repair Cost
  - (b) Cost of loss of effectiveness
  - (c) Cost of damage
  - (d) Loss of income due to failure

(iv) Cost of preventive maintenance (C_m)

When the product is put to use, the customer has to spend money every year on items (ii), (iii) & (iv). If the product is used for, say n years, then the *present value* of the money that the user has to spend for all the years can be calculated as follows:

$$V_1 = C_i + \sum_{j=1}^{n} (C_{oj} + C_{fj} + C_{mj})[1/1 + i]^j \qquad (12.21)$$

where i is the annual interest rate (expressed as a fraction) and C_{oj}, C_{fj} and C_{mj} are the respective costs incurred in the jth year and assumed to be paid at the end of that year.

If, at the end of the nth year the scrap value of the product is V_s, then the present value of the n-year-old product is

$$V_2 = V_s\,[1/1+i)]^n \tag{12.22}$$

Then the *present Cost* of the product is

$$C_p = V_1 - V_2$$

$$= C_i + \sum_{j=1}^{n} [1/1+i)]^j\,[C_{oj} + C_{fj} + C_{mj}] - V_s\,[1/(1+i)]^n$$

$$= C_i + \sum_{j=1}^{n} [1/(1+i)]^j\,(C_{yj}) - V_s\,[1/(1+i)]^n \tag{12.23}$$

Where C_y is the yearly cost. The product having lowest C_p should be choice of the customer. However, while making decisions he has to keep in mind other factors such as availability of spares, possible increase in costs in future, etc.

Example 12.2

Let us consider that a customer has to make a choice between product A and B whose costs are shown in Table 12.1. The annual interest rate is 10%.

TABLE 12.1

Cost	C_i	C_{y1}	C_{y2}	C_{y3}	V_s
Product A	20,000	1000	1600	2200	15,000
Product B	15,000	1500	1800	2000	10,000

Solution

The present costs of products A and B are calculated as follows:

$$C_{pA} = 20000 + 1000[1/1.1] + 1600[1/1.1]^2 + 2200[1/1.1]^3 - 15000[1/1.1]^3$$
$$= 11551$$

$$C_{pB} = 15000 + 1500[1/1.1] + 1800[1/1.1]^2 + 2000[1/1.1]^3 - 10000[1/1.1]^3$$
$$= 11840$$

It is clear from the above calculations that in spite of a higher initial cost, product A is more economical. This is due to low failure and maintenance costs as a result of its higher reliability.

* * *

12.51 Depreciation-Cost Models

After a product has operated for a period of time, either it is considered as scrap and sold at scrap value, or its value is considered to be much lower than the initial cost due to the ageing and wear. If V_s is the value of the product after n years of operation, the difference between the initial cost C_i and V_s is a cost on the part of the user. This cost is known as the depreciation cost and is to be spent by the user in n years of use. If D_j is the portion of the total depreciation cost for the jth year of operation, then the usage cost during the jth year can be represented as

$$U_j = D_j + C_{oj} + C_{fj} + C_{mj} \tag{12.24}$$

For a straight line depreciation model,

$$D_j = (C_i - V_s) / n \tag{12.25}$$

and is constant for all the years.

However, usually depreciation is high initially and dereases as the product value decreases. In this case, D_j is the value of the product at the end of $(j-1)^{th}$ year multiplied by rate of depreciation.

$$\text{or,}\quad D_j = d\, C_i\, (1-d)^{j-1} \tag{12.26}$$

where d is the rate of depreciation. The factor d can be found as follows:

The product value at the end of n years is:

$$V_s = C_i\, (1-d)^n \tag{12.27}$$

Therefore,

$$d = 1 - [V_s/C_i]^{1/n} \tag{12.28}$$

$$\text{then}\quad D_j = C_i\, [V_s/C_i]^{(j-1)/n} \times [1-(V_s/C_i)^{1/n}]$$

$$= C_i[V_s/C_i]^{j/n} \times [(C_i/V_s)^{1/n} - 1] \tag{12.29}$$

Example 12.3

Consider an electronic instrument initially costing \$1000. Assume that its resale value after five years of use is about \$500. Compare the two depreciation models.

Solution

For the linear depreciation model,

$$d = \frac{1000 - 500}{5 \times 1000} = 0.1 \text{(per dollar)}$$

and the annual depreciation is \$100.

For the non-linear declining model,

$$d = 1 - [500/1000]^{1/5} = 0.129 \text{ (per dollar)}$$

The depreciation for all the years has been given in Table 12.2 and a graphic comparison of both the models has been shown in Fig. 12.4.

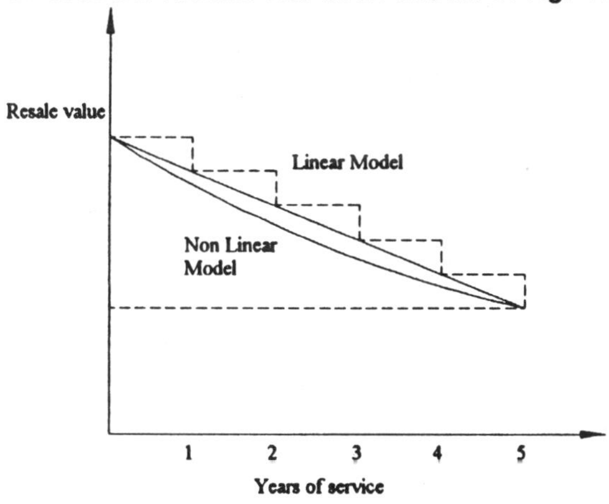

Fig. 12.4 Depreciation Models.

TABLE 12.2

Year(j)	Initial Cost	Depreciation	Cost at the end of the year
1	1000	129	871
2	871	112.4	758.6
3	758.6	97.9	660.7
4	660.7	85.2	757.5
5	575.5	74.2	501.3

* * *

12.6 AVAILABILITY COST MODELS FOR PARALLEL SYSTEMS

As already explained, sometimes it is more economical to put more than one product or equipment in parallel to achieve higher reliability. In such a case, an optimum number of units would minimize the cost due to the operation maintenance of the entire system. Consider a system composed of m units operating in parallel with the following specifications:

C_1 = Operation and maintenance cost per unit time (per equipment)
C_2 = penalty cost(due to down time) per unit time
D = System down-time
U = System up-time

Then the total system cost due to operation, maintenance and failures per unit time will be

$$C_s = \frac{C_1 mU + C_2 D}{U + D} \qquad (12.30)$$

The system down time for a parallel system with m units is

$$D = U[B_s/A_s] \qquad (12.31)$$

where A_s is the system availability and $B_s = 1 - A_s$ the system unavailability.

Also, $B_s = (B)^m$ (12.32)

where B is the unavailability of each unit. then

$$\begin{aligned} C_s &= C_1 m A_s + C_2 B_s \\ &= C_1 m + (C_2 - C_1 m) B^m \end{aligned} \qquad (12.33)$$

It is clear that as m increases the first term increases and the second term decreases and therefore there exists a value of m for which C_s is minimum. This can be found by solving the equation

$$\frac{dC_s}{dm} = 0 \qquad (12.34)$$

Example 12.4

Consider an equipment with steady state unavailability 0.3. Its operating and maintenance cost is \$1000 per equipment per hour and down time cost is

\$15,000 per hour. The number of equipments to be operated in parallel in order to minimize the total system time cost is to be determined.

Solution

Here, Unavailability of equipment B = 0.3
C_1 = \$1000, and C_2 = \$15,000

Let m be the number of equipments to be operated in parallel in order to minimise C_s. The system cost C_s can be obtained as a function of m as follows:

$$C_s = 1000\,m + (15000 - 1000\,m)(0.3)^m$$

Therefore, $dC_s/dm = 1000 + (0.3)^m[-19060 + 1204\,m] = 0$

On solving this equation, we find the value of m lies between 2 and 3. Now

$$(C_s)_{m=2} = \$3170 \text{ and } (C_s)_{m=3} = \$3324$$

Thus m=2 gives optimum value of C_s and hence the number of equipments to be operated in parallel to minimise the operation and down-time cost is two in this case.

* * *

12.7 COST EFFECTIVE CHOICE OF SUBSYSTEMS

In most situations a system designer is permitted to utilize a fixed amount of money to design a system and for each of the components several options are available. The aim of the system designer is to meet the requirement of reliability level within the range of the available resources. He can do so by selecting each component of such a category so that the system may have the required reliability level. The following situations are very likely to occur:

1. There will always be a configuration which will have the lowest cost amongst all possible configurations.
2. Same reliability level may be achieved for different costs.
3. System may have different reliability levels for the same cost in two or more configurations.
4. The reliability level can be higher for a combination of components which results in lower system cost i.e. system reliability need not be a monotonically increasing function of cost
5. Also, there will exist a configuration having the highest reliability level amongst all the possible component groups.

It is desirable to know which configurations result in an increment of system reliability corresponding to some increment in the system cost. It helps the system designer in designing a system with appropriate cost and reliability.

TABLE 12.3

	Component 1			Component 2			Component 3	
A_1	0.90	10	B_1	0.80	5	C_1	0.95	40
A_2	0.95	30	B_2	0.90	20	C_2	0.98	100
A_3	0.98	100	B_3	0.95	50			

For instance, suppose an engineer has to design a system which has three components connected in series. The number of options with their cost and reliability corresponding to each component are given in Table 12.3.

TABLE 12.4

	R	C		R	C		R	C
$A_1B_1C_1$	0.684	55*	$A_3B_1C_1$	0.745	145	$A_3B_1C_2$	0.768	245
$A_1B_2C_1$	0.770	70*	$A_3B_2C_1$	0.838	160	$A_3B_2C_2$	0.864	220
$A_1B_3C_1$	0.812	100	$A_3B_3C_1$	0.884	190	$A_3B_3C_2$	0.912	250*
$A_2B_1C_1$	0.722	75	$A_1B_1C_2$	0.706	115	$A_2B_1C_2$	0.837	135
$A_2B_2C_1$	0.812	90*	$A_1B_2C_2$	0.794	130	$A_2B_2C_2$	0.838	150
$A_2B_3C_1$	0.857	120*	$A_1B_3C_2$	0.838	160	$A_2B_3C_2$	0.884	180*

The component groups categorized by the various degrees of reliability yield 18 combinations shown in Table 12.4.

The six expected desirable configurations can now be analysed from Table 12.4. These configurations shown in this table are also exhibited graphically in Fig.12.5.

Now, the problem arises how to generate only these six optimum configurations mathematically so that the system designer may get maximum benefit of his resources without wasting much time and without the fear of choosing a configuration which has less reliability than possible for the given cost.

The situation may arise in which the minimum reliability requirement and the maximum cost permitted is predecided. In such a case one has to see only those optimum configurations which satisfy both the required conditions.

A method for the solution of this problem is presented in the form of an algorithm. We assume that component j has M_j options available with different reliabilities and costs. The reliabilities and costs corresponding to all possible options of components have been given in the following set:

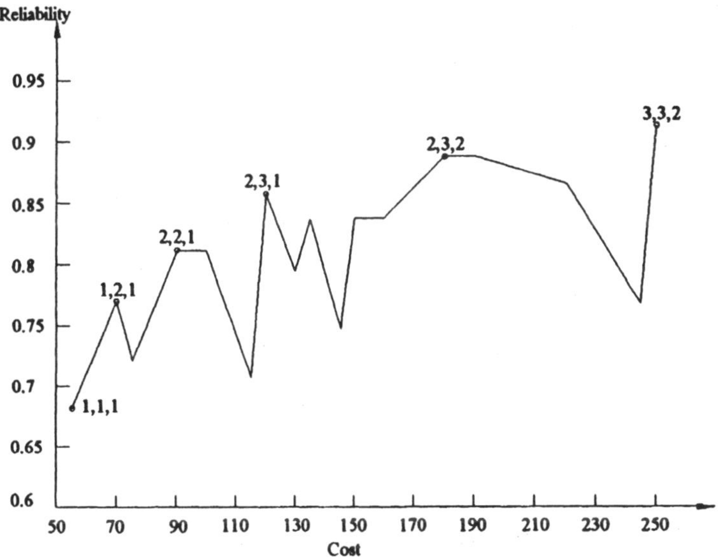

Fig.12.5 Reliability cost combinations.

$(R_{i,j}, C_{i,j})$ for j = 1,2,... ,N and for each j,i = 1,2,...,M_j. The total number of comnbinations of choices is thus

$$\prod_{j=1}^{N} M_j$$

For a fixed cost constraint, this problem could be solved by dynamic programming, but when cost varies, everytime it gives rise to a new problem and hence it becomes very difficult and time consuming to handle the problem by dynamic programming particularly at the design stage. The following heuristic method is introduced to deal with such problems of varying cost. The method is simple and very useful for the system designer. The number of optimum configurations directly obtained by this method is only

$$[\sum_{j=1}^{N} (M_j-1) + 1]$$

12.71 Algorithm

0. $i=1,\ l=1$ (Initialize)

1. Calculate

$$R_l = \prod_{j=1}^{N} R_{i,j}$$

$$C_l = \sum_{j=1}^{N} C_{i,j} \qquad (12.35)$$

Where l corresponds to the number of times step 1 is performed

2. Calculate $A_l = [F(R_{i,j}, C_{i,j})],\ j=1,2,\ldots,N$ (12.36)

   Where $F(R_{i,j}, C_{i,j}) = \Delta R_{i,j} / \Delta C_{i,j}$ (12.37)

   and $\Delta R_{i,j} = R_{i+1,j} - R_{i,j}$ (12.38)

   $\Delta C_{i,j} = C_{i+1,j} - C_{i,j}$ (12.39)

3. Choose j_1, as that values of j which corresponds to the largest element of vector A_l

   $F(R_{i,j1}, C_{i,j1}) \geq F(R_{i,k}, C_{i,k}),\ k = 1,2,\ldots,j_1-1, j_1+1,\ldots,N$

4. Replace $F(R_{i,j1}, C_{i,j1})$ in A_l by $F(R_{i+1,j1}, C_{i+1,j1})$

   and $R_{i,j1}$, $C_{i,j1}$, in R_l and C_l by $R_{i+1,j1}$, $C_{i+1,j1}$ respectively.

5. When $l \rightarrow \sum_{j=1}^{N} (M_j-1)$; stop.

   Otherwise increment l and go to step 1.

12.8 REPLACMENT POLICIES

Replacement theory is concerned with situations in which the efficiency of a system that has deteriorated over time, can be restored to its previous level of efficiency by some kind of remedial action. The problem with which we are concerned is to determine the times at which such remedial actions should be taken in order to optimise the appropriate measure of effectiveness. The measure of effectiveness may be efficiency, age or economic value. In general, we have to strike a trade off between increasing and decreasing cost functions. The increasing cost function is due to the decreasing efficiency of the system due to aging or wear. This favours the decision to replace the system at an early age to reduce the cost due to operating and maintenance. In contrast, the decreasing cost function is due to the depreciation of the original system. In other words, spreading the capital cost over longer time periods results in a lower average cost. This favours the decision not to replace the system. Minimum cost is obtained by summing both the increasing and decreasing costs and determining the minimum total cost.

For deteriorating items, the problem consists of balancing the cost of replacing old items with new items against the cost of maintaining the old items effieciently. In the case of replacement of items that fail, the problem is that of determining which items to replace and how frequently to replace them, so as to balance the wasted life of the items replaced earlier against the cost of down time of the system due to the item's failure in service or use.

The following assumptions will be followed throughout this section :

(a) Items are completely effective until they fail, after which they are completly ineffective.
(b) Queuing problems (arising because of several items failing simultaneously) are ignored since it is assumed that maintenance/repair crew size is unlimited or sufficient to carry out maintenance/repairs.
(c) Failed items are replaced with identical items, that is , the replaced item has the same life time distribution as that of the failed item.
(d) The replacement time is negligible.

12.81 Mathematical Model

Let C = purchasing cost of equipment
S = scrap value
r(t) = operating cost at a time t

Total cost in running the equipment =

Captial cost + Operating cost - Scrap value of the equipment

$$= C + r(t) - S \tag{12.40}$$

If the equipment is used for T years, then the total running cost incurred will be

$$K(T) = \int_0^T r(t)\,dt \tag{12.41}$$

Thus,
Total cost incurred on the equipment in T years
= Capital cost + Total running cost in T years - Scrap value
= C + K (T) - S (12.42)

The average cost per year incurred by the equipment is given by

$$A(T) = \frac{C + K(T) - S}{T} \quad (12.43)$$

We want to find that value of T for which A(T) is minimum. Hence, we differentiate Eqn.(12.43) with respect to T, and make it equal to zero.

$$dA(T)/dT = [-(C-S)/T^2] + [r(T)/T] - [(1/T^2)\int_0^T r(t)\, dt]$$

Thus,

$$r(T) = \frac{C - S + K(T)}{T} = A(T) \quad (12.44)$$

From Eqn.(12.44) ,we can conclude that we should replace the equipment when the average annual cost reaches the minimum.

Example 12.5

The cost of a machine is $15000 and its scrap value is $1000. The maintenance costs of the machine (as found from the records) are as follows:

Year	Cost of Maintenance
1	200
2	300
3	500
4	650
5	800
6	1000
7	1600
8	2100
9	2700

When should the machine be replaced ?

Solution

In this problem, r(t) is discrete, hence

$$K(T) = \sum_{i=1}^{T} r(t).$$

We wish to find the minimum value of A(T). Here, C = 15000, S = 1000. The value of A(T) has been calculated for different years and is given in Table 12.5.

TABLE 12.5
Calculations for A(T)

Years t	r(t)	K(T)	C-S+K(T)	A(T)
1	200	200	14200	14200
2	300	500	14500	7250
3	500	1000	15000	5000
4	650	1650	15650	3912
5	800	2450	16450	3290
6	1000	3450	17450	2908
7	1600	5050	19050	2721
8	2100	7150	21150	2643
9	2700	9850	23850	2650

From Table 12.5, it may be seen that A(T) is minimum in the eighth year. Thus, the machine should be replaced at the end of eighth year, otherwise the average annual cost will again increase.

* * *

Example 12.6

A lorry fleet owner finds from his past records the cost per year of running a lorry and its resale value, as given in Table 12.6. the purchase price of the lorry is $25000. At what stage should the lorry be replaced ?

Solution

From Table 12.6, it may be noted that the scrap value is a decreasing function of time. We now wish to minimise A(T). The analysis of the problem is given in Table 12.7

TABLE 12.6
Cost Data for Running a Lorry
(purchase price of the lorry: $ 25000)

Year of operation	Resale price at end of year	Annual operating cost	Annual Maintenance cost
1	15000	6300	300
2	13500	7000	500
3	12000	7700	1000
4	9000	9500	1500
5	8000	11500	2500
6	7500	13000	3500
7	7000	14300	4500

TABLE 12.7
Analysis of Example 12.6

Year of operation	Resale price at end of year, S(t)	Investment cost C-S(t)	Annual operating cost r(t)	Cumulalative of r(t), K(t)	Total annual cost C-S(t) + K(t)	Average cost A(t)
1	15000	10000	6300	6300	16300	16300
2	13500	11500	7000	13300	24800	12400
3	12000	13000	7700	21000	34000	11333
4	9000	16000	9500	30500	46500	11625
5	8000	17000	11500	42000	59000	11800
6	7500	17500	13000	55000	72500	12083
7	7000	18000	14300	69300	87300	12471

Table 12.7 indicates that the value of A(T) is minimum in the third year. Hence, the lorry should be replaced after every three years which results in the lowest average annual cost of $11333.

* * *

13
RELIABILITY MANAGEMENT

13.1 INTRODUCTION

Reliability is no more a subject of interest confined to only academicians and scientists. It has become a serious concern for practising engineers and manufacturers, sales managers and customers, economists and government leaders. The reliability of a product is directly influenced by every aspect of design and manufacturing, quality engineering and control, commissioning and subsequent maintenance, and feedback of field-performance data. The relationships between these activities are shown in Fig.13.1. A well-planned

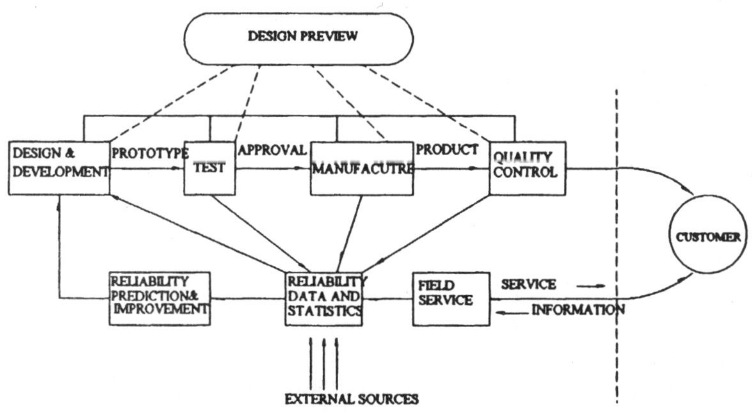

Fig.13.1 Reliability and product life-cycle.

and efficiently managed reliability programme makes possible a more effective use of resources and results in an increase in productivity and decrease in wastage of money, material, and manpower. As organizations grow more and more complex, communication and coordination between various activities become less and less effective. The cost of ineffective communication can be dangerously expensive in terms of both time and money. Moreover reliability achievement needs, in addition to proper coordination of information, a specialized knowledge of each and all of the interrelated components in a system. This places a great emphasis on the creation of an independent group which could not only coordinate between different departments but also carry out all reliability activities of the organization.

The managing of reliability and quality control areas under the impact of today's organized world competition is a highly complex and challenging task. Management's reliability and quality control ingenuity in surmounting the technological developments required for plant equipment, process controls, and manufactured hardware requires a close working relationship between all producer-and user-organization elements concerned.

The techniques and applications of reliability and quality control are rapidly advancing and changing on an international basis. Industry views the use of higher performance and reliability standards as scientific management tools for securing major advantage over their competition. The application of these modern sciences to military equipment, space systems, and commercial products offers both challenge and opportunity to those responsible for organization effectiveness. The use of intensified reliability and quality programs as a means to improving product designs, proving hardware capability, and reducing costs offers far reaching opportunity for innovations in organization and methods.

The effects of the increasing complexity, reliability, schedule, and cost competition on the reliability and quality control organization have required that all top management be aware of the most logical cost-saving areas and be assured that the product is as dependable as possible under the allowable conditions of contract or competition.

To manufacture an excellent quality product with a very high numerical reliability sometimes requires much more money than a customer is willing to pay. Therefore, since high reliability and acceptable product costs are often initially difficult to achieve, it becomes necessary that timely management decisions be made regarding reliability, schedule, and cost trade-offs. These decisions require the use of very exacting and cautiously selected information and careful organization of implementing action in order to obtain the most value for the money expended.

13.2 MANAGEMENT OBJECTIVES

The management objectives in organizing the reliability and quality control department should be to design and develop an organizational plan that will provide the controls necessary to assure that the services and products of the parent organization meet contractual requirements. These management objectives may be stated in many different ways, but in essence they probably control and reliability department is to assure that competitively proved services and hardware that meet or exceed the customer's requirements are provided.

Of course, there must be an optimum balance between the quality and reliability aspects of a product and its cost; otherwise, the industry may price itself out of the range that the customer is willing or has the ability to pay. Also, in some instances the customer may deliberately elect to sacrifice some reliability assurance for schedule reasons. Deliberate actions are required of management in order to accomplish its planned objectives for a program effectively and to assure that any trade-offs affecting product reliability and maintenance are clearly understood by the producer and customer.

Management is responsible for the business enterprise showing a profit. It is in this area that quality control and reliability have the responsibility to assist top management by assuring that planned actions are met in the design, manufacture, and use phases of the hardware. The company that develops a reputation for the manufacture of reliable products within budget will usually grow and prosper. Certainly a manufacturing or service enterprise of high integrity and enthusiasm will increase the prosperity and security of the organization and employees, as well as contribute to the social well-being of the community and nation.

Management of each organization element must be flexible and able to react quickly to meet the demands of any possible competition or new customer requirement. The ability to react quickly, objectively, and effectively to quality and reliability challenges and to anticipate these needs before difficulties arise is an organization characteristic most desired. Quality control and reliability departments have a responsibility to minimize warranty and customer service complaints by planned preventive actions as well as timely corrective-action coordinations. A satisfied customer is a most important contributing factor to the continuance of the manufacturing enterprise and the achievement of management objectives.

The reliability requirements should be clearly stated at the design and development stage itself. While setting reliability objectives it is worth considering the following objectives of the organization:

1. Maximize output,
2. Optimize reliability,
3. Minimize waste,
4. Maximize customer satisfaction and reputation,
5. Optimize job satisfaction, and
6. Minimize discontent.

All concerned should participate in deciding specific objectives and agree for the ways and means of achieving them. Management by objectives approach places greater emphasis on the importance of the basic decisions made during design and development cycle in terms of reliability and how well it satisfies the needs for which it is intended.

All objectives, whether requirement specifications or design instructions, are essentially a means of communicating information to others. Therefore they should be:

1. Clearly understandable,
2. Unambiguous, and
3. Realistic in terms of resources available.

A reliability specification format can be prepared for each type of product. Even though the content may vary considerably from one type to another, the typical contents may include:

1. The type and source of component failure data.
2. Reliability assessment methods to be employed.
3. Confidence levels required for reliability predictions
4. Mode of reliability specification:

    (a) MTTF (mean time to failure) for nonrepairable items,
    (b) MTBF (mean time between failures) for repairable items,
    (c) Probability of success for *one-shot* devices whose operation is limited to a single operation cycle,
    (d) Failure rate, and
    (e) Mean number of operations before an item fails (for devices such as switches, connectors, relays, circuit breakers, etc.)

5. Maximum acceptable down time and mean time to repair (maintainability characteristics).

6. Maintenance policy:

    (a) Repair plan,

(b) Availability of spares,
(c) Maintenance personnel requirements, and
(d) Test facilities.

7. Details of environmental conditions and methods of operation

13.3 TOP MANAGEMENT'S ROLE IN RELIABILITY AND QUALITY CONTROL PROGRAMS

Management must provide the controls needed to assure that all quality attributes affecting reliability, maintainability, safety, and cost comply with commitments and satisfy the customer's requirements. Tersely stated, management must have well-planned policies, effective program planning, timely scheduling, and technical training. Management must clearly state and support its objectives and policies for accomplishing the product quality and reliability and assign responsibility for accomplishment to appropriate functions throughout the organization.

Top management's basic objective is to provide and maintain quality and reliability organizations capable of efficiently accomplishing the necessary inspection, test, and analytical laboratory services to assure that all products satisfy the specified requirements of quality and reliability. The quality control organization must support these objectives in a timely, objective, and helpful manner. Improved product performance and lower costs must be continually emphasized, and the results must be made visible to management.

Fig.13.2 depicts a typical top-management organization which shows the responsible management of the combined quality control and reliability control departments. This arrangement provides for the entire function to be headed by a director, with the quality control and reliability control functions headed by managers. In this manner the necessary coordination, services, and assurances at the equally important policy setting operating levels of the various programs are kept on the policy course and not allowed to *drift off* to the detriment of any one aspect. Advantages of this combined quality control and reliability organization are that top management has one point of communication and the overhead costs of combined R&QC organization may be lower than for separate organizations.

13.31 Time-phase Planning, Scheduling, and Implementation

The importance of reliability and quality control management control through detailed scheduling of each item of the reliability and quality task must be emphasized. Care must be exercised to sequence reliability and quality program elements to coincide with related total program plans. For example, it would not be practical to request a major change in existing procedures when the contract is nearing completion and the return will not justify the effort expended. Nor would it be practical to expect the accomplishment of tests in nonessential areas of operation when the cost of the test equipment would not be justified by the service the equipment would provide. However, the purchase and installation of equipment for assurance may more than justify itself when compared with the potential impact of equipment failure in customer operations.

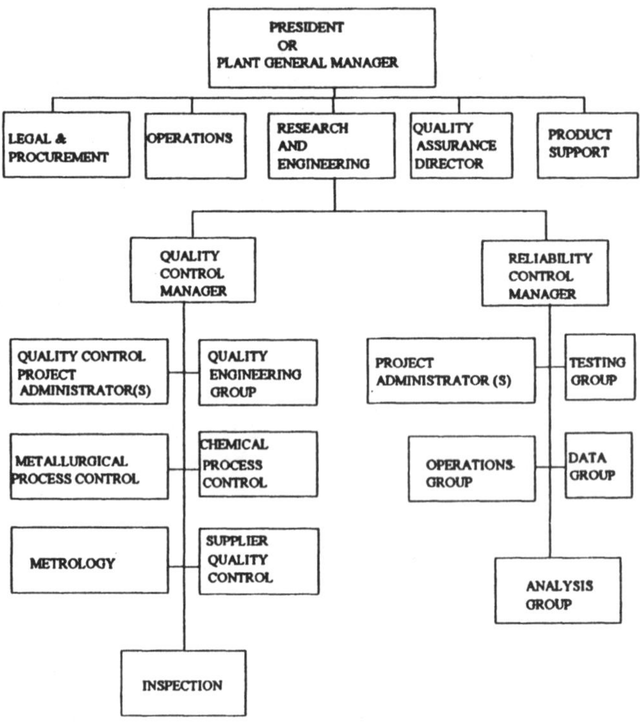

Fig.13.2 Top-management organisation.

Management follow-up and evaluation of reliability and quality program progress should be accomplished by use of audits and simple reports that are specifically designed for the purpose. These management reports serve as decision-making tools and forewarn management in the event progress becomes static. Timely management action must be readily available and applied as needed to many areas of the manufacturing sequences to maintain a good, smooth-flowing, low-cost operation.

13.32 Management Selection of Key Personnel

Management must recognize and choose the type of persons that are needed to fill the key positions in the reliability and quality control organization. Management must know that these selected people will be able to work closely with and motivate others to accomplish their respective tasks. Top management philosophy establishes the element for employee motivation throughout the enterprise.

Top management must be organizationally situated to apprise, counsel, and instruct the middle management that reports to them. All levels of management must maintain clear two-way communications and motivate others without destroying initiative and creativity.

When top management can report improvements in progress, whether it be in implementing a new program or during the actual manufacturing process, the chances are good that the operations of the particular departments are contributing effectively to assuring a fair profit for the business enterprise.

13.4 COST EFFECTIVENESS CONSIDERATIONS

13.41 Organization Responsibility

Responsibility for costs within the reliability and quality control organizations can be most effectively accomplished when specific, capable individuals are charged with coordinating all matters relating to cost analysis and budget control. However, the assignment of coordination responsibility to these individuals must not be allowed to detract from the duty of each member of the reliability and quality control organization to maintain a high level of cost effectiveness.

The cost control function within the reliability and quality control organization is most frequently located within the quality control Administrative Group, the Quality Control Systems Group, or the Quality Control Engineering Group. Regardless of which group is given the responsibility, the director of reliability and quality control and his department managers must maintain very close and continuing communications with the responsible individuals. Timely analysis of trends and decisions and guidance should be provided frequently.

13.42 Timely Cost Planning

The reliability and quality control management team has value to the total organization that is related directly to its favourable impact on product reliability, performance, and costs. Its contribution to the organized task is of greatest value when performance, reliability, and maintainability of the product are optimized with total program costs.

Although many individuals cooperatively contribute to the overall performance schedule-cost profit objective, it is necessary that the executive authority of R&QC management enter into the cycle whenever the desired voluntary cooperation in other branches of the organization falters or the need for new ground rules and policy decisions becomes evident.

Product quality assurance is most economically secured when the conditions which might lead to loss of sale, customer rejection, or excessive warrantly cost are predicted,prevented, or corrected at the earliest possible time.

13.43 Incentive Contracts

The abrupt deemphasis of cost plus fixed fee military contracting has focused attention upon the incentive contract as a means for assuring effective management interest in achieving product reliability and maintenance commitments. With this medium, a specified scale of incentive- and sometimes penalty is applied as a factor in the total contract price. Penalty scales are usually applied at lower rates than incentive scales and may be omitted in competitive fixed price contracts.

13.44 Cost Analysis and Budgeting

Every product merits an analysis of the total tasks to be performed with the allowed costs. The estimation of costs for every function must be quite close to the final actual costs of the specific function if effective results are to be achieved. It is apparent that the general readjustment (usually arbitrary cuts) of budgetary estimates by top management will be in those areas where the departmental estimates and accounting reports of past performance on similar programs are in obvious disagreement.

13.45 Equipment and Facility Costs

Cost estimation of the equipment and facilities required for standards and calibration, process control, inspection and test is another essential task for reliability and quality control engineers. Applicable staff and line personnel should be given the opportunity to take part in the planning of all equipment and facilities expansion, retirement, or replacement.

Great care must be exercised to determine that adequate justification exists for the addition or replacement of facilities. Improved product reliability and lower costs must be tangible and measurable. Savings predicted should offset the cost of new equipment and facilities within a period prescribed by top management.

13.46 Cost Records

Reliability and quality control organizations have the responsibility for generating and maintaining the important segments of product records of rework and scrap costs, testing costs, warranty costs, etc., upon which pricing structures, company procedures, redesign, and even critical litigation have been founded. The cost of these record-keeping and data processing activities must certainly be compared with their worth to the company. The responsibility for this falls upon those who implement and make the system work.

Cost estimation for this requirement must include the consideration of savings through the use of automated data processing equipment, the ever increasing cost of records storage and data retrieval, the nature of any contractual requirement for data reproduction and translation, participating in data centers.

13.47 Quality and Reliability Cost Control

To control cost in the quality and reliability programs, careful long range planning must be exercised by management. This planning must be accomplished by those to whom top management has delegated the responsibility and who will be held accountable for the implementation of the plans. The controlling of these long range plans at the time of implementation is one of the basic principles of cost control.

Sturdy programs, research and development programs, production programs, prevention, assessment, rework, and scrap cost estimates should all be made in the long range plans whereby proper budgeting may be forecast and arrangements made.

13.5 THE MANAGEMENT MATRIX

The adroitness of a company to remain competitive and maintain its profit level requires more than the ability to engineer and produce products in quantity. The matrix technique applied to decision making provides an objective means for solving various management problems. Quality assurance of a product or system is a significant factor in the growth pattern of a company. The departmental functions, policies and responsibilities dictate the type of organizational structure which can best fulfill the objectives of the consumer and the company. At the top management level, the matrix technique is useful in determining the organisation structure based upon the responsibilities delegated to each department and as a basis for penetrating new market areas. In all cases, the effectiveness of the management process is directly related to profitability through consumer assurance that product performance and quality are maximized within the negotiated cost structure.

Management of a department responsible for administration of the quality assurance program in a division of a company primarily oriented to research, development and production of diversified products and systems requires special planning, techniques and philosophy. The management must have the capability to continually maintain the proper level of customer satisfaction and evaluate product performance even though the products and systems are usually required to perform at limits bounded by the state of the art. In general, each product or system has performance requirements in scope and magnitude such that the product assurance requirements specified are as diverse as the product line, depending upon the customer documents or procurement agency involved in the contract.

The solution, to the stated conditions must be one of dynamic planning of the steps in organizing to accomplish the department objectives. Elements of the matrix can then be sequentially incorporated into the organizational structure in logically phased steps. The matrix planning is always an evolutionary process to eliminate the administrative stresses associated with revolutionary changes due to new business and profound requirements. A continual audit of the structure, and contract requirements should be conducted to validate the effectiveness of the organization in cost and performance and its applicability with program demands.

A study of programs determine the need for an operational analysis since the interface relations between the sections for each contract would have to be established during the proposal stage. Each new program is placed in the organization after a decision has been made as to the need for establishing it as a project. Several factors are considered and the methodology of decision theory is applied. The following factors are considered as the most heavily weighted.

1. *Customer Requirement*

Certain programs are of such magnitude that management and communications must extend in an unbroken line through all levels of procurement. The need for a specific organizational structure is a customer requirement. This does not assure that all activities will be performed by the project but that authority and responsibility for compliance with requirements is maintained by the project.

2. *Special Requirements*

The product or system and/or contractual requirements are so specific and different that existing procedures cannot suffice.

3. *Schedule*

This objective requires special attention. A tight schedule requires appropriate manpower to evaluate acceptability of the production flow. In some cases, the personnel performing acceptance must be certified in special ways or have specific talents.

4. *Product Complexity and Skill Levels*

Product complexity (processes, test techniques, production fabrication) and skill levels are such that the product is significantly different from related products.

5. *Dollar Volume as a Function of Time*

The ratio of program cost/time is high. This implies a concentrated program effort is required.

6. *Manpower Availability*

The program requirements for specialized manpower are such that this factor is considered. This objective is not heavily weighted since it is related to attainment of other objectives.

These objectives are weighted in terms of the various courses of action using the matrix approach to establish a decision. This approach has a basic purpose of analyzing the array of actions and depicting the decision in mathematical terms.

The management function then utilizes this tool for planning and action in performance of its activities. The organization matrix provides the mechanism for management in an expeditious manner and efficient departmental control commensurate with this company's products and philosophies.

The placement of quality and reliability assurance in the overall organizational structure should be considered on the basis of optimum product control and assurance which minimizes the total program costs.

The effective use of the matrix technique in decision making can be a useful tool which focuses attention upon all program requirements and allows the decision maker to efficiently trade-off or heavily weight those sections which contribute the greatest to program overall success. The technique is adaptable through all levels of management and provides a documented analysis for the decision maker to use in re-evaluating his original decision in the light of new information.

13.6 RELIABILITY AND QUALITY CONTROL FACILITIES AND EQUIPMENT

The nature of the reliability and quality control activity imposes an added burden upon the planning which must precede the provision of facilities and equipment. The managers of plant engineering and facilities functions are under constant pressure to hold down the costs of space, equipment, and material, as well as the cost of personnel. In the natural optimism for self confidence in the organization and its product, quality and reliability methods and equipment requirements are sometimes taken for granted.

To anticipate the necessary provisions for product assurance in advance of the final (production) design and manufacturing places reliability and quality in superposition with profits. Advance planning of all such costs is necessary if management is contractually responsible for reliability and quality performance, and certainly to whatever modicum the company feels ethically bound in the absence of a specification.

It is desirable that the provisions for reliability and quality control facilities and equipment be made in close cooperation with the company's engineering design group; if feasible, the planning should be made during the concept and preliminary design phase of the product, and certainly in conjunction with plans for new plant locations or structural additions to the existing plant. It is important that any particular requirements for test equipment be given to management so that they can be provided in the planning layout of new facilities.

Coordination of reliability and quality control with design engineering results in knowledge of what the product is intended to do. This information and the contract specifications will allow the setting up of economical quality control sampling plans and appropriate reliability demonstration test levels, thereby determining the appropriate facilities and test equipment.

This liaison enhances the compatibility of test tolerances at all stages of product inspection and permits an orderly expansion of generic tolerances from part supplier to assembly producer to consumer.

13.61 Funding and Schedules

The critical demands of advance planning for reliability and quality control equipment appear in the funding and scheduling of the production master plan. Equipments which require long lead procurement must be included within the master schedule to minimize the terms of loan capital provided for this purpose. Similarly, the funding requirements for facilities must be evaluated, for these will include such considerations as inspection area lighting, temperature, humidity, air conditioning, clean room, air control and flow distribution, special disposal and sanitation installations, personnel safety provisions, and mobile access into all such areas.

13.62 Equipment Specifications

To unify the management of reliability and quality control organizations, it is to the advantage of each that test equipment procurement specifications be generated within the organization. In this way no other operating group can establish the boundaries of test and inspection by indirection and reliability quality tests can be established over the full design spectrum of the product. Also, for companies with multiple product lines, reliability and quality control man management can see the entire test picture and advise the purchase of equipments compatible with any tests which may be required. In this manner fewer equipments of greater capability may involve less capital expenditure than more equipments of limited and singular capability. If the company procurement policy does not allow the generation of reliability and quality control specifications, the management of those functions will certainly elect to advise the procuring group of its judgment through appropriate intracompany communication.

13.63 Reliability and Quality Control Design of Test Equipment

In some organizations the reliability and quality control groups have been given the responsibility for test equipment design. This requires that very capable engineers be made responsible for this effort. When adequately staffed, certain advantages may accrue through this organization policy. These advantages include improved timeliness and effectiveness of test equipment, greater emphasis on automation, improved supplier coordination, improved integration of all test functions, and optimum emphasis on nondestructive inspection and test methods.

Disadvantages may develop if capable equipment design personnel are not available to staff the equipment design function. This frequently leads to the use of alternate or makeshift types of equipment, which do not provide optimum cost effectiveness. It must be recognized that an additional level of coordination with product engineering is required in order to assure maximum compatibility of the test equipment with a product. This is very important, particularly for products which have frequent model changes.

13.7 RELIABILITY DATA

Data acquisition includes identifications and description of the system and, subsequently, collection of operating and maintenance experience data, and test description and results. Data analysis involves organization of data for specific components or subsystems and identification of their failure events and rates.

13.71 Planning a Data-Collection Programme

The primary advantage of acquisition and accumulation of data is that these can be used to predict the reliability of a component (or system) when it is operated under the conditions which these data represent. In planning a programme to collect data, due consideration should be given to the several factors that are important to the success of the programme:

1. A complete set of clearly stated technical objectives must be established.
2. The methods by which the required data will be collected and processed should be stated.
3. A detailed written document which is in effect a specification for the work to be done during the study must be prepared. This specification should normally contain:

    (a) A brief and factual account of the development and objectives of the reliability programme,
    (b) explicit definition of terms that are of interest to the study and that are used throughout the specification,
    (c) data requirements, such as item of data, criteria, unit of measurements, etc.,
    (d) a complete and detailed technical inventory of the product to be evaluated, and
    (e) materials and facilities needed for the evaluation.

13.72 Data-acquisition Methods

Two methods are usually employed in collecting the required data, depending upon the relative importance of accuracy vs. cost.

The *first method* is to supply the operational, maintenance, and production personnel with what are know as *data forms* or *failure forms* containing blanks for the desired information, and ask that forms be completed as directed. Forms containing the raw data are returned to a central collection point (reliability group) for processing. This method has the advantage of low cost but the data so collected are invariably of questionable accuracy and completeness. The operational, maintenance, and production personnel, in general, tend to look upon data collection as mere paper work, and in the pressure of more urgent responsibilities they tend to neglect it.

The *second method* is to employ technical personnel who have the assigned responsibility for carrying out the measurement programme. This method has numerous advantages. A few important ones are enumerated below:

1. Personnel concerned can be given a thorough understanding of the objectives of the study.

2. A high interest in the study can be maintained at the source of the data.

3. As a result of (1) and (2), the evaluation personnel can make the necessary decisions to keep the study on the right course.

4. Data supplied under the conditions of close monitoring and recheck require a negligible amount of rework and interpretation before final processing.

5. Selective attention can be given to developing details or trends that are pertinent to evaluation.

6. Inconsistencies and errors in the data can be detected through cursory checks and analyses.

13.73 Use of Samples

Since it is seldom feasible to make measurements on the entire population, the use of statistical techniques is necessary. Such techniques permit the extrapolation of the results obtained from a sample of the population as a whole and therefore to other similar populations.

The use of samples in the measurement of reliability requires that the final result be presented as an estimated value with the confidence limits to indicate the probable range within which the population mean will fall. The larger the size of the sample, the narrower will be the confidence interval.

13.74 Analysis and Reporting

As failure forms are received they are reviewed, and completed by a member of the reliability group. This person should have a good understanding of the complete system so that he is able to judge the consequences of a failure with respect to the system and establish it critically. He should also be able to initiate the necessary corrective action.

Reporting represents one of the reliability groups greatest responsibilities. It furnishes all levels of engineering activities and management with information relevant to their needs. Consequently, the possibility of misinterpretation must be minimized. Answers to questions such as the following should be available in the reporting:

1. Were the data taken from the development tests, field tests, component tests, system tests?
2. What were the environmental conditions?
3. Were the data homogeneous and representative?
4. How large was the sample size?
5. What assumptions were made concerning the shape of the failure distribution?

13.75 Data Management

A data management system needs to be established for the purpose of collection and evaluation of reliability data from equipment manufacturers and users. The important tasks of the data management would be:

1. Collection and analysis of input or field data, test data, and manufacturer's data, and

2. Classification of the collected data by equipment and event types in order to facilitate evaluation and correlation of data.

A breakdown of functions required in data collection and classification is shown in Fig. 13.3.

13.76 Data Bank

A reliability data bank is an integral part of a reliability group. It usually consists of:

1. An event store, and
2. a reliability data store.

It serves the following two main purposes:

1. It provides information to its contributors regarding the performance

(availability, reliability, etc.) of their own plant, and

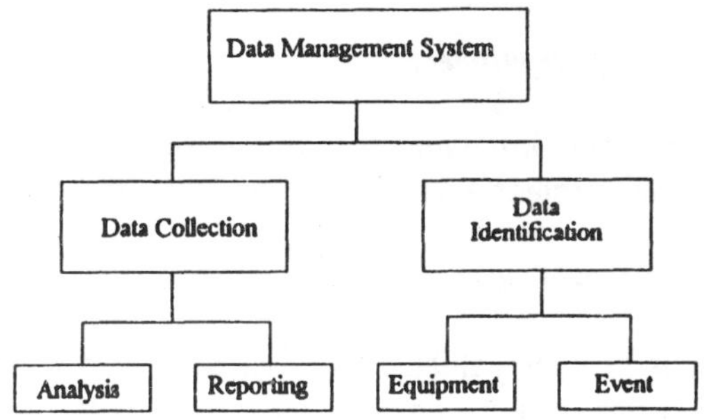

Fig. 13.3 Data management system.

2. It provides the generic reliability data required by the project analysis section of the reliability group.

The functional relationships of the data-bank system are depicted in Fig. 13.4.

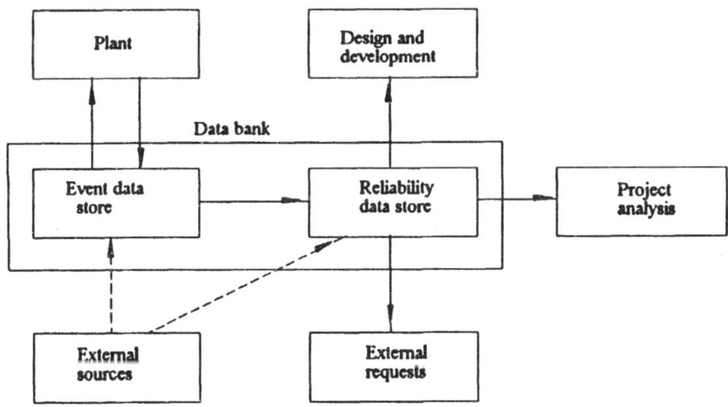

Fig. 13.4 Data bank.

13.761 Internal Data Sources

Positive controls are necessary for a reliability department to assure flow of all relevant data into their files. Periodic check by reliability personnel is not satisfactory; formalized document routing to reliability and/or sign-off provisions are required.

The following areas often generate information vital to reliability control and should be periodically monitored to establish that no new data sources are bypassing the reliability files.

1. *Purchasing/ Subcontracting Department*

Look for major subcontracts involving test requirements and individual tests subcontracted directly at project engineering request.

2. *Library Acquisition Lists and Document Control Inventories*

A large company has much valuable data generated from *one time only* sources, libraries can serve as checkpoints which often turn up these occasional inputs.

3. *Contracts Department*

Often plans, proposals, or performances bearing on reliability are forwarded to the customer without the reliability department being notified. Screening or receipt of all documents is not proposed; only a positive check off arrangement within contract department to assure transmittal of relevant data is suggested.

4. *Field service Engineering*

Too often this department is isolated physically from design reliability engineering and runs its own failure analyses and quick fixes on customer-accepted assemblies. Full copies of all issuances should go to reliability department.

5. *Environmental Test Laboratory*

Laboratories usually compile schedules, plans, status reports, etc. on a regular basis. The need for reliability department to utilize these is determined by the degree of centralization of the test control function and its integration within the reliability organization.

13.762 Structuring of Data at Original Source

Tabulating, storing, or analyzing data is made difficult or impossible unless the data are structured (i.e., placed on standard format, coded, descriptors identified, etc.) at the source to the degree of rigidity appropriate to the volume and anticipated complexity of search.

1. *Failure Reports*

Control on failure reports will vary with the volume of reports to be handled.

A small quantity can reasonably be tabulated, and the trends analyzed and studied, by using manual methods and by working from the original narrative descriptions. As the quantity of reports grows, the necessity of conventional coding and *restricted English* terms increases if the information is to be handled on a mass basis. A computer search is possible only when each *field* or box (by which a search might be made) is restricted to a stipulated selection of terms or figures on the original report. The trends thus revealed naturally require subsequent engineering interpretation of significance.

2. *Test Reports*

The format is basically designed for reports on component parts, although it is applicable to tests of assemblies of greater size and complexity. Specific adaptations can be made easily if the testing in question is consistently on a particular type of product. However, the principles of utilizing a standardized format to facilitate rapid interpretation, coding, and retrieval still apply. The common requirements for date, full identification of the part, tabular description of the tests and results, plus a narrative summary of conclusions, constitute a universal disciplinary framework to guide the reporting of any methodical testing.

13.8 TRAINING

The performance of personnel who define, design, procure, manufacture, construct, test, repair, and operate equipment is inherently variable because of inequalities in skills, knowledge, personality, education, and training. This performance variability affects the quality of equipment and manufactured products. Advance planning for training is an activity that should coincide with the advance product goals set by the management of the company.

The plan of action by management for the advance planning of the goals rests and is dependent on the company's resources such as facilities, tools, raw materials, personnel, productive capacity, sales outlets, etc. Because business is subject to change, it is rather difficult to predetermine definite training courses during the early product planning stage. But when a product becomes firm business and specifications are known, training plans must be activated on a time phased basis.

13.81 Reliability and Quality Control Training

The purpose of reliability and quality control training is to communicate skills, methods, ideas, objectives and attitudes to all personnel levels in an organization. Effective training incorporates the identifying, measuring, and supplying of the training needs that develop day by day in various activities. Reliability and quality control management should assure the accomplishment of education programs to indoctrinate all personnel whose work relates to the product's reliability. The assigned personnel must understand the value of their individual contributions to the product and be motivated to provide excellent results.

The need for additional specialized training can be evaluated by consulting the sources of information concerning any new task. The program plan certainly should indicate the various operations which require accomplishment. With the various operations and responsibilities known, the job performance and qualification requirements relating to the task should be explored.

One of the duties of the quality assurance engineer should be to ensure that supervisory personnel become aware of the training needs of their workers and to make certain that means are devised and used to determine exactly what, when, and how training is to be implemented and made effective (Fig.13.5).

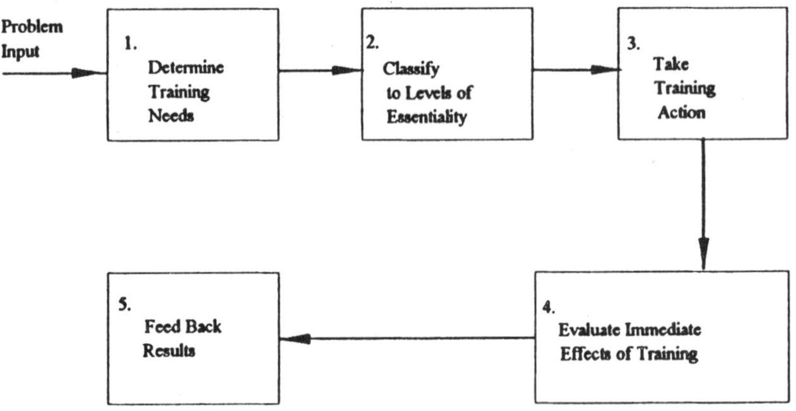

Fig.13.5 Systematic planning and training cycle.

13.82 Specific Training Needed

In order to measure the employee's knowledge and determine what specific training is needed, we can use what might be called job knowledge quotient. Job knowledge quotient is a series of test questions designed to be answered by employees. Different sets of questions can be made applicable to specific areas required of job knowledge.

An employee's experience and background provide management with an indication of the needs of training that can be expected. Once the information is gathered and analyzed, there should be an understanding of how much and what kind of training the employee needs. Training needs comprise the skills, knowledge, information, and attitudes which individuals require to meet reliability and quality specifications.

Changing demands often alter requirements of skill and knowledge and increase training needs.

Subjective measurement devices and techniques are available to identify and gauge these needs. These techniques and devices are:

1. Job or activity analysis
2. Tests or examinations
3. Questionnaire or improvement-checklist surveys
4. Purposeful observations and consultations based on history of errors
5. Reliability and quality control reports and audits
6. State of the art surveillance and review for changes
7. Merit and performance ratings.

The primary objectives of reliability and quality control training and indoctrination are to:

1. Promote reliability and quality control consciouness in all personnel engaged on the project.

2. Emphasize to personnel in engineering, manufacturing, reliability, quality control, purchasing, etc., the specific effects of their particular jobs in contributing to or detracting from system reliability.

3. Afford all personnel with sufficient knowledge and understanding of the specific and general factors affecting system reliability to assure the incorporation of good reliability techniques into the design and manufacture of equipment.

4. Assure that all reliability and quality control personnel are capable of performing their tasks effectively and efficiently.

5. Concentrate attention on those areas of activity considered to be particularly amenable to a reliability improvement effort.

13.83 Informal Training

Informal training (on the job) occurs throughout industry when any member of management gives instructions to his subordinates. Skill in such communication is important in achieving desired actions. Motivation for quality and reliability is a daily task and is the result of organized effort. It requires the measurement of progress and gives frequent feedback to employees of the quality of job they are doing. Control charts provide a scoreboard of personnel performance. This feedback of information, when coupled with plans for corrective-action patterns, will promote desired motivation.

13.84 Formal Training

Formal training occurs when skills, experience, ideas, and information are organized into a classroom curriculum to achieve desired levels of skills and understanding. The objectives in training programs must be stated, and they must be realistic. The applicable subject matter must be organized and accurate, and methods must be suited to subject matter. Instructors must be qualified and experienced, and proper evaluation and feedback for curriculum improvement must be provided. Schedules must be realistic and planned to have personnel trained as the task is implemented.

13.85 Evaluation of Training

Evaluation of training is necessary to determine whether trainees have or have not reached predetermined goals. The basis of effective evaluation is the observation and measurement of same performance before planned training and after training.

A comparison of the results evaluates training. Evaluation is based upon a record of all available evidence which shows the degree to which training objectives were or were not realized, the improvements affected, and the ultimate effects on production activities. Training which involves measurement of errors, defects, failures, waste or speed and productivity can be evaluated and measured objectively with the *before and after* approach. Training which involves mental skills and long term development will involve subjective measurement.

The following factors can be used to evaluate training for both mental and physical skills:

1. Statistical measurement of before and after performance recorded on control charts
2. Checklist enumeration of improvements before and after performance
3. Recorded changes in job-performance ratings by supervisor
4. Written tests and examinations
5. Tabulation and analysis of quality control and reliability reports

6. Comparison with stimulated control groups
7. Comparison with personnel case histories
8. Number of hours spent in training

13.86 Guidelines for Effective Evaluation

1. Evaluation must seek out successes as well as failures.
2. Evaluation must start with specific skill objectives to be achieved.
3. Evaluation must be built around a systematic long term, continuous plan as required.
4. Evaluation must determine the degree to which training resulted in sufficient learning.
5. Evaluation should be made immediately before there are significant losses from other sources.
6. Evaluations tailored to one's own activities are better than the use of ready made ones by outsiders.

Training records should be maintained in a manner similar to production - or inventory record maintenance. These records, when accumulated over a period of time, should represent an inventory of skills and a distribution of variability in both professional and trade skills existing in the facility.

14
RELIABILITY APPLICATIONS

In this chapter, we discuss two typical applications of Reliability Engineering. The first,, *Reliability analysis of instrument landing systems*, concerns the application of reliability techniques to a safety system and was carried out by the author himself. The second, *Reliability analysis of banking system*, deals with both hardware and software reliability with emphasis on the later and has been taken from Musa's book on Software Reliability. These two case histories are expected to clarify many of the concepts discussed in the previous chapters of the book.

14.1 INSTRUMENT LANDING SYSTEM

The ability to land an aircraft under low or zero visibility conditions is probably the most vital factor, which determines the reliability of air travel. An electronic system now in use provides a solution to this problem. This system is known as Instrument Landing system (ILS). The function of Instrument Landing system is to enable the pilot of an aircraft to make a safe approach to, and landing on, a runway even under adverse conditions of weather and visibility. It is essential that our system should also be reliable enough so as to make the landing safe. That is why reliability analysis is important and considerable efforts are made to obtain a high system reliability.

14.11 System Details

The function of an ILS is accomplished by the provision of azimuth guidance, elevation guidance and distance from threshold information. The essential elements of the Instrument Landing System, illustrated in Fig. 14.1, consist of a runway localizer for azimuth guidance, glidepath equipment for elevation guidance, and marker beacons.

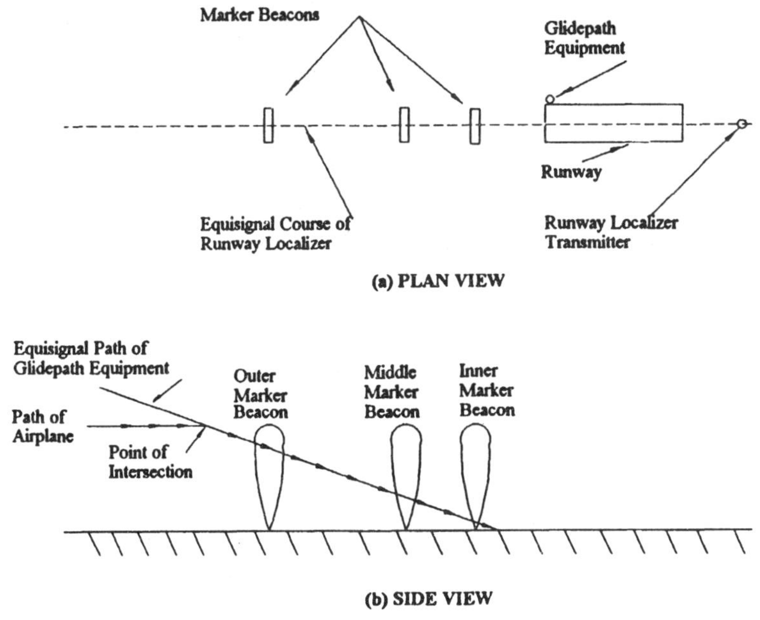

Fig. 14.1 Schematic diagram showing essential features of the Instrument Landing System.

The runway localizer provides the lateral or azimuth guidance that enables the airplane to approach the runway from the proper direction. Signals carrying azimuth guidance information are produced by a VHF Localizer equipment. The glidepath equipment provides an equisignal path type of guidance in the vertical plane analogous to the guidance in azimuth provided by the equisignal path of the localizer.

The combination of runway localizer and glidepath equipment provides the pilot with sufficient information to approach the runway in the correct direction, and to bring the aircraft down to earth along a glidepath that will provide a safe landing. In addition three VHF marker beacons are sited along the extended centre-line of the runway to provide distance from threshold information to an aircraft making an instrument approach. These three beacons are referred to as the *outer*, *middle* and *inner* markers, and are located nominally at 3.9 nautical miles, 1050 meters and 75-480 meters respectively from the landing threshold. The marker beacons provide indications to the pilot of an aircraft that these specific points along the approach path have been reached. The marker beacons radiate a horizontally polarized signal at 75 MHz. This carrier signal is modulated by a coded audio tone, the frequency of the audio tone and the code depending on the position of the marker beacon. The outer marker modulation is a tone of 40 Hz coded two dashes per second. The middle marker modulation is a tone of 1300 Hz coded alternate dots and dashes, the dashes keyed at the rate of two dashes per second and the dots at the rate of six dots per second. The inner marker modulation is a tone of 3000 Hz coded six dots per second.

The vertical radiation pattern of each marker beacon is adjusted to provide coverage over the following distances measured on the ILS glidepath and localizer course line.

(i) Inner marker : 150 $\pm$ 50 Metres.
(ii) Middle marker : 300 $\pm$ 100 Metres.
(iii) Outer marker : 600 $\pm$ 200 Metres.

In this study, we consider the reliability analysis of localizer and glidepth equipment only as beacons can be assumed failure free with reasonably good degree of confidence. Also we assume the installation of category II (Cat II) level equipment for highlighting few salient points in reliability analysis.

14.12 Reliability Analysis

Reliability Analysis of equipment is necessary to have a quantitative knowledge of the reliability so that further improvement in reliability can be made. This study deals with reliability and MTBF calculations for ILS ground equipment -Localizer, and Glidepath .

Reliability is the probability that an equipment functions satisfactorily for a given period of time under specified operating and environmental conditions. It can be expressed in terms of the failure rate and MTBF as follows:

$$R = e^{-\lambda t} = e^{-t/m} \quad (14.1)$$

Where,

R Reliability of the system
λ Failure rate of the system.
t Time
m MTBF of the system.

Several methods of reliability and MTBF prediction have already been discussed and each one of them depends upon the degree of theoretical and technical data available and the degree of accuracy required for prediction.

In this study, Parts Count Method has been followed which involves the counting of each generic type of component such as fixed film resistors, ceramic capacitors, silicon NPN transistors, etc. The counted number is multiplied by a generic failure rate for each part type and then these products are added to obtain the failure rates of each functional block of the system. The failure rates for this study were taken from MIL-HDBK-217.

14.13 Localizer

The runway localizer radiates a horizontally polarized composite field pattern modulated by 90 Hz and 150 Hz tones. (The radiated wave consists of a

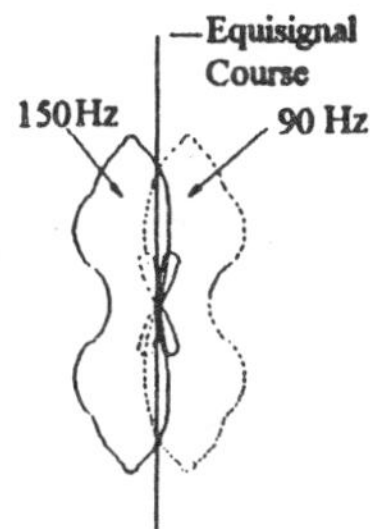

Fig.14.2 Directional pattern used in runway Localizer of Instrument Landing System.

single carrier wave which is simultaneously associated with two sets of continuously radiated amplitude -modulated sidebands represented by modulated frequencies of 90 Hz and 150 Hz, respectively). The composite field pattern, shown in Fig.14.2, comprises of two patterns which correspond to the relative strengths of the 90 Hz and 150 Hz sidebands as a function of direction. Further, the equisignal course directions are indicated by equality in the strength of the two modulations which are separated by suitable filters in receiver output, separately rectified and then applied with opposite polarity to a zero centermeter. This radiation pattern produces a coarse sector, about the extended centre line of the runway, in which the 90 Hz tone predominates on the left hand side of the approach to the runway and the 150 Hz tone predominates on the right. The difference in the depth of modulation (DDM) of the two tones is zero along the course line. The localizer carrier frequency is within the band 108 MHz to 112 MHz.

The Localizer is a static installation which provides the azimuth guidance to an aircraft. The localizer aerial system is sited on the extended centre line of the runway at a distance which is consistent with safe obstruction clearance practice from the stop end of the runway. The localizer transmitter cabinet and ancillary equipment are housed in a building at the rear of, and within a radius of 90 meters from the centre of the aerial system. A remote control facility is provided in the airfield control building. In category-II, one transmitter provides power to the aerial, while the other transmitter remains off. The radiated signals are continuously checked by a duplicated monitoring system.

A Cat II system has two channels, each consisting of main transmitter Unit, Motor Drive Unit and Mechanical Modulator in addition to Coaxial Distribution Unit, Aerial Distribution Unit, Localizer Aerial Arrays, Monitor Aerials and Associated Equipment, Control Unit (local) and Control Unit (remote).

14.14 Glidepath

Signals carrying elevation guidance information are produced by UHF glidepath equipment which produces a horizontally polarized composite field pattern modulated by 90 Hz and 150 Hz tones. This radiation field pattern produces a glidepath sector ,about a straight line descent path in a vertical plane containing the centre line of the runway, in which the 90 Hz tone predominates above the descent path and the 150 Hz tone predominates below. As a result of the predominance of one tone above and one tone below the descent path, difference in the depths of modulation are apparent in the glidepath sector. The DDM is zero along a straight line descent path, providing the glidepath and increases with increasing vertical displacement from the descent path. The glide path transmitter carrier frequency is within the band 328 MHz to 336 MHz. The glidepath equipment is static installation which provides the elevation guidance of the Instrument Landing System.

The glidepath equipment provides a range of at least 10 nautical miles (18.5 km) within the sectors, 8° either side of the glidepath course line, bounded by upper and lower planes 1.75θ and 0.30θ respectively where θ is the glidepath angle.

In order to ensure that there will be only one equisignal glidepath, the lower antenna is so excited that its lobe maximum is larger than the maximum of the upper antenna and is so placed that its pattern has a maximum that is at a relatively large angle above the horizon as shown in Fig.14.3. Different side band frequencies are radiated from these antennas in the same manner as indicated for localizer in Fig.14.2. The proper glidepath is in the range of 2 degree to 5 degree. Since the glidepath equipment must be placed at the side of the runway so that it will not present a hazard, the antenna patterns in the horizontal plane must be carefully controlled so that the glidepath will have the correct slope along the azimuth course defined by the localizer. The category-II equipment is identical to that of localizer equipment.

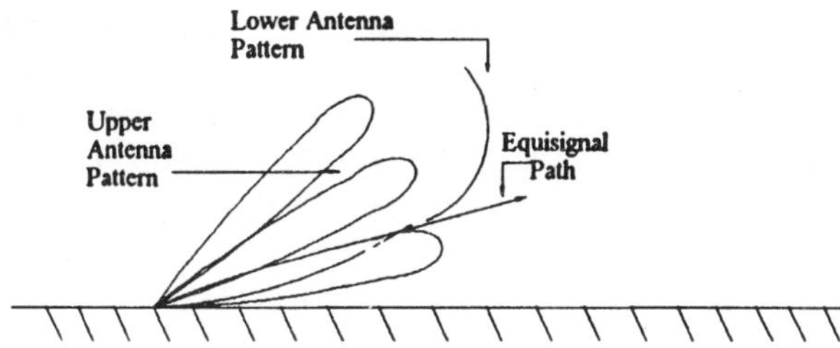

Fig.14.3 Antenna patterns producing equisignal Glidepath for Instrument Landing System.

14.15 Localizer Reliability Calculations

The functional performance of the localizer equipment of the ILS has been described. Based on this functional performance we obtain the Reliability Logic Diagram (RLD) for Cat II system which has been shown as RLD -1 in

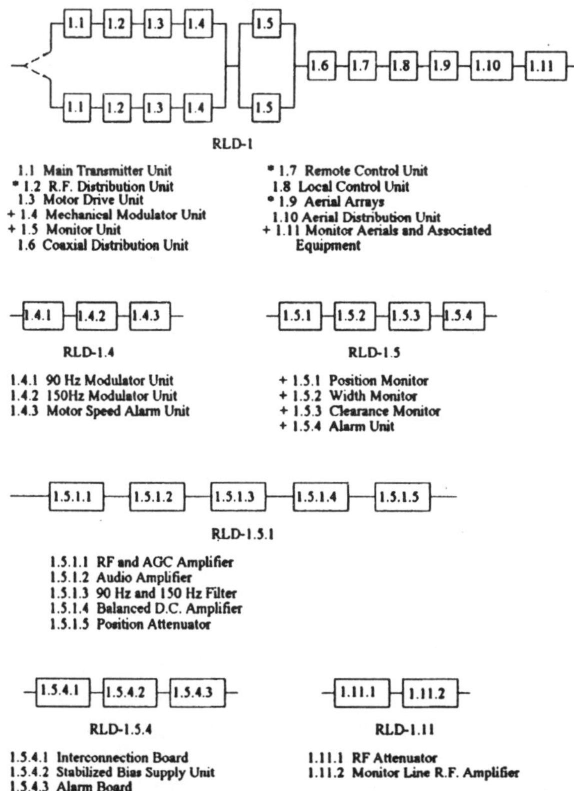

Fig.14.4 RLD for Localizer.

Fig.14.4. Some of the blocks (*) namely 1.2, 1.7 and 1.9 do not contribute to the failure of the equipment and are therefore not analyzed further. Some other blocks, namely 1.1, 1.3, 1.6, 1.8 and 1.10 are simple and their failure rates can be directly estimated by finding out the failure rates of the constituent components. Blocks such as 1.4, 1.5 and 1.11 require further decomposition in separate sub-blocks and are indicated by (+). The numbering of the blocks has been done in such a way that it clearly shows that this is the sub block of which particular block. The following points have been kept in view while analyzing Fig.14.4.

(i) RF distribution unit and localizer aerial arrays do not contribute to the failure rate of the system.

(ii) The components of the coaxial distribution unit have not been included in the analysis (based on experience) except for four switchover relays.

(iii) The remote control unit has only some switches and all other functions are confined to the local control unit only. Therefore, remote control unit is not considered in the reliability analysis.

(iv) In the local control unit , identity tone detectors have not been considered into reliability analysis as their failure does not result into the failure of the equipment.

(v) The failure rates of Aperture Monitor Combining unit in the Aerial Distribution Equipment and of the monitor dipoles in Monitor Aerials and Associated equipments have been taken as equal to zero.

(vi) In cat II system the stand by channel B comes into operation when the main channel A fails. In practice channel A is operated for some time, then channel B is operated for some time, then channel A and so on. Therefore, the effective failure rate of each channel would be the half of each channel's failure rate, calculated on the assumption of continuous operation.

(vii) Nearfield I monitor & Nearfield II monitor are in active parallel redundancy. Each monitor unit comprises of position monitor, width monitor, clearance monitor and alarm unit as shown in RLD - 1.5. Each of these three monitors comprises of five units as shown in RLD -1.5.1. This figure is drawn for position monitor. For other monitors, the first four units will remain the same and the fifth one will be replaced by appropriate attenuator (position/width/ clearance).

The failure rate calculations for localizer are shown in the respctive tables.

The failure rate given in these tables refer to the values per million parts and therefore are to be multiplied by 10^{-6}. They have been taken from Mil-HDK-217. The following notes will be helpful in understanding these tables.

(i) Reference Note No. has been included in the tables for each entry. Its significance is:

(a) Note No. 1 indicates that the value has been estimated using the Handbook.
(b) Note No. 2 indicates that the value has been estimated by referring to Part Stress Method in the Handbook.
(c) Note No. 3 indicates that the failure rate for this item has been calculated in another table. The numbers of the tables and the numbers in the Reliability Logic Diagrams are self explanatory.

(ii) Ground fixed environment (G_F) has been assumed for failure rate calculations.

(iii) Resistors are of carbon composition type. They have been assumed classified according to a style with 2 letters. For resistors and capacitors, commercial, non-mil quality has been assumed and the value of quality factor π_Q is taken as 3.

(iv) Diodes have been divided into two categories - General purpose (silicon) and Zener & Avalanche. Both these and transistors are assumed to be of non-mil hermetic type with $\pi_Q=5$.

(v) Connections of PCBs with coaxial cable are taken to fall in the category of coaxial connectors. Control panel with coaxial cable fall in the category of circular, rack & panel arrangement. Wiring Board connectors fall in the category of printed wiring Board. Sockets, Plugs, etc. are considered similar to coaxial type of connectors for failure rate estimation. Transformers are categorized into two types-Audio transformers and RF transformers. They are assumed to be of non-mil type and $\pi_Q=3$. Inductors are also assumed to have $\pi_Q=3$.

(vi) Switches are assumed to be of non-mil type . Key switch and jack switch are assumed to have the failure rate equal to that of push button type. π_Q is taken to be as 20. Master switch is assumed to be of toggle type and $\pi_Q=20$. For rotary switches $\pi_Q = 50$, for general purpose relays $\pi_Q = 6$.

(vii) Quartz crystal, fuses, lamps (neon and incandescent) are assumed to be of MIL-C-3098 specification and meters are assumed to be of MIL-M-10304 specification.

(viii) Warning devices, batteries and all the elements of Aerial Distribution unit except the resistors and capacitors are assumed to have zero failure rate.

14.151 Reliability expression and MTBF determination

Table 14.1 summarizes the failure rates of all the constituent units of localizer. These failure rates have been obtained as shown in the subsequent tables. The actual values for all components could not be reproduced for obvious reasons.

TABLE 14.1

Failure Rates for Units of Localizer

Sr.No	Name of the Component	Failure Rate
1.	Main Transmitter Unit	f_1
2.	R.F. Distribution Unit	f_2
3.	Motor Drive Unit	f_3
4.	Mechanical Modulator Unit	f_4
5.	Monitor Unit	f_5
6.	Coaxial Distribution Unit	f_6
7.	Remote Control Unit	f_7
8.	Local Control Unit	f_8
9.	Aerial Arrays	f_9
10.	Aerial Distribution Equipment	f_{10}
11.	Monitor Aerials and Associated Equipment	f_{11}

The block diagram is shown in Fig. 14.4. Let R_1 be the reliability for both the channels, each comprising of blocks 1.1 to 1.4. Let R_2 be the reliability for parallel combination of blocks 1.5. Let R_3 be the reliability of blocks 1.6 to 1.11 in series.

Then, the localizer reliability R_L is given by

$$R_L = R_1 * R_2 * R_3$$

$$R_1 = (1 + \lambda_1 * t) \exp(-\lambda_1 * t)$$

As each channel works for only about half the time,

$$\lambda_1 = (1/2)(f_1 + f_2 + f_3 + f_4)$$

Also,

$$R_2 = 2\exp(-f_5 * t) - \exp(-2f_5 * t)$$

TABLE 1.1
Main Transmitter Unit

Sr. No	Name of the Component	Ref. Note No	Qty.	Generic failure	Π_Q	Failure rate
1	Fixed resistor	1	112	0.0110	3	3.6960
2	Variable resistor	1	17	0.1400	3	7.1400
3	Fixed paper capacitor	1	11	0.0260	3	0.8580
4	Fixed ceramic capacitor	1	37	0.0180	3	1.9980
5	Fixed tantalum capacitor	1	22	0.2800	3	18.480
6	Variable air trimmer capacitor	1	13	1.9000	3	74.100
7	Silicon npn transistor	1	21	0.0160	5	1.6800
8	Silicon pnp transistor	1	8	0.0240	5	0.9600
9	Fixed inductor	1	21	0.0110	3	0.6930
10	Variable inductor	1	5	0.0230	3	0.3450
11	General purpose diode	1	15	0.0031	5	0.2325
12	Zener diode	1	6	0.0120	5	0.3600
13	R.F. Transformer	1	1	0.1500	3	0.4500
14	Quartz crystal	2	1	0.2000	-	0.2000
15	Thyristor	1	4	0.1000	5	2.0000
16	Rotary switch	1	2	0.9600	50	96.000
17	Socket	1	3	0.0170	3	0.1530
18	Plug	1	3	0.0170	3	0.1530
	Total					209.4985

The total failure rate for blocks 1.6 to 1.11 in series is given by :

$$\lambda_3 = f_6 + f_7 + f_8 + f_9 + f_{10} + f_{11}$$

Therefore,

$$R_3 = \exp(-\lambda_3 * t).$$

As the reliability expression is not a simple exponent, we derive the MTBF by integrating this expression from 0 to ∞. Therefore, MTBF for the localizer m_L is given by :

$$m_L = \int_0^\infty R_L \, dt$$

TABLE 1.3

Motor Drive Unit

Sr. No	Name of the Component	Ref. Note No	Qty.	Generic failure rate	Π_Q	Failure rate
1.	Fixed resistor	1	37	0.0110	3	1.2210
2.	Fixed ceramic capacitor	1	11	0.0180	3	0.5940
3.	Fixed tantalum capacitor	1	15	0.2800	3	12.600
4.	Silicon npn transistor	1	2	0.0160	5	0.1600
5.	Fixed inductor	1	1	0.0110	3	0.0330
6.	General purpose diode	1	16	0.0031	5	0.2480
7.	Zener diode	1	2	0.0120	5	0.1200
8.	R.F. Transformer	1	1	0.1500	3	0.4500
9.	Thyristor	1	6	0.1000	5	3.0000
10.	Jack switch	1	1	0.0029	20	0.0580
11.	Meter	2	1	10.000	-	10.000
12.	Lamp (Incandescent)	2	1	1.0000	-	1.0000
13.	General purpose relay	1	1	0.3300	6	1.9800
14.	Plug	1	3	0.0170	3	0.1530
	Total					31.5150

TABLE 1.4

Mechanical Modulator Unit

Sr. No	Name of the Component	Ref. Note No	Qty.	Generic failure rate	Π_Q	Failure rate
1.	90 Hz Modulator Unit	3	1	18.249	-	18.2490
2	150 Hz Modulator Unit	3	1	18.249	-	18.2490
3	Motor Speed Alarm Unit	3	1	11.760	-	11.7600
	Total					48.2580

TABLE 1.4.1

90 Hz Modulator Unit

Sr. No	Name of the Component	Ref. Note No	Qty.	Generic failure rate	Π_Q	Failure rate
1.	Fixed paper capacitor	1	7	0.0260	3	0.5460
2	Variable air trimmer capacitor	1	3	1.9000	3	17.100
3	R.F. Transformer	1	1	0.1500	3	0.4500
4	Socket	1	3	0.0170	3	0.1530
	Total					18.2490

14.16 Glidepath reliability calculations

The functional performance of the glidepath equipment of the ILS has already been described. Based on the functional performance we obtain the Reliability Logic Diagram for the Cat II system which is shown as RLD-2 in the Fig. 14.5.

The blocks in this diagram are numbered as 2.1 to 2.11. Some of the blocks (*) namely 2.2, 2.7,2.9 and 2.10 do not contribute to the failure of the equipment and are therefore not analyzed further. Some other blocks namely 2.1, 2.3, 2.6, 2.8, and 2.11 are simple and their failure rates can be directly estimated by using the failure rates of the constituent components. Blocks such as 2.4 and 2.5 are decomposed into various sub-blocks and it is indicated by(+).

The failure rate evaluation of the glidepath equipment has been carried out assuming the points as indicated in the case of localizer except for the following:-

(i) The number of switch over relays in the coaxial distribution unit is now 3 instead of 4.

(ii) In the local control unit, identity tone detectors are not used in this case.

(iii) Based on experience the aerial distribution unit is assumed failure free and therefore not included in the reliability analysis.

(iv) All the associated units except the RF amplifier in the monitor aerials and associated equipments have zero failure rate.

14.161 Reliability expression and MTBF determination

Table 14.2 summarizes the failure rates of all the constituent units of the glidepath equipment. These failure rates have been obtained as shown in the subsequent tables.

The block diagram is shown in Fig.14.5. Let R_1 be the reliability for both the channels, each comprising of blocks 2.1 to 2.4. Let R_2 be the reliability for parallel combination of blocks 2.5. Let R_3 be the reliability of blocks 2.6 to 2.11 in series . Then the glidepath reliability R_G is given by

$$R_G = R_1 * R_2 * R_3$$

$$R_1 = (1 + \lambda_3 * t) \exp(- \lambda_3 * t)$$

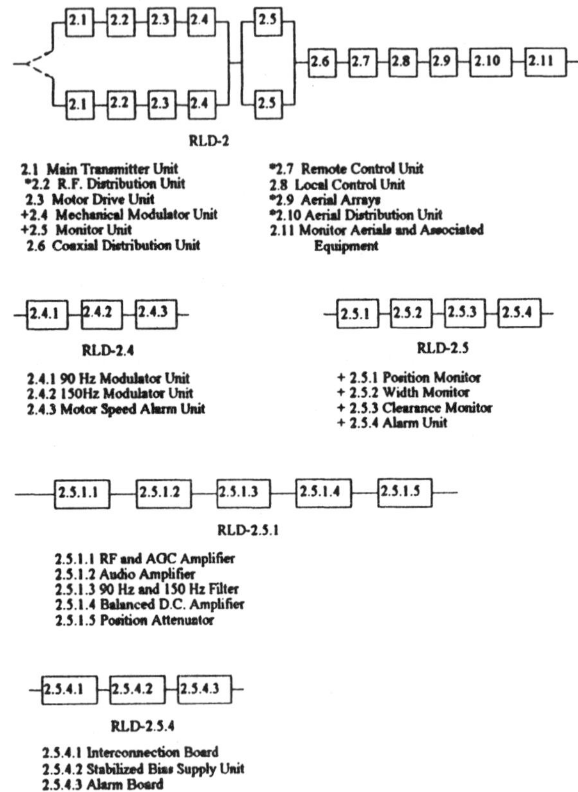

Fig.14.5 RLD for Glidepath.

As each channel works for only about half the time,

$$\lambda_3 = (1/2)\,(g_1 + g_2 + g_3 + g_4)$$

Also,

$$R_2 = 2 \exp(-g_5 * t) - \exp(-2\, g_5 * t)$$

The total failure rate for blocks 2.6 to 2.11 in series is given by :

$$\lambda_4 = g_6 + g_7 + g_8 + g_9 + g_{10} + g_{11}$$

Therefore,

$$R_3 = \exp(-\lambda_4 * t)$$

TABLE 14.2
Failure Rates for Units of Glidepath

Sr.No	Name of the Component	Failure Rate
1.	Main Transmitter Unit	g_1
2.	R.F. Distribution Unit	g_2
3.	Motor Drive Unit	g_3
4.	Mechanical Modulator Unit	g_4
5.	Monitor Unit	g_5
6.	Coaxial Distribution Unit	g_6
7.	Remote Control Unit	g_7
8.	Local Control Unit	g_8
9.	Aerial Arrays	g_9
10.	Aerial Distribution Equipment	g_{10}
11.	Monitor Aerials and Associated Equipment	g_{11}

As the reliability expression is not a simple exponent, we derive the MTBF by integrating this expression from o to ∞. Therefore, the MTBF for the glidepath m_G is given by

$$m_G = \int_o^\infty R_G \, dt$$

It may be observed that the localizer as well as the glidepath make use of active parallel as well as standby redundancy in some subsystems. Therefore, the failure rate will be the function of time. Hence, talking about a consolidated failure rate for these units is meaningless. Therefore, we have evaluated the reliability expressions and used them to evaluate the MTBF for these units.

14.2 BANKING SYSTEM

Software reliability measurement techniques are of great value in the evaluation of software engineering technology and in operational software management. Most of the project management applications relate to subsystem or system test phases. The fact that software reliability measurement is better developed for the later stages of the software life cycle is not a problem. On many projects system test represents 40 to 50 percent of pre- release costs. Improved decision making in this period can indeed have an impact! We discuss a typical case study for a banking system.

A bank desiring to set up the data network shown in Fig.14.6 hires an outside firm to design the hardware for the front end processor and system controller. In addition the firm is also contracted to develop the necessary software that will run on both of these processors. Wishing to monitor their

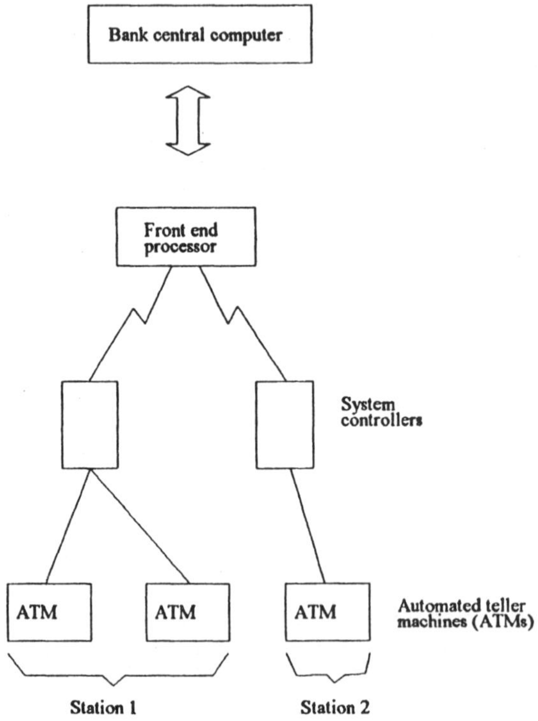

Fig.14.6 Bank data network.

progress towards a failure intensity objective for the system controller software, the firm decided to use a software reliability model. They will collect failure interval data during system test. In practice the firm would also monitor the reliability of the front end processor software. Combining the failure interval data with estimates for the amount of available resources and some project parameters, a report similar to that shown in Fig.14.7 can be generated. The report shows the most likely value of the indicated quantity in the center. The lower and upper confidence bounds (for various confidence limits) are shown sandwiched around it. For example, we are 75 percent confident that the present failure intensity is between 0.079 and 0.125 failure/CPU hr. The *completion date* refers to the date of meeting the failure intensity objective. It is in month-day-year format.

The 75 percent confidence interval has been found from experience to be a good compromise between a higher confidence and the resultant larger interval.

SOFTWARE RELIABILITY PREDICTION
BASIC MODEL
DATCOM EXAMPLE

BASED ON SAMPLE OF 180 TEST FAILURES
EXECUTION TIME IS 450.57 HR
FAILURE INTENSITY OBJECTIVE IS 0.20e-03 FAILURES/CPU HR
CALENDAR TIME TO DATE IS 180 DAYS
PRESENT DATE 9/1/86

	CONF. LIMITS				MOST	CONF. LIMITS			
	95%	90%	75%	50%	LIKELY	50%	75%	90%	95%
TOTAL FAILURES	190	191	193	196	199	204	208	213	217
FAILURE INTENSITIES (FAILURES/1000 CPU HR)									
INITIAL F.I	850.1	878.6	924.1	968.8	1033	1099	1147	1197	1230
PRESENT F.I	66.9	71.5	79.1	87.2	99.8	114.0	125.0	137.4	145.8
ADDITIONAL REQUIREMENTS TO MEET FAILURE INTENSITY OBJECTIVE									
FAILURES	10	11	13	15	19	24	28	33	37
EXEC. TIME (CPU HR)	899.6	939.5	1007	1080	1197	1336	1450	1586	1684
CAL. TIME (DAYS)	36.0	37.9	41.4	45.3	51.8	59.8	66.7	75.2	81.5
COMPLETION DATE	100786	100986	101086	101786	102386	103186	110786	111686	112286

Fig. 14.7 Sample project status report.

A software reliability model can be used to help answer managerial questions relating to project status and scheduling. Three such questions are:

1. Is this software ready for release?
2. When will it be ready for release?
3. Should we regress to a previous version from present?

To answer the first question, compare the present failure intensity with the project's failure intensity objective. The question regarding when the software will be ready for release can be answered by observing the completion date line in Fig.14.7. We can determine whether we should regress to a previous version by tracking present failure intensity for each version. If the new version is not meeting the failure intensity objective and the old one is, and the difference between the failure intensities is substantial, it will probably be worth the effort to regress.

The model can help the manager, through simulation, reach trade-off decisions among schedules, costs, resources, and reliability and can assist in determining resource allocations. One chooses several values of each parameter that is to be varied, applies the model to compute the effects, examines the results, and iterates this procedure as required.

To illustrate, consider the DATCOM project system test period report (Fig.14.7) for September 1,1986. The project had a test team of 3 and a debugging team of 5 people. A failure intensity objective of 0.0002 failure/CPU hr for the system controller software was set originally. The estimated completion date was expected to be October 23, with a 75 percent confidence interval of October 13 to November 7. Assuming this date to be unsatisfactory, the effect of several different managerial actions on the schedule can be studied:

1. increasing the failure intensity objective,
2. working up to 50% overtime (as required to expedite the project),
3. increasing the size of the test team,
4. increasing the size of the debugging team, or
5. making more computer time available.

We will present the results of the studies here to show their usefulness. It is assumed that the increases of actions 3 and 4 are made by reallocating experienced people from other parts of the project so that negligible training time is involved.

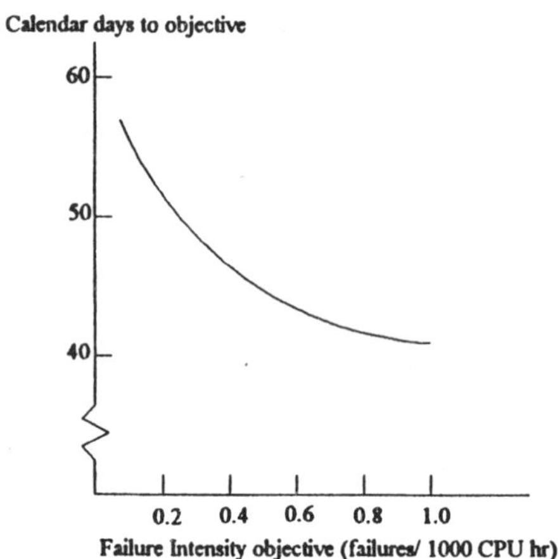

Fig.14.8 Effect of failure intensity objective on predicted completion date for bank project.

The effect of increasing the failure intensity objective on the schedule is illustrated in Fig.14.8. The effect of varying the failure intensity objective on additional cost is illustrated in Fig.14.9. *Additional cost* represents the testing and debugging cost required to reach the objective.

The effects of resource changes are indicated in Fig.14.10. Overtime and additional personnel have negligible effect on schedules for this project at this point in its history. Making more computer time available has a substantial effect. This indicates that this project is essentially though not completely limited by the computer time available. If the manager can make both more computer time available and back off on the failure intensity objective, it may be possible to change the estimated completion date even further. Note that we cannot generalize these conclusions to other projects. Effects of resource changes are specific to a particular project and time.

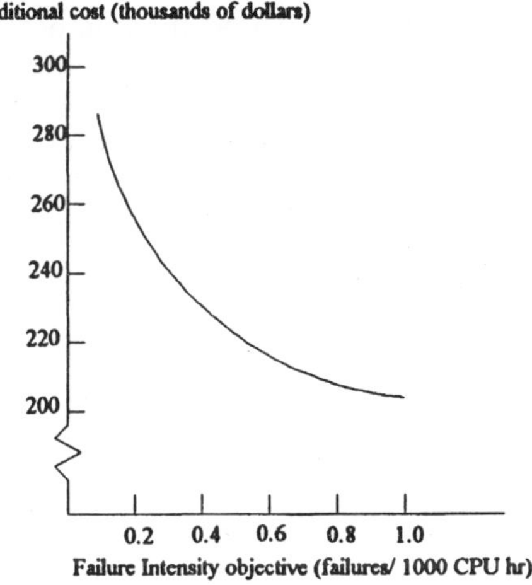

Fig. 14.9 Effect of failure intensity objective on additional cost for bank project.

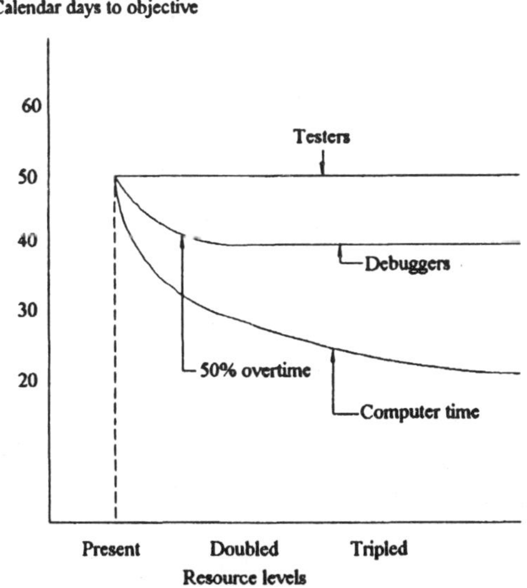

Fig. 14.10 Effect of resource levels on predicted completion date for bank project.

We have the capability to combine component reliabilities to determine system reliability. Consider the event diagram shown in Fig.14.11 for our bank data network system. It presents the view of a user at station 1 of how the functioning of components affects the functioning of the system. The view from station 2 is similar except that there is only one ATM. The bank's central computer is assumed to have a hardware reliability and software reliability of 1 for simplicity in presenting this example. The ATMs have a reliability of 0.995 for a 24-hr period. The front end processor and system controller hardware have 24-hr reliabilities of 0.99 and 0.995, respectively. The front end processor and system controller software each have a failure intensity in execution time of 0.004 failure/CPU hr. The front end processor utilization is 0.95. The system controller utilization is 0.5. We wish to find the reliability of the system for a 24-hr period as seen from station 1 and from station 2.

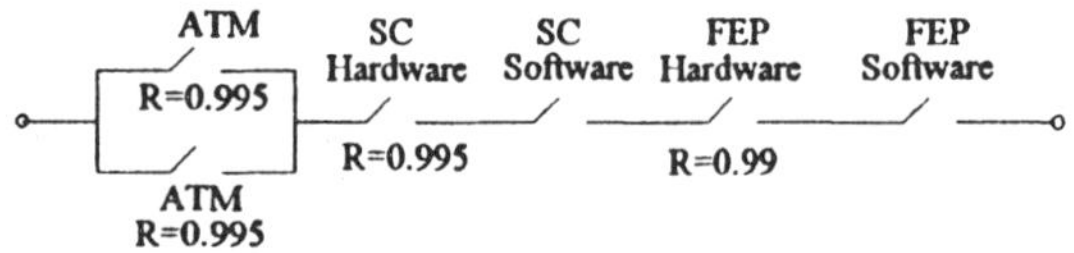

Fig.14.11 Bank data network failure event diagram (as seen from station 1).

The calendar time failure intensities for the front end processor and system controller software will be 0.0038 failure/hr and 0.002 failure/hr, respectively. The 24-hr reliabilities can be calculated, using a standard formula for relating failure intensity and reliability, as 0.913 and 0.953. The overall 24-hr period reliability as seen from station 1 is calculated to be 0.857 and that from station 2 turns out to be 0.853. If the bank considers this unacceptable, improvements should be made first in the front end processor software and then in the system controller software.

PROBLEMS

1. An equipment obeying exponential law of reliability has 97 percent probability of survival in first 100 hours of operation. What is the probability of its survival in:

   (a) First 200 hours operation?
   (b) Post 100 hours of operation provided it has survived for the 900 hours of the 1000 hours of useful life?

2. An engine shaft has a failure rate of 0.5×10^{-7}/hr. The shield used with the shaft has a failure rate of 2.5×10^{-7} /hr. If a given company has 5000 engines with these shafts and shields and each engine operates for 350 days of useful life. Estimate the number of shafts and shields that must be replaced annually.

3. The reliability R(t) of an item is assumed to be exponentially decreasing function :

   $R(t) = \exp(-t/10^{-4} \text{ days})$

   What is the probability that the item will still be functioning without failure at t = 300 days, given that the unit functioned without failure at t = 100 days ?

4. It is required to produce a device having a reliability of at least 95 % over a period of 500 hr. Estimate the maximum permissible failure rate and minimum MTBF.

5. A home computer manufacturer determines that his machine has a

constant failure rate of $\lambda = 0.4$/year in normal use. For how long should the warranty be set if no more than 5% of the computers are to be returned to the manufacturer for repair?

6. A device has a constant failure rate of 0.7/ year.

   (a) What is the probability that the device will fail during the second year of operation?
   (b) If upon failure the device is immediately replaced, what is the probability that there will be more than one failure in 3 years of operation?

7. Somebody wants to take a 1000 km trip by car. The car has a constant failure rate of $\lambda = 10^{-4}$ per kilometer travelled. What is the probability that the destination is reached without the car breaking down?

8. The weather radar system of an airliner has an MTTF of 1140 hours. Assuming that the failure rate is constant, solve the following problems:

   (a) What is the probability of failure during a 4-hour flight?
   (b) What is the maximum duration of a flight such that the reliability may not drop below 0.99?

9. The failure rate for a certain type of component is $\lambda(t) = \lambda_0 t$ where $\lambda_0 >> 0$ and is constant. Find its reliability, mortality and MTBF.

10. Two types of components with identical electrical characteristics have different failure rates. The failure rate of component A is 1%/ 1000 hrs. and that of B is $10^{-6}t$, where t is in hours. Which component is more reliable for an operating time of (i) 100 hrs (ii) 100 mts ?

11. An engineer approximates the reliability of a cutting assembly by

$$R(t) = \begin{cases} (1-t/t_0)^2 & , 0 \leq t < t_0 \\ 0 & , t \geq t_0 \end{cases}$$

   (a) Determine the failure rate.
   (b) Does the failure rate increase or decrease with time?
   (c) Determine the MTTF.

12. Define failure rate and express it in terms of reliability. The reliability expression for a system is given as:

$$R(t) = \exp[-(a+bt)t]$$

Determine the expression for $\lambda(t)$.

13. The PDF for the time to failure of an appliance is

$$f(t) = \frac{32}{(t+4)^3}, \quad t > 0,$$

where t is in years.

(a) Find the reliability R(t),
(b) Find the failure rate $\lambda(t)$,
(c) Find the MTTF.

14. A sample of 150 electronic components is subjected to testing (presumably in useful life). Three failures are found at the end of 400 hours, four more at the end of 800 hours, two more at the end of 1200 hours, four more at the end of 1800 hours and no further failures are found when the test is terminated at 2500 hours.

(a) Estimate the MTBF if failed components are replaced when found.
(b) Estimate MTBF if no replacements are made.

15. For the reliability analysis, 300 diodes were placed for a life test. After 1500 hr, 16 diodes had failed and test was stopped. The times at which failures occurred are: 115, 120, 205, 370, 459, 607, 714, 840, 990, 1160, 1188, 1300, 1380, 1414, 1449 and 1497 hrs. Determine the failure rate of the diodes.

16. A sample of 200 resistors is subjected to testing (presumably in the useful life period). Five failures are found at the end of 500 hours. Six more at the end of 800 hours, 2 more at the end of 1200 hours, 6 more at the end of 1800 hours and no further failures are found when the test is terminated at the end of 2400 hours.

(a) Estimate the MTBF if failed resistors are replaced when found.
(b) Estimate the MTBF if no replacements are made.

17. Twenty small generators were put under test for a period of 1500 hours. One generator failed at 400 hours and was replaced by new one. A second failed at 500 hours and was also replaced. A third and fourth failed at 550 and 600 hours, respectively, and were removed

from testing, but were not replaced. A fifth malfunctioned at 700 hours was immediately repaired, and was put back into test. A sixth malfunctioned at 800 hours but was kept in test. Later analysis showed this failure was due to governor malfunction. Estimate the failure rate of the generators.

18. Ten units are placed on life test, and the failure times are 9, 19, 27, 35, 40, 46, 50, 55, 56, 60 hr. Plot f(t), λ(t), Q(t) and R(t).

19. In the life-testing of 100 specimens of a particular device, the number of failures during each time interval of twenty hours is shown in Table below. Estimate the MTTF for these specimens.

TABLE

Time Interval Hours	Number of failures during the interval
$T \leq 1000$	0
$1000 < T \leq 1020$	25
$1020 < T \leq 1040$	40
$1040 < T \leq 1060$	20
$1060 < T \leq 1080$	10
$1080 < T \leq 1100$	5

20. In a well-shuffled deck of 52 playing cards, what is the probability that the top card is:

(a) A diamond (b) a black card, and (c) a nine ?

21. The PDF of the lifetime of an appliance is given by

$f(t) = 0.25t\, e^{-0.5t} \quad , t \geq 0,$

where t is in years.

(a) What is the probability of failure during the first year?
(b) What is the probability of the appliance's lasting at least 5 years?
(c) If no more than 5% of the appliances are to require warranty services, what is the maximum number of months for which the appliance can be warranted?

22. A device is put into service on a Monday and operates seven days each week. Each day there is a 10% chance that the device will break down. (This includes the first day of operation). The maintenance crew is not available on weekends, and so the manager

hopes that the first breakdown does not occur on a weekend. What is the probability that the first breakdown will occur on a weekend?

23. A man and his wife appear for an interview for two posts. The probability of husband's selection is 1/7 and that of the wife's selection is 1/5. What is the probability that only one of them will be selected ?

24. A president is to be elected from the membership of a political organization which has 100 members. If the ratio of male to female is 4:1 and half of both men and women are married, what is the probability that

(a) the president is a man,
(b) the president is a married woman, and
(c) the president is a married man or married woman?

25. Consider the following if statement in a program:

if B then s_1 else s_2

The random experiment consists of 'observing' two successive executions of the if statement. The sample space consists of four possible outcomes:

$$S = [(s_1,s_1), (s_1,s_2), (s_2,s_1), (s_2,s_2)] = [t_1, t_2, t_3, t_4]$$

Assume the following probability assignment:

$$Pr(t_1) = 0.34, Pr(t_2) = 0.26, Pr(t_3) = 0.26, Pr(t_4) = 0.14$$

Determine the probability of the following events:

(a) At least one execution of the statement s_1.
(b) Statement s_2 is executed the first time.

26. A company producing electric light bulbs has an annual inspected output of 7.8 million bulbs and its inspection department is assessed as having a reliability of 0.9. A particular customer buys a batch of 4500 light bulbs from this company in which he finds that 9 are faulty. On the basis of these data, what is the estimate of the average number of bulbs which the company rejects each year in the inspection department?

27. A binary communication channel carries data as one of two types of

signals, i.e. ones or zeros. A transmitted zero is sometimes received as a one and a transmitted one is sometimes received as a zero because of noise. For a given channel, assume a probability of 0.94 that a received zero is a transmitted zero and a probability of 0.91 that a received one is a transmitted one. Further assume a probability of 0.45 of transmitting a zero. If a single signal is sent, determine:

(a) Probability that one is received,
(b) Probability that a zero is received,
(c) Probability that a one was transmitted if a one was received.

28. A cinema house gets electric power from a generator run by diesel engine. On any day, the probability that the generator is down (event A) is 0.025 and the probability that the diesel engine is down (event B) is 0.04. What is the probability that the cinema house will have power on any given day? Assume that occurrence of event A and event B are independent of each other.

29. A has one share in a lottery in which there is one prize and two blanks ; B has three shares in a lottery in which there are three prizes and 6 blanks; compare the probability of A's success to that of B's success.

30. Four persons are chosen at random from a group containing 3 men, 2 women and 4 children. Calculate the chances that exactly two of them will be children.

31. A manufacturing concern specializing in high-pressure relief valves subjects every valve to a particular acceptance test before certifying it as fit for use. Over a period of time, it is observed that 95% of all valves manufactured pass the test. However, the acceptance test adopted is found to be only 98% reliable. Consequently, a valve certified as fit for use has a probability of 0.02 of being faulty. What is the probability that a satisfactory valve will pass the test?

32. A certain firm has plants A, B and C producing respectively, 35%, 15% and 50% of the total output. The probabilities of a non defective product are 0.75, 0.95, and 0.85 respectively. A customer receives a defective product. What is the probability that it came from plant C ?.

33. In a survival test involving mechanical valves, 1000 valves are tested. On the average, 822 valves survive 400 cycles of operation, and out of these, 411 valves survive 700 cycles of operation. What is the probability of a valve surviving 400 cycles as well as 700

cycles of operation? What is the probability that a valve will survive 700 cycles of operation if it has already survived 400 cycles of operation?

34. A given lot of small devices is 98 percent good and 2 percent defective. To be certain of using a good device, each device is tested before installation. The tester itself is not totally reliable since it has the following conditional probabilities:

P(says good/actually good) = 0.95
P(says bad/actually bad) = 0.95

A device is tested with the tester indicating the device is bad. What is the probability the device is actually bad?

35. An anti-aircraft gun can take a minimum of four shots at an enemy plane moving away from it. The probability of hitting the plane at first, second, third, and fourth shots are 0.4, 0.3, 0.2, and 0.1 respectively. What is the probability that the gun hits the plane ?

36. A device has a random failure rate of 20 failures/10^5 hour for an operating period of 300 hours, what is the probability of

(a) No failure
(b) One failure
(c) Two failures
(d) Two failures or less
(e) More than two failures.

37. Four identical electronic units are connected in parallel. Each has a reliability of 0.9. Estimate the probability of 0, 1, 2, 3, and 4 of these units remaining operative.

38. It is found that the number of system breakdowns occurring in a given length of time follows a Poisson distribution with a mean value of a 2 break-downs. What are the probabilities, in the same length of the time, of the system having

(a) no break-down
(b) 1 break-down
(c) 2 break-downs
(d) 10 break-downs
(e) Less then 3 break-downs
(f) three or more than 3 break-downs.

39. An illuminated mimic diagram in a plant control room has 150 nominally identical bulbs which are required to be permanently illuminated. If the probability of any one bulb being out at any one time is 0.01, what is the probability of

(a) at least 5 bulbs being out,
(b) not more than 3 bulbs being out,
(c) exactly 4 bulbs being out?

40. Verify that the function defined by $f(t) = 0.1e^{-.25t} + 0.06e^{-0.1t}$ for all number $t>0$, with $f(t) = 0$ for $t<0$, is a density function and find the expected value of a random variable having such a density function.

41. The time (measured in years), X, required to complete a software project has a pdf of the form:

$$f_X(x) = \begin{cases} kx^2(1-x^3), & 0 \le x \le 1 \\ 0 & \text{otherwise} \end{cases}$$

Determine the probability that the project will be completed in less than 4 months.

42. A device has a lifetime which is known to be an exponential random variable X with $E(X) = 10$ years. Find the value of t_o for which the probability is exactly 1/2 that the device lasts at least t_o years; that is, $P(X>t_o) = 1/2$.

43. A simple resistive element of fixed value 10 ohms is known to obey Ohm's law. The current flowing through this element is randomly distributed according to a rectangular distribution which has lower and upper limits of 4 A and 6 A respectively. What is the probability that the voltage developed across the element meets the requirement of being at least 45V?

44. A room is lit by five nominally identical lamps. All lamps are switched on together and left switched on. It is known that the times to lamp failures after they have been switched on is rectangularly distributed between a lower limit of 8000 hr and an upper limit of 12,000 hr. What is the mean time to the room being in darkness? How would this mean time be affected if the number of lamps was increased to a total of 15?

45. A delicate electronic clock, which is circular in shape, is to be housed in a box with a diametrical clearance of not less than 2.0 mm and

not greater than 7.0 mm. This clearance is provided for cooling purposes. The radius of the clock is a random variable following a normal probability law with a mean of 20.0 cm and a coefficient of variation of 1%. The manufacturing process adopted to produce the housing results in making the inner radius of the box also a random variable following a normal probability law with a mean of 20.2 cm and a coefficient of variation of 2%. Evaluate the probability that the specified clearance will be met for a clock and its housing.

46. An electronic amplifier, when normally functioning, is found to have random variations in power output from all causes which follow a rectangular distribution between the limits of 45 mW and 55 mW. In addition, the amplifier has a probability at any time of 10^{-2} of being in the catastrophic or completely unavailable state where the power output is effectively zero. What is the reliability of the amplifier in meeting a requirement for the power output to be greater than 47 mW?

47. A control system has a power output measured in watts, W, which as a result of variations in the elements within the system, is randomly distributed with respect to time according to the p.d.f. fw(W) where:

$$\begin{aligned}
fw(W) &= 0 && \text{for } 0 \le W \le 42.5 \\
fw(W) &= 0.032W - 1.36 && \text{for } 42.5 \le W \le 45 \\
fw(W) &= 0.08 && \text{for } 45 \le W \le 55 \\
fw(W) &= 1.84 - 0.032W && \text{for } 55 \le W \le 57.5 \\
fw(W) &= 0 && \text{for } 57.5 \le W < \infty.
\end{aligned}$$

Draw the shape of this p.d.f. and calculate the reliability of the control system if the requirement for the power output at a particular time is (a) that it should be between 45 W and 57 W, (b) that it should be between 43 W and 57 W and (c) that it should be less than 55 W.

48. A given component has an MTBF of 10^6 hr, what is the reliability for an operating period of 10 hr for 5 such components in series ?

49. A regulated power supply consists of a step down transformer, rectifier, filter and a regulator. The constant failure rates of these components are :

Transformer	:	1.56% failures/1000 hours
Rectifier	:	2.00% failures/1000 hours
Filter	:	1.70% failures/1000 hours

Regulator : 1.40% failures/1000 hours

Determine the reliability of this supply if it is required to operate for (1) 500 hours (2) 1000 hours (3) 1500 hours. Comment on reliability vs hours of operation . What is the failure rate of total supply unit ?

50. A manufacturer of 16K byte memory boards finds that the reliability of the manufactured boards is 0.98. Assume that the defects are independent.

(a) What is the probability of a single byte of memory being defective?
(b) If no changes are made in design or manufacture, what reliability may be expected from 128K byte boards?

51. An electronic amplifier is made up of 50 resistors, 20 capacitors, 10 transistors, 10 diodes and 10 variable resistors. Every component must be in the working state for the amplifier to be working. Each component has times to failure which follow an exponential distribution with mean values of 5×10^6 hr, 2×10^6 hr, 10^6 hr, 10^7 hr and 5×10^5 hr for resistors, capacitors, transistors, diodes and variable resistors respectively. What is:

(a) the mean time to failure of the amplifier and
(b) the probability that the amplifier has failed by a time of 100 hr?

52. A certain component has a failure rate of 4×10^{-8}/hr in the on- state and a failure rate of 4×10^{-9}/hr in the off-state. On average, over the life of this component, it is only 25% of the time in the on-state. What is the effective failure rate of this component?

53. A measurement system consists of a sensor unit and an indicator unit. Starting from time zero, the times to failure for each unit are exponentially distributed. The mean time to failure for the sensor is 6000 hr and that for the indicator is 3000 hr. If either unit fails the system remains in the failed state. What is:

(a) the mean time to system failure,
(b) the probability of the system being in the failed state after 1000 hr, and
(c) the probability of the system being in the successful state after 4000 hr ?

54. A system is composed of 5 identical independent elements in parallel. What should be the reliability of each element to achieve a

system reliability of 0.96 ?

55. Four capacitances of 25 μF each are connected in parallel to act as a single capacitance of 100 μF required for the successful operation of a unit. If the reliabilities of these capacitors are 0.6, 0.7, 0.8, and 0.9 respectively, find the reliability of the unit.

56. A solid fuel booster engine has been test fired 2760 times. On 414 occasions the engine failed to ignite. If a projectile is fitted with three identical and independent booster engines of this type, what is the chance on launching of the projectile that,

(a) all three engines fail to ignite,
(b) at least one of the engines fails to ignite ?

57. The reliability function for a relay is $R(t) = \exp(-\lambda K)$ where K is the number of cycles and $\lambda = 10^{-4}$/cycle. A logic circuit uses 10 relays. The specific logic circuit used is unknown. What range should K have for the system reliability to be 0.95 ?

58. A 10kW power supply system is to be designed. The following alternatives are available:

1. One single generator of 10kW rating with a failure rate of 0.20 per 1000hr,
2. two generators each rated for 10kW and with a failure rate of 0.25 per 1000 hr, and
3. three generators each rated for 5kW and with a failure rate of 0.20 per 1000 hr.

Which of the alternatives would you select ? Why ? Assume that the redundant units are statistically independent.

59. Two circuit breakers of the same design each have a failure-to- open-on-demand probability of 0.02. The breakers are placed in series so that both must fail to open in order for the circuit breaker system to fail.

What is the probability of system failure

(a) If the failures are independent, and

(b) If the probability of a second failure is 0.1, given the failure of the first ?

(c) In part (a) what is the probability of one or more breaker failures on demand ?

(d) In part (b) what is the probability of one or more failures on demand ?

60. A microprocessor system consists of the following units:

Unit	Number	Failure rate
Processor	1	λ_P
Main Memory Unit	3	λ_{MM}
Disk Controller	1	λ_{DC}
Disk Drive	4	λ_{DD}
Video Terminal	1	λ_{VT}

each with a constant failure rate. The system configuration is shown in Fig. For the system to operate, the processor, terminal and disk controller must function together with two of the memory units and three of the disk drives. Obtain an expression for the reliability of the system and the system MTBF.

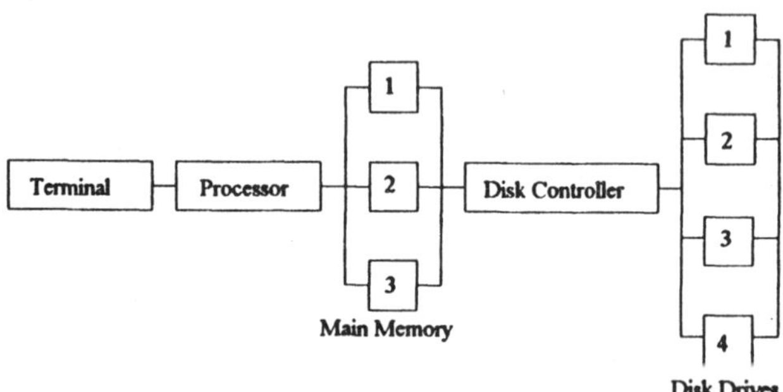

61. An equipment consists of 100 tubes. Twenty of these are connected functionally in series (branch A). This branch A is in turn connected in series to a parallel combination of branches B and C. The branch B and C contain 20 and 60 tubes respectively, connected functionally in series. The reliability of each tube in branch A, B and C respectively is $p_a = 0.95$, $p_b = 0.93$, and $p_c = 0.96$. Calculate the overall reliability of the equipment.

62. Three generators, whose data is given below, are connected in parallel. Determine the reliability of the system if the required load is 100kW.

Generator	Capacity	Reliability
1	50kW	0.98
2	100kW	0.97
3	50kW	0.99

63. A critical measuring instrument consists of two sub-systems connected in series. Sub-systems A and B have reliabilities 0.90 and 0.92, respectively, for a certain operating time. It is necessary that the reliability of the instrument be raised to a minimum value of 0.917 by using parallel sub-systems of A alone. Determine how many units of A should be used with one B to get a minimum reliability value of 0.98. What is the actual reliability value obtained?

64. A PC/XT has the following units with their respective failure rates in (%/1000 hrs.) as indicated:

i	CPU(incl. RAM and interfaces)	1.0
ii	Co-processor	2.0
iii	Key Board	0.8
iv	VDU	2.5
v	Hard Disc	3.0
vi	Floppy Drive 1	1.5
vii	Floppy Drive 2	1.5
viii	Printer	3.5

(a) Determine the reliability of each unit for 2,000 hrs. of operation.

(b) Determine the reliability of the system and MTBF if only one floppy drive is sufficient.

(c) How is the reliability of the system and MTBF modified if we consider that the Co-processor is used only 40% of the time, and printer is used only 20% of the time.

65. The circuit in the following picture shows a battery, a light, and two switches for redundancy. The two switches are operated by different people, and for each person there is a probability of 0.9 that the person will remember to turn on the switch. The battery and the light have reliability 0.99. Assuming that the battery, the light, and the two people all function independently, what is the probability that the light will actually turn on?

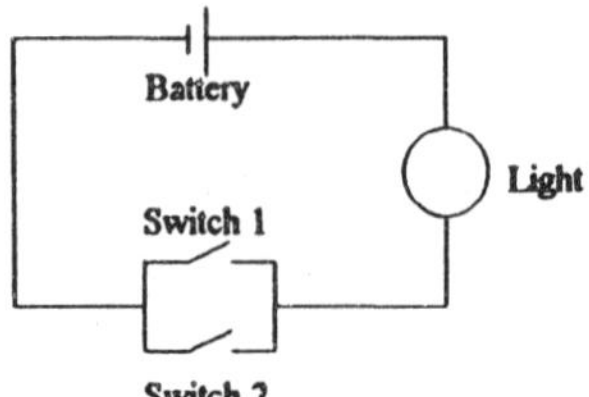

66. A computer system has three units as shown in Fig. Their reliabilities are as follows:

Card reader	= 0.89
Central processing unit(CPU)	= 0.98
Line printer	= 0.85

Determine the system reliability. If you want the system reliability to be not less than 0.95, what steps would you take? Draw the improved system diagram and calculate its actual reliability.

67. A system consists of three components in series, each with a reliability of 0.96. A second set of three components is purchased and a redundant system is built. What is the reliability of the redundant system (a) with high-level redundancy, (b) with low-level redundancy?

68. Given the following component reliabilities, calculate the reliability of the two systems.

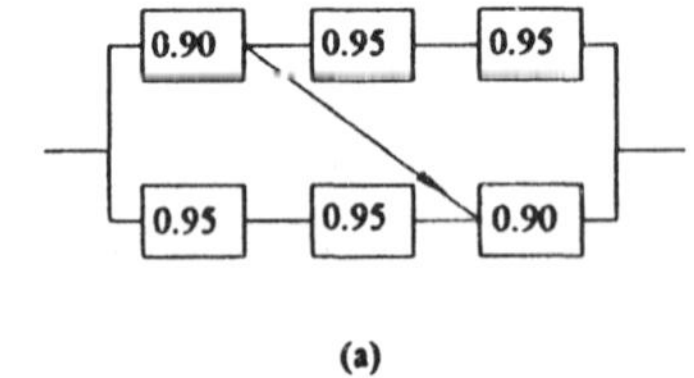

(a)

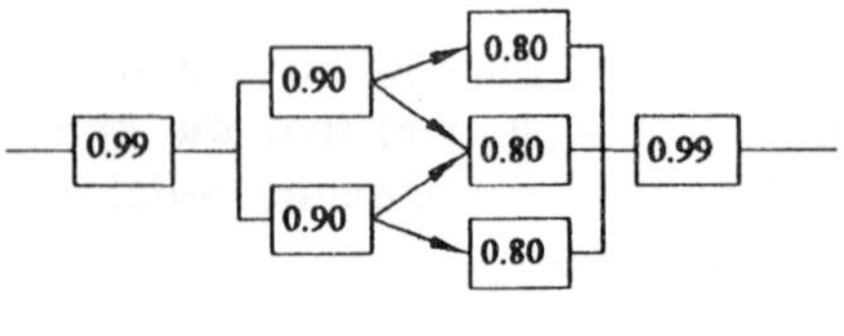

(b)

69. Four elements of a system each have a constant probability of 0.1 of being in the failed state at any time. What is the system probability of being in the failed state if the elements are so connected that system successes is achieved when :

(a) any 1 or more of the 4 elements are successful,
(b) any 2 or more of the 4 elements are successful,
(c) any 3 or more of the 4 elements are successful, and
(d) only all 4 elements are successful?

70. An electronic equipment comprises five active, independent, and identical units. The equipment will only operate successfully if at least three of the units are operating normally. Each unit has a constant failure rate, $\lambda = 0.004$ failure/hr. Calculate the system mean time to failure.

71. A 10-elements system is constructed of independent identical components so that 5 out of the 10-elements are necessary for system success. If the system reliability R must be 0.99, how good must the components be ?

72. Assume a designer has the freedom to use as many elements in parallel as he wishes. If an element has a reliability of 0.6 over a fixed time interval, determine the minimum number of parallel elements he must use to achieve a unit reliability of at least 0.95 for the following two cases:

(a) Successful unit operation will result if at least one element operates.
(b) If at least two parallel elements must operate.

73. If the level of stress changes during a mission, then the failure rate also changes. At take off, for example, an aircraft engine has to generate a greater torque to get the higher engine thrust required. At cruising altitude and speed, torque requirements are reduced. Assume the stress profile of an aircraft flight is as shown:

(a) Find an expression for reliability of a single engine for one flight.

(b) Assume a four engine aircraft. If all four engines are required for takeoff and climb, but only two out of four are required for completing the flight, determine the entire system reliability for one flight.

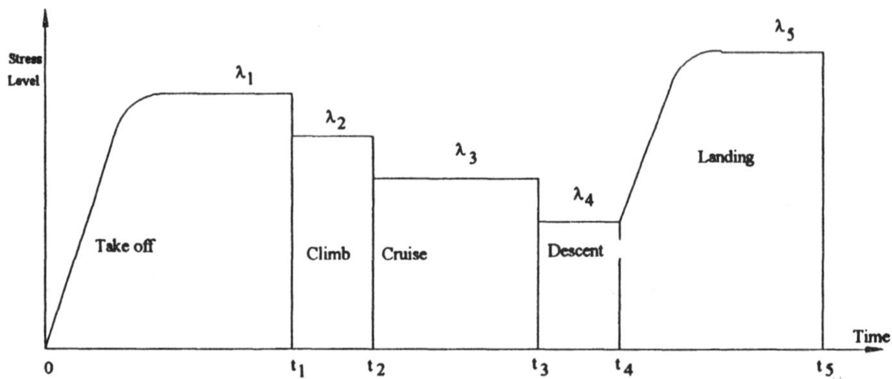

74. A pipeline carrying fluid has two valves as shown below. Draw the reliability logic diagram if

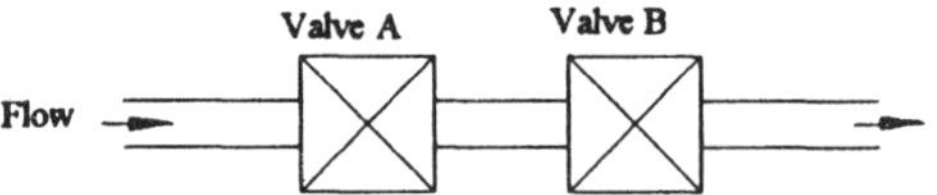

(a) both of them are normally closed and expected to open when required to permit flow, and
(b) both of them are normally open and expected to close to block the flow.

75. In an electronic circuit a diode function is necessary. To increase the reliability of this diode function one wants to apply active redundancy. One can afford, however, no more than three diodes. The diodes used may exhibit both open failures and short-circuit failures, the associated probabilities are:

-Open failure p_o = 0.02,
-Short circuit failure p_s = 0.01

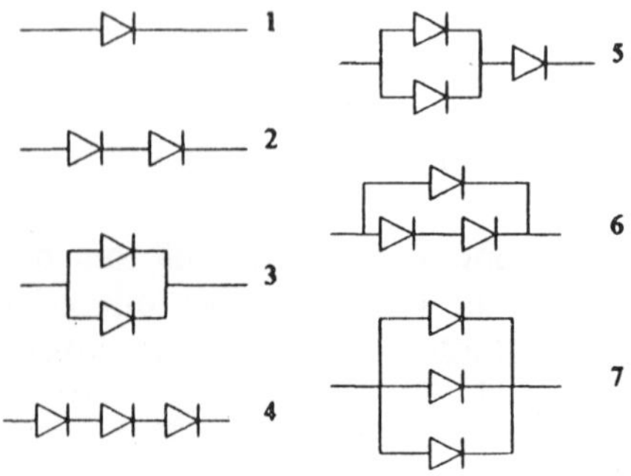

The diodes fail stochastically independent. Indicate for which of the circuits below the reliability is maximal and motivate your answer.

76. An electronic system can fail in two mutually exclusive failure modes, i.e., type I (open mode) and type II (short mode). The open and short modes constant failure rates are $\lambda_1 = 0.002$ and $\lambda_2 = 0.004$ failures/ hour, respectively. Calculate the value of the following items for a 100 hr mission:

1. System reliability
2. Open mode failure probability
3. Short mode failure probability

77. The failure probabilities of a diode subject to double failure are related as follows:

$$q_o + q_s = 0.2$$
$$q_o/q_s = 0.3$$

What type of redundancy would you apply in order to increase the reliability ? What is the maximum reliability possible and number of diodes required ? If the ratio $q_o/q_s = 1.5$, how would it affect the redundancy design ? Determine the maximum system reliability and the number of diodes required.

78. The identical components of the system below have fail-to-danger probabilities of $p_d = 10^{-2}$ and fail-safe probabilities of $p_s = 10^{-1}$.

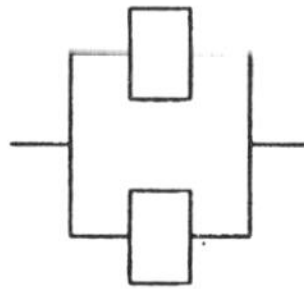

(a) What is the system fail-to-danger probability?
(b) What is the system fail-safe probability?

79. A small nuclear research reactor has three absorber rods which are suspended above the reactor and are designed to drop into the reactor core and shut the reactor down in the event of any untoward incident. The three rods are designated A, B and C and it has been found that the probability of each of these rods failing to drop on demand is $p_a = 0.005$, $p_b = 0.01$ and $p_c = 0.001$. If it is known that any two or more of three rods entering the reactor core will

safely shut the reactor down, what is the probability of failing to shut the reactor down when called upon to do so?

80. A system has MTBF of 200 hrs. Calculate the 100 hr. reliability of a system in which one such unit is operative and two identical units are standing by.

81. Two components each with an MTBF of 1000 hours are connected in (1) active (2) standby redundancy. Determine the overall system MTBF.

82. The failure rate of a device is constant equal to 0.06×10^{-3} per hr. How many standby devices are required to achieve a reliability of more than 0.985 for an operating period of 10,000 hrs? What is the MTTF of the resulting system ?

83. A d.c. generator has a failure rate of 0.0002 failures/hour. In case of its failure, a battery is used as a standby whose failure rate is 0.001 failure /hour when in operation. Find the reliability and MTBF for this system for a mission time of 10 hours assuming perfect sensing and switching. If the sensing and switching device has a 99 percent reliability for a switching function, how are the results modified ?

84. Calculate the reliability of 10 hrs operating period of a parallel system with two units, each having a failure rate of 0.01 failures/hour. Do likewise for a two unit standby system using the same units and assuming 100 percent reliability of sensing, switching and idling. Compare the two on the basis of reliability and MTBF.

85. The power supply to the operating unit of a hospital is provided by a generator whose failure rate follows an exponential distribution law with parameter $\lambda_1 = 0.005$ per hour. A standby battery unit is coupled through a decision switch which has a reliability $r_d = 0.90$. Calculate the reliability of the power supply system for a mission time of 10 hours if the battery failure rate follows a distribution law with parameter $\lambda_2 = 0.001$ per hour.

86. For the diagram shown in figure, determine the reliability expression, if each branch has identical probability of success of 0.80.

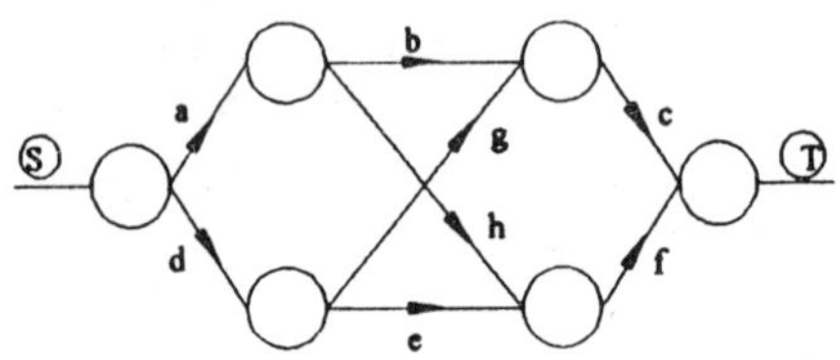

87. The graph shown in fig represents a four-station communication network. The four nodes represent the four stations and the six branches represent two way communication links between the pair of stations. Find all minimal tie-sets for transmission between a and b and derive an expression for reliability of communication between these nodes.

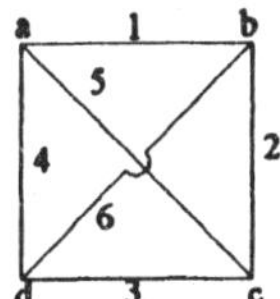

88. Two towns are connected by a network of communication channels. The probability of a channel's failure-free operation is R, and channel failures are independent. Minimal level of communication between towns can be guaranteed provided at least one path containing properly functioning channels exists. Given the network below, determine the probability that the two towns will be able to communicate. Here -***- denotes a communication channel.

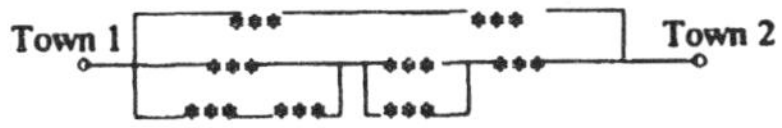

A network of communication channels.

89. Six computer systems are interconnected by means of the network below. Each connecting line of the network can be used in both directions and has an independent failure probability $p_0 = 0.1$. Calculate the probability that a successful information exchange can occur between the computers 1 and 4.

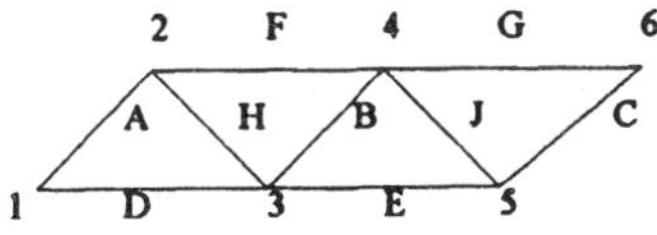

90. An information processing station R can receive information from four transmission stations T1, T2, T3 and T4. The four transmission stations are connected by two-way links as shown in Fig. What is

the probability that R will fail to receive information from T1 if the probability of failure of transmitting links are identical and equal to q.

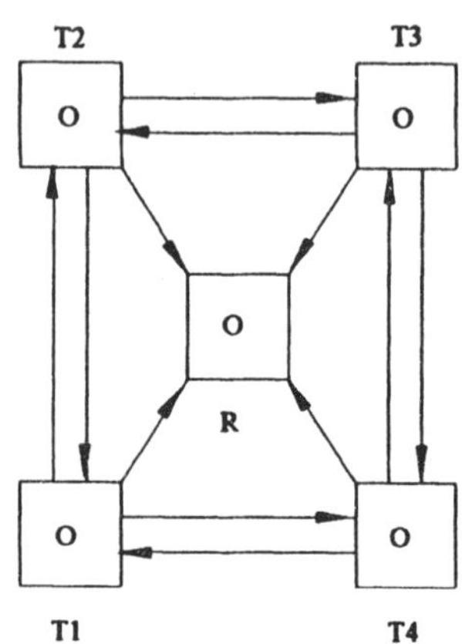

91. A five component system is connected as shown in fig. Derive an expression for system reliability using Baye's theorem. If all the components are identical and independent with a reliability of 0.8, determine the system reliability.

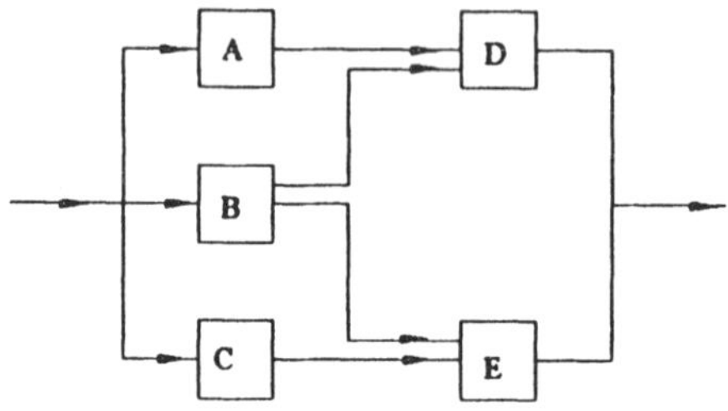

92. A vibration monitoring system consists of six sub-systems, all connected in series. The predicted reliabilities as obtained from an analysis are $R_1 = 0.993$, $R_2 = 0.996$, $R_3 = 0.998$, $R_4 = 0.997$, $R_5 = 0.987$, and $R_6 = 0.989$. Calculate the system reliability. If it is desired to increase the reliability by 3.33%, determine the percentage by which the reliability of each sub-system should be increased. Assume an exponential model for each sub-system.

93. Four units are connected in series, with reliabilities $R_1 = 0.85$, $R_2 = 0.9$, $R_3 = 0.8$ and $R_4 = 0.95$. Calculate the system reliability. If the reliability is to be increased to a value of 0.65, how should this be apportioned among the four units according to the minimum effort method ?

94. A system having three serial units is to be designed to have an overall reliability of 0.95. The complexity of third serial unit is expected to be twice as much as that of the second unit and complexity of the second unit is expected to be three times as high as that of the first unit. While first and third units are to operate all the time, second unit is to operate only for 50% of the total time. Allocate unit reliabilities to three units.

95. Three units of a system have predicted reliabilities 0.75, 0.85 and 0.95 respectively. It is desired to allocate the reliabilities such that the system reliability is not less than 0.75. Obtain the necessary solution by

(a) Basic Method
(b) Minimum Effort Method

If the cost of a unit with reliability r is $k\tan^2(\pi r/2)$, calculate the % age increase in cost for both the solutions.

96. Devise Hamming code consisting of data digits and check digits to encode the 5-bit data word 10101. Show how one error can be detected and corrected.

97. A message coded in Hamming Code is transmitted through a noisy channel.

The received message is

1011001 0111001 0011011 1110110

Decode the message assuming that at most a single error can occur in a word.

98. Consider a system having 5 components in series whose data are :

Component No.	Reliability	Cost	Weight
1	0.75	35	77
2	0.91	75	22
3	0.98	58	63
4	0.78	39	52
5	0.95	90	18

Find the optimum allocation of redundancies for maximizing system reliability if $C_S < 500$ and $W_S < 400$.

99. Consider a system having 5 components in series whose data are :

Component	P_i	C_i	W_i	V_i
1	0.785	5.2	77.9	34.9
2	0.916	75.1	22.2	28.6
3	0.986	58.5	63.3	87.3
4	0.755	36.1	52.7	45.9
5	0.949	90.8	8.0	58.6

Find the optimum allocation of redundancies for maximizing system reliability if
$C_s < 500$, $W_s < 589.6$ and $V_s < 686.2$

100. Determine the optimum number of redundancies to maximise reliability in the case of a system having 6 components with the following details:

Component	Reliability	Cost, $	Weight, Kg
1	0.80	10	2
2	0.90	20	3
3	0.85	20	4
4	0.75	15	6
5	0.70	10	2
6	0.90	15	7

Component No.4 is digital in nature. Component No.3 can permit the use of standby redundancy only but not parallel redundancy. Component 1 and 2 cannot be procured more than a total of 2 units each.

The total system cost should not exceed 200 Dollars and total system weight should not exceed 60 Kg.

101. An electrical supply system is subject to failure which causes loss of supply to the consumer. The mean time between such failures is known to be 398 hr and the meantime to repair the failures and restore the supply is known to be 2 hr. What is the average value of the availability of the supply to the consumer over a long period of time?

102. An engine is to be designed to have a minimum reliability of 0.7 and a minimum availability of 0.95 over a period of 1000 hrs. Determine the mean repair time and the frequency of failures of the engine.

103. A computer has an MTTF = 34 hr and an MTTR = 2.5 hr.

(a) What is the availability?
(b) If the MTTR is reduced to 1.5 hr, what MTTF can be tolerated without decreasing the availability of the computer?

104. For a computer unit, a suitable air-conditioning system has to be designed. It should have a minimum reliability value of 0.95 for an operation of 800 hours. The minimum availability value over the same period of time is required to be 0.98. Assuming constant hazards for failure and repair, estimate the time to failure and the mean repair time.

105. If a transmitter is to be designed to have a reliability greater than 0.90 over 1000 hr and a minimum availability of 0.99 over that period, determine the values of MTTF and MTTR.

106. A large office block has a fire detection and alarm system which is subject to a mean constant failure-rate of two failures per year (1 year = 8760 hr) and each failure that occurs takes, on average, 4 hr to detect and repair. The system is also subject to a quarterly routine inspection and test on which occasions it is out of action for a fixed time of 1 hr. If the expected probability of fire occurrence in the building over a period of time is 0.073, what is the probability of an undetected fire by the alarm system over the same period of time?

107. In a workshop a minimum of two lathes are required to operate continuously to meet the demand. The ratio of the repair rate to the failure rate of each lathe is 4. If the workshop has three identical lathes, determine the probability that at least two of them are available in the long run.

108. You are given a system with n components. The MTBF for each component is 100 hrs. and the MTTR is 5 hrs. Each component has its own repair facility. Find the limiting availability of the system when:

(1) All components are required for the system to function.
(2) At least one component is required for the system to function.

109. A two component parallel system uses both identical components each with $\lambda = 10^{-5}$/hr and $\mu = 10^{-2}$/hr. Calculate the %age increase in mean time to failure with the use of repair facilities if the system uses:

(a) Active Parallel Redundancy
(b) Standby Redundancy.

110. Three generators each of 20KW have different values of availability figures as supplied by different manufactures. The availability values are 0.96,0.98 and 0.92 respectively. Determine the overall system availability if the required load is 40 KW.

111. A system consists of two units in active redundancy. The units have a constant failure rate λ of 10^{-3} per hour and fail stochastically independent.

(a) How large is the MTTFF if no corrective maintenance at unit level is performed?
(b) How large will the MTTFF be if there are repairs allowed at the unit level? The repair rate μ is 10^{-1} per hour.
(c) What is the addressing frequency of the repair channel in case(b) if it may be assumed that $\lambda << \mu$?
(d) The repair costs of a unit are $500 per event. However, if the system goes down the costs, mainly because of the loss of production, are $5,000 per event. Determine, based on the outcome of (a), (b) and (c) whether it is economically sound to perform corrective maintenance at the unit level.

112. The following three units are in operation in parallel in the reliability sense.

Unit	Failure/rate hr	Repair rate/hr
A	0.004	0.10
B	0.005	0.15
C	0.003	0.06

If the system is operating as a one-out-of-three parallel system, determine

1. System availability,
2. Frequency of system failure,
3. Mean down- time, and
4. Mean up- time.

113. In testing certain systems whose operating time upto failure was normally distributed, we obtain ten realisations of the operating time upto failure (in hours): 115, 75, 80, 150, 75, 100, 120, 95, 70, 100. Find the confidence bounds for the mean of the operating time upto failure with a level of confidence of 95%.

114. Twenty identical items were tested for 200 hr. Nine of the total items failed during the test period. Their failure times are specified in table below. The failed items were never replaced. Determine whether the failure data represent the exponential distribution.

Failure number	1	2	3	4	5	6	7	8	9
Failure times (hr)	10	15	2	45	60	85	13	16	20

115. A relatively large number of nominally identical pumps are installed on a process plant and it is known that the times to failure for the pumps follow an exponential distribution. Over a 6-month period it is found that 5 pump failures have occurred in a sample of 20 of the pumps. If an estimate of the population mean failure-rate is made from this sample, what are the symmetrical 90% confidence limits on the estimate?

116. A non-replacement reliability test is carried out on 20 high- speed pumps to estimate the value of the failure rate. In order to eliminate wear failures, it is decided to terminate the test after half of the pumps have failed. The times of the first 10 failures(in hours) are

33.7, 36.9, 46.8, 56.6, 62.1, 63.6, 78.4, 79.0, 101.5, 110.2

(a) Estimate the MTTF.
(b) Determine the 90% confidence interval for the MTTF.

117. One wants to determine the MTTF of a new monolithic digital-to-analog converter at 25°C/ 77°F/ 298K. For that purpose 60 converters are operated for 1000 hours at 100°C/ 212°F/ 373K and 60 converters for 1000 hours at 85°C/ 185°F/ 358K.

At 100°C, the MTTF turned out to be 6.5×10^3 hours. At 85°C this was 2.4×10^4 hours. Assume that the failure process behaves as a chemical process with a reaction rate:

$$Q = Q_0 \exp(-E_A/kT)$$

What is the MTTF of this converter at 25°C?

118. Suppose that a sample of 20 units passes an acceptance test if no more than 2 units fail. Suppose that the producer guarantees the units for a failure probability of 0.05. The buyer considers 0.15 to be the maximum acceptable failure probability.

(a) What is the producer's risk?
(b) What is the buyer's risk?

119. The same data have been fit with both the basic and logarithmic poisson models. The parameters obtained are :

Basic	Logarithmic poisson
λ_0 = 20 failures/cpu hr	λ_0 = 50 failures /cpu hr
v_0 = 120 failures	θ = 0.025/ failures

Note that the logarithmic poisson usually has higher initial failure intensity. At first, recall that this falls off more rapidly than the failure intensity for the basic model, but finally it falls off more slowly. We require to examine some consequences of this behavior.

First, determine the additional failures and additional execution time required to reach a failure intensity objective of 10 failures /cpu hr, using both models. Then repeat this for an objective of 1 failure /cpu hr. Assume in both cases that you start from the initial failure intensity.

120. A computing center has a reliability objective of 0.90 for an 8- hr shift for its interactive service. The system requirement is simply that service be provided, regardless of the response time involved. All reliabilities are measured with respect to this shift. It has a dual processor configuration fed by a front-end processor, as shown in fig below. The front-end processor has a reliability of 0.99 and its operating system, 0.95. The reliability of each mainframe processor is 0.98. What must the reliability of the mainframe operating system be to meet the overall reliability objective?

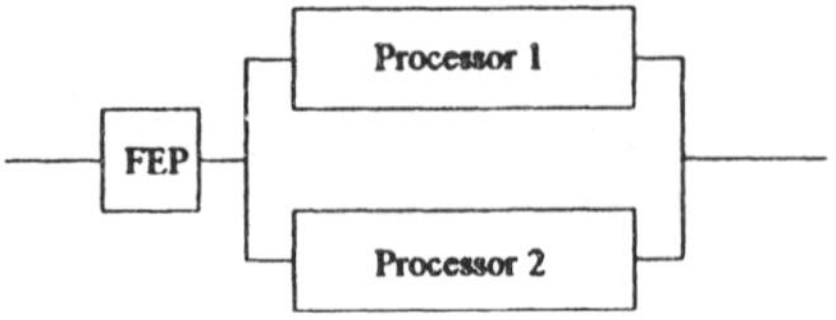

Computing Centre Configuration.

121. A program with 50,000 source instructions and a source to object expansion ratio of 4 will be executed on a machine with an average instruction execution rate of 333,333 instructions /cpu sec. On similar projects, a fault exposure ratio of $K = 1.67 \times 10^{-7}$ has been experienced, along with a fault density of 6 faults per 1000 source instructions. The fault reduction factor $B = 1$. Estimate the initial

failure intensity at the start of the system test.

122. The Soft Landing software service company has won a service contract to provide recovery service for a patient control and billing system. The service is provided to doctors in a region who own personal computers. It has a failure intensity of 1 failure /100 cpu hr. The average system runs 10 cpu hr /week and there are 600 doctors to be serviced. The average service person can make 2 service calls daily, including travel and allowing for enough spare personnel to prevent excessive service backlogs from building up.

How many service personnel do we need ? Assuming a cost of $200/call, what annual contract fee must we charge each doctor to achieve 20 % profit on sales ?

123. A program has an initial failure intensity of 10 failures/ cpu hr. We want to test and debug this program until a failure intensity of 1 failure/10 cpu hr is achieved. Assume the following resource usage parameters.

Resource usage	Per hr	Per failure
Failure identification effort	3 person hr	2 person hr
Failure correction effort	0	6 person hr
Computer time	1.5 cpu hr	1 cpu hr

(a) What resources must be expended to achieve the reliability improvement required ? Use the logarithmic Poisson execution time model. Assume a failure intensity decay parameter of 0.05.
(b) If the failure intensity objective is cut to half, are the resources required doubled?

124. A change to 2000 source instructions is being made in a program of 100,000 source instructions. The fault density of base program at the start of the system test was 5 faults /1000 source instructions. Assume that the fault reduction factor B = 1. The initial failure intensity of the base program was 2 failures /cpu hr. It was released with failure intensity of 6 failures /100 cpu hr. Fixes of faults are not made during the operational phase. What is the expected impact on failure intensity of the operational program?

125. Which of the following systems you will recommend as economical from reliability point of view (the system having least cost/ reliability ratio) ? The reliability and cost per component in configuration A B and C respectively are as below:

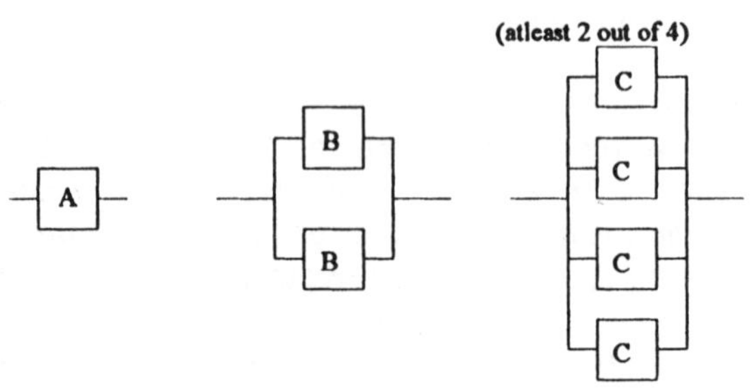

$P_a = 0.90$ $P_b = 0.70$ $P_c = 0.60$

$C_a = \$ 1000/-$ $C_b = \$ 500/-$ $C_c = \$ 200/-$

ANSWERS TO ODD NUMBERED PROBLEMS

1. (a) 0.9409 (b) 0.97

3. 0.9802

5. 47 days

7. 0.905

9. $R(t) = \exp(-\lambda_o t^2/2)$, $f(t) = \lambda_o t \exp(-\lambda_o t^2/2)$, $m = (\pi/2\lambda_o)^{1/2}$

11. (a) $\lambda(t) = \dfrac{2}{t_o[1-(t/t_o)]}$; $0 \le t \le t_o$.

    (b) The failure rate increases from $2/t_o$ at $t=0$ to infinity at $t=t_o$.

    (c) $m = t_o/3$

13. (a) $R(t) = 16/(t+4)^2$ (b) $\lambda(t) = 2/(t+4)$

    (c) $m = 4$ years

15. 0.000682/ hr

17. 1.7762×10^{-4} /hr

19. 1046 hrs

21. (a) 0.0902 (b) 0.2873 (c) 8 months

23. 2/7

25. (a) 0.86 (b) 0.40

27. (a) 0.5275 (b) 0.4725 (c) 0.949

29. 7 : 16

31. 0.999

33. 0.411, 0.500

35. 0.6976

37. 0.0001, 0.0036, 0.0486, 0.2916, 0.6561

39. (a) 0.018 (b) 0.935 (c) 0.047

41. 53/729

43. 0.75

45. 0.216

47. (a) 0.896 (b) 0.992 (c) 0.900

49. $\lambda = 6.66 \times 10^{-5}$ /hr, R = 0.9672, 0.9355, 0.9049

51. (a) 19,610 hrs (b) 0.0051

53. (a) 2000 hrs (b) 0.3935 (c) 0.1353

55. 0.3024

57. $51 < K < 13{,}514$

59. (a) 0.0004 (b) 0.002
    (c) 0.0396 (d) 0.038

61. 0.10765

63. 3 A's, R = 0.9191

65. 0.9703

67. (a) 0.9867 (b) 0.9952

69. (a) 0.0001 (b) 0.0037

(c) 0.0523 (d) 0.3439

71. 0.885

73. (a) $R = \exp(-\lambda_1 t_1) \cdot \exp[-\lambda_2(t_2-t_1)] \cdot \exp[-\lambda_3(t_3-t_2)]$

$\cdot \exp[-\lambda_4(t_4-t_3)] \cdot \exp[-\lambda_5(t_5-t_4)]$

(b) $R = \exp(-4\lambda_1 t_1) \cdot \exp[-4\lambda_2(t_2-t_1)]\ [6R'^2 (1-R')^2$

$+ 4R'^3 (1-R') + 4R'^4]$

where, $R' = \exp[-\lambda_3(t_3-t_2)] \cdot \exp[-\lambda_4(t_4-t_3)] \cdot \exp[-\lambda_5(t_5-t_4)]$

75. Circuit in fig.6 is optimal.

77. Series configuration of 2 diodes and R = 0.886,

Parallel configuration of 2 diodes and R = 0.832.

79. 6.5×10^{-5}

81. (i) 1500 hrs (ii) 2000 hrs

83. Reliability of 0.99999 and MTBF of 6000 hrs,

Reliability of 0.99997 and MTBF of 5990 hrs.

85. 0.9949

87. $R = p_{ab} + q_{ab} p_{ac} p_{bc} + q_{ab} p_{ad} p_{bd} q_{ac} + q_{ab} p_{ad} p_{bd} p_{ac} q_{bc}$

$+ p_{ad} p_{cd} p_{bc} q_{ab} q_{ac} q_{bd} + p_{ac} p_{cd} p_{bd} q_{ab} q_{ad} q_{bc}$

89. 0.988

91. 0.94208

93. 0.881, 0.881, 0.881, 0.95

95. 0.8478, 0.9109, 0.9710; 196.898%;

0.8885, 0.8885, 0.95; 22.05%

97. 9996

99. 3,2,2,3,1

101. $\lambda = 1/398$, A = 0.995

103. (a) 0.9315 (b) 20.4 hrs

105. 9491 hrs, 96 hrs

107. 0.896

109. (a) 3.33×10^4 % (b) 5×10^4 %

111. (a) 1500 hrs (b) 50,000 hrs

(c) 2×10^{-3} /hr (d) \$ 1.10 /hr

113. 79.10 hr - 116.9 hr

115. 0.197 faults /yr - 1.05 faults /yr

117. 1.66×10^7 hrs

119. 60 failures and 4.16 CPU hr, 64 failures and 3.2 CPU hr;

114 failures and 18 CPU hr, 156 failures and 39.2 CPU hr

121. 0.3 failures /CPU hr

123. (a) χ_I = 778 per-hr, χ_F = 552 per-hr, χ_C = 389 CPU hr

(b) No, Somewhat less

125. System "C" is optimal.

REFERENCES

BOOKS

1. Amendola A. and Bustamante A., ***Reliability Engineering,*** Kluwer Academic Publishers, Netherlands, 1988.

2. Amstadter B.L., ***Reliability Mathematics (Fundamentals; Practices; Procedures),*** McGraw-Hill Book Co., Inc., New York, 1971.

3. Apostolakis G., Garribba S. and Volta G., ***Synthesis and Analysis Methods for Safety and Reliability Studies,*** Plenum Publishing Corporation, New York, 1980.

4. Balagurusamy E., ***Reliability Engineering,*** Tata McGraw-Hill Publishing Company Limited, New Delhi, 1984.

5. Barlow R.E. and Proschan F., ***Mathematical Theory of Reliability,*** John Wiley & Sons, Inc., New York, 1965.

6. Bazovsky I., ***Reliability Theory and Practice,*** Prentice-Hall Inc., Englewood Cliffs, New Jersey, 1961.

7. Billinton R. and Allan R.N., ***Reliability Evaluation of Engineering Systems: Concepts and Techniques,*** Plenum Press, New York, 1983.

8. Breipohl A.M., ***Probabilistic Systems Analysis,*** John Wiley & Sons, Inc., NewYork, 1970.

9. Chorafas D.N., ***Statistical Processes and Reliability Engineering,*** D.Van Nostrand Co., Inc., New Jersey, 1960.

10. Colombo A.G. and Keller A.Z., ***Reliability Modelling and Applications,*** D.Reidel Publishing Co., Holland, 1987.

11. Deo N., ***Graph Theory with Applications to Engineering and Computer Science,*** Prentice -Hall Inc., Englewood Cliffs, New Jersey, 1974.

12. Dhillon B.S., ***Quality Control, Reliability, and Engineering Design,*** Marcel Dekker, Inc., New York, 1985.

13. Dhillon B.S. and Singh C., ***Engineering Reliability: New Techniques and Applications,*** Wiley-Interscience, John Wiley & Sons, Inc., New York, 1981.

14. Dummer G.W.A. and Griffin N., ***Electronic Equipment Reliability,*** John Wiley & Sons, Inc., New York, 1960.

15. Feller W., ***An Introduction to Probability Theory and its Applications,*** Volume-1, Wiley Eastern Pvt. Ltd., New Delhi, 1968.

16. Gnedenko B.V., ***The Theory of Probability,*** MIR Publications, Moscow, 1969.

17. Govil A.K., ***Reliability Engineering,*** Tata McGraw Hill Publishing Company Limited, New Delhi, 1983.

18. Green A.E., ***Safety Systems Reliability,*** John Wiley & Sons Ltd., New York, 1983.

19. Ireson W.G., ***Reliability Hand Book,*** McGraw-Hill, Inc., New York, 1966.

20. Ireson W.G. and Coombs C.F., Jr. (Editors), ***Handbook of Reliability Engineering and Management,*** McGraw-Hill Book Co., Inc., New York, 1988.

21. Klaassen K.B. and Jack C.L.van Peppen, ***System Reliability,*** Chapman and Hall, Inc., New York, 1989.

22. Llyod D.K. and Lipow M., ***Reliability: Management, Methods and Mathematics,*** Prentice-Hall, Inc., Englewood Cliffs, New Jersey, 1962.

23. Locks M.O., ***Reliability, Maintainability and Availability Assessment,*** Hayden Book Co., Inc., New Jersey, 1973.

24. Mann N.R., Schafer R.E. and Singpurwalla N.D., ***Methods for Statistical Analysis of Reliability and Life Data,*** John Wiley & Sons, Inc., New York, 1974.

25. Misra K.B., ***Reliability Analysis and Prediction,*** Elsevier Science Publishers, Netherlands, 1992.

26. Musa J.D., Iannino A. and Okumoto K., ***Software Reliability,*** McGraw Hill Book Co., Singapore, 1987.

27. Myers G.J., ***Software Reliability: Principles and Practices,*** John Wiley & Sons, Inc., New York, 1976.

28. Page L.B., ***Probability for Engineering,*** Computer Science Press, New York, 1989.

29. Papoulis A., ***Probability, Random Variables and Stochastic Processes,*** McGraw-Hall Kogakusha Ltd.,Tokyo,1965.

30. Pieruschka E., ***Principles of Reliability,*** Prentice-Hall, Inc., Englewood Cliffs, New Jersey, 1963.

31. Polovko A. M., ***Fundamentals of Reliability Theory,*** Academic Press, Inc., New York, 1968.

32. Rau J.G., ***Optimization and Probability in Systems Engineering,*** Van Nostrand Reinhold Co., New York, 1970.

33. Sandler G.H., ***System Reliability Engineering,*** Prentice-Hall, Inc., Englewood Cliffs, New Jersey, 1963

34. Shooman M.L., ***Probabilistic Reliability: An Engineering Approach*** McGraw-Hill, New York, 1968.

35. Shrinath L.S., ***Reliability Engineering,*** Affiliated East-West Press Pvt. Ltd., New Delhi, 1985.

36. Sinha S.K., ***Reliability and Life testing,*** Wiley Eastern Limited, New Delhi, 1986.

37. Smith O.C., ***Introduction to Reliability in Design,*** McGraw- Hill Inc., New York, 1976.

38. Tillman FA., Hwang C.L. and Kuo W., ***Optimization of Systems Reliability,*** Marcel Dekker, Inc., New York, 1980.

39. Trivedi K.S., ***Probability and Statistics with Reliability, Queuing and Computer Science Application,*** Prentice-Hall, Inc., Englewood Cliffs, New Jersey, 1982.

40. Von Alven W.H. (Editor), ***Reliability Engineering,*** Prentice-Hall, Inc., Englewood Cliffs, New Jersey, 1964.

RESEARCH PUBLICATIONS

1. Abraham J.A., ***An Improved Algorithm for Network Reliability,*** IEEE Trans. Reliability, Vol. 28, pp 58-61, April 1979.

2. Aggarwal K.K., ***Redundancy Optimization in General Systems,*** IEEE Trans. Reliability, Vol.R-25, pp 330-332, December 1976.

3. Aggarwal K.K., ***A New Concept in the Reliability Modelling,*** Annual Reliability and Maintainability Symposium, Atlanta, U.S.A., 1989.

4. Aggarwal K.K., ***Economical Design of Reliable System- Some Practical Solutions,*** International Journal of Quality and Reliability, Vol.8, pp 349-354, 1992.

5. Aggarwal K.K., ***Optimum Redundancy Allocation in Non-Series Parallel Systems Using Boolean Differences,*** IEEE Trans. Reliability, Vol.28, pp 79-80, 1979.

6. Aggarwal K.K., ***Integration of Reliability and Capacity in Performance Measure of a Telecommunication Network,*** IEEE Trans. Reliability, Vol.34, pp 184-186, 1985.

7. Aggarwal K.K., Gupta J.S. and Misra K.B., ***A New Heuristic Criterion for Solving a Redundancy Optimization Problem,*** IEEE Trans. Reliability, Vol. R-24, pp 86-87, April 1975.

8. Aggarwal K.K., Misra K.B. and Gupta J.S., ***A Simple Method for Reliability Evaluation of a Communication System,*** IEEE Trans. Communication, Vol. Com-23, pp 563-565, May 1975.

9. Aggarwal K.K., Misra K.B. and Gupta J.S., ***A Fast Algorithm for Reliability Evaluation,*** IEEE Trans. Reliability, Vol. R-24, pp 83-85, April 1975.

10. Aggarwal K.K., Misra K.B. and Gupta J.S., ***Reliability Evaluation: A Comparative Study of Different Techniques,*** Microelectronics and Reliability, Vol.14, pp 49-56, 1975.

11. Aggarwal K.K. and Gupta J.S., ***On Minimizing the Cost of Reliable Systems,*** IEEE Trans. Reliability, Vol.24, pp 205-208, 1975.

12. Aggarwal K.K. and Rai S., ***Reliability Evaluation in Computer Communication Networks,*** IEEE Trans. Reliability, Vol.R-30, pp 32-36, April 1981.

13. Aggarwal K.K. and Rai S., ***Symbolic Reliability Evaluation Using Logical Signal Relations,*** IEEE Trans. Reliability, Vol. R-27, pp 202-205, August 1978.

14. Aggarwal K.K., Chopra Y.C. and Bajwa J.S., ***Modification of Cut Sets for Reliability Evaluation of Communication Systems,*** Microelectronics and Reliability , Vol.22, pp 337-340, 1982.

15. Aggarwal K.K., Chopra Y.C. and Bajwa J.S., ***Topological layout of Links for Optimizing the s-t Reliability in a Computer Communication Network,*** Microelectronics and Reliability, Vol.22, pp 341-345, 1982.

16. Aggarwal K.K., Chopra Y.C. and Bajwa J.S., ***Capacity Consideration in Reliability Analysis of Communication Systems,*** IEEE Trans. Reliability, Vol.31, pp 171-181, 1982.

17. Aggarwal K.K., Chopra Y.C. and Bajwa J.S., ***Reliability Evaluation by Network Decomposition,*** IEEE Trans. Reliability, Vol.31, pp355- 358, 1982.

18. Agrawal A. and Barlow R., ***A Survey of Network Reliability and Domination Theory,*** Operations Research, Vol.32, pp 478-492, 1984.

19. Anderson R.T., ***Reliability Design Hand Book,*** IIT Research Institute, April 1979.

20. Ashrafi N. and Berman O., ***Optimization Models for Selection of Programs, Considering Cost and Reliability,*** IEEE Trans. Reliability, Vol.41, pp 281-287, June 1992.

21. Balagurusamy E. and Misra K.B., ***Failure Rate Derating Chart for Parallel Redundant Units with Dependent Failures,*** IEEE Trans. Reliability, Vol.25, pp 122, June 1976.

22. Balagurusamy E. and Misra K.B., ***Reliability of a Parallel System with Non-Identical Units,*** IEEE Trans.Reliability, Vol.R-24, pp 340- 341, December 1975.

23. Banerjee S.K. and Rajamani K., ***Closed form Solutions for Delta-Star and***

Star-Delta Conversions for Reliability Networks, IEEE Trans. Reliability, Vol.R-25, pp 115-118, June 1976.

24. Bennets R.G., *On the Analysis of Fault Trees,* IEEE Trans. Reliability, Vol.R-24, pp 175-185, August 1975.

25. Brijendra Singh, *A Study of Network Reliability,* D.Phil Thesis, University of Allahabad, Allahabad, India, 1991.

26. Brown D.B., *A Computerized Algorithm for Determining the Reliability of Redundant Configuration,* IEEE Trans. Reliability, Vol.R-20, pp 121-124, August 1971.

27. Buzacott J.A., *Network Approaches to Finding the Reliability of Repairable Systems,* IEEE Trans. on Reliability, Vol.R-19, pp 140- 146, November 1970.

28. Chopra Y.C., *Reliability Analysis and Optimization in Communication Systems,* Ph.D Thesis, Punjab University, Chandigarh, India, 1982.

29. Deo N. and Medidi M., *Parallel Algorithms for Terminal- Pair Reliability,* IEEE Trans. Reliability, Vol.41, pp 201-209, June 1992.

30. Downs T. and Garrone P., *Some New Models of Software Testing with Performance Comparisons,* IEEE Trans. Reliability, Vol.40, pp 322- 328, August 1991.

31. Dugan J.B., *Automated Analysis of Phased-Mission Reliability,* IEEE Trans. Reliability, Vol.40, pp 45-52, April 1991.

32. Dugan J.B. and Trivedi K.S., *Coverage Modeling for Dependability Analysis of Fault Tolerant Systems,* IEEE Trans. Computers, Vol.38, pp 775-787, June 1989.

33. Elperin T., Gretsbakh I. and Lomonosov M., *Estimation of Network Reliability using Graph Evaluation Models,* IEEE Trans. Reliability, Vol.R-40, pp 572-581, December 1991.

34. Evans M.G.K., Parry G.W. and Wreathall J., *On the Treatment of Common -Cause Failures in the System Analysis,* Reliability Engineering, Vol.39, pp 107-115, 1984.

35. Fratta L. and Montanari U.G., *Synthesis of Available Networks,* IEEE Trans. Reliability , Vol.R-25, pp 81-87, June 1976.

36. Fratta L. and Montanari U.G., ***A Boolean Algebra Method for Computing the Terminal Reliability of a Communication Network,*** IEEE Trans. Circuit Theory, Vol.CT-20, pp 203-211, May 1973.

37. Fratta L. and Montanari U.G., ***A Recursive Method Based on Case Analysis for Computing Network Terminal Reliability,*** IEEE Trans. Communication, Vol. COM-26, pp 1166-1176, August 1978.

38. Gopal K., Aggarwal K.K. and Gupta J.S., ***Reliability Evaluation in Complex Systems with many Failure Modes,*** International Journal of Systems Science, Vol.7, pp 1387-1392, 1976.

39. Gopal K., Aggarwal K.K. and Gupta J.S., ***A New Method for Reliability Optimization,*** Microelectronics and Reliability, Vol.17, pp 419- 422, 1978.

40. Gopal K., Aggarwal K.K. and Gupta J.S., ***A New Method for Solving Reliability Optimization Problems,*** IEEE Trans. Reliability, Vol.29, pp 36-37, 1980.

41. Gopal K., Aggarwal K.K. and Gupta J.S., ***On Optimal Redundancy Allocation,*** IEEE Trans. Reliability, Vol.27, pp 325-328, 1978.

42. Gopal K., Aggarwal K.K. and Gupta J.S., ***Reliability Analysis of Multistate Device Networks,*** IEEE Trans. Reliability, Vol. R-27, pp 233-235, August 1978.

43. Gopal K., Aggarwal K.K. and Gupta J.S., ***A New Approach to Reliability Optimization in GMR Systems,*** Microelectronics and Reliability, Vol.18, pp 419-422, 1978.

44. Gopal K., Aggarwal K.K. and Gupta J.S., ***An Event Expension Algorithm for Reliability Evaluation in Complex Systems,*** International Journal of Systems Science, Vol.10, pp 363-371, 1979.

45. Gopal K., ***Reliability Analysis of Complex Networks and Systems,*** Ph.D Thesis, Kurukshetra University, Kurukshetra, India, 1978.

46. Gupta H. and Sharma J., ***A Delta-Star Transformation Approach for Reliability Evaluation,*** IEEE Trans. Reliability, Vol R-27, pp 212-214, August 1978.

47. Hansler E., ***A Fast Recursive Algorithm to Calculate the Reliability of a***

Communication Network, IEEE Trans. Communication Vol.COM-20, pp 637-640, June 1972.

48. Hansler E., McAulifee G.K. and Wilkov R.S., ***Exact Calculation of Computer Network Reliability, Networks,*** Vol. 4, pp 95-112, 1974.

49. Heidtmann K.D., ***Smaller Sums of Disjoint Products by Subproduct Inversion,*** IEEE Trans. Reliability, Vol.38,pp 305-311, Aug.1989.

50. Heidtmann K.D., ***Improved Method of Inclusion- Exclusion Applied to k-out-of-n Systems,*** IEEE Trans. Reliability, Vol.R-31, pp 36-40, April 1982.

51. Hurley R.B., ***Probability Maps,*** IEEE Trans. Reliability, Vol.R-12, pp 39-44, September 1963.

52. Jasman G.B. and Kai O.S., ***A New Technique in Minimal Path and Cutset Evaluation,*** IEEE Trans. Reliability, Vol.34, pp 136-143, 1985.

53. Jensen P.A. and Bellmore M., ***An Algorithm to Determine the Reliability of Complex Systems,*** IEEE Trans. Reliability, Vol.R-18, pp 169-174, November 1969.

54. Lee S.H., ***Reliability Evaluation of Flow Network,*** IEEE Trans. Reliability, Vol.R-29, pp 24-26 April 1980.

55. Li D. and Haimes Y.Y., ***A Decomposition Method for Optimization of Large- System Reliability,*** IEEE Trans. Reliability, Vol.41, pp 183-189, June 1992.

56. Lin P.M., Leaon B.J. and Huang T.C., ***A New Algorithm for Symbolic System Reliability Analysis,*** IEEE Trans. Reliability, Vol. R-25, pp 2-15, April 1976.

57. Locks M.O. and Biegel J.E., ***Relationship Between Minimal Path-Sets and Cut-Sets,*** IEEE Trans. Reliability, Vol.R-27, pp 106-107, June 1978.

58. Locks M.O., ***Inverting and Minimizing Path-Sets and Cut-Sets,*** IEEE Trans. Reliability, Vol R-27, pp 106, June 1978.

59. McLeavey D.W. and McLeavy J.A., ***Optimization of a System by Branch-and -Bound,*** IEEE Trans. Reliability, Vol. R-25, pp 327-329, December 1976.

60. Mercado J.de, Spyratos N. and Bowen B.A., ***A Method for Calculation of Network Reliability,*** IEEE Trans. Reliability, Vol.R-25, pp 71-77, June 1976.

61. Misra K.B., ***Optimum Reliability Design of a System Containing Mixed Redundancies,*** IEEE Trans. Power Apparatus and Systems, Vol.PAS-94, pp 983-993, May 1975.

62. Misra K.B., ***A Method of Solving Redundancy Optimization Problems,*** IEEE Trans. Reliability, Vol. R-20, pp 117-120, August 1971.

63. Misra K.B., ***An Algorithm for Reliability Evaluation of Redundant Network,*** IEEE Trans. Reliability, Vol.R-19, pp146-151, November 1970.

64. Misra K.B. and Sharma U., ***An Efficient Algorithm to Solve Integer-Programming Problems Arising in System- Reliability Design,*** IEEE Trans. Reliability, Vol.40, pp 81-91, April 1991.

65. Nakagawa Y., ***Studies on Optimal Design of High Reliable System: Single and Multiple Objective Nonlinear Integer Programming,*** Ph.D Thesis, Kyoto University, Japan, December 1978.

66. Nakagawa Y., Nakashima K. and Hattori Y., ***Optimal Reliability Allocation by branch- and- bound Technique,*** Vol.R-27, pp 31-38, April 1978.

67. Nakagawa Y. and Nakashima K., ***A Heuristic Method for Determining Optimal Reliability Allocation,*** IEEE Trans. Reliability, Vol.R- 26, pp 156-161, August 1977.

68. Nakazawa H., ***Decomposition Methods for Computing the Reliability of Complex Networks,*** IEEE Trans. Reliability, Vol-30, pp 289-292, December 1981.

69. Page L.B. and Perry J.E., ***A Model for System Reliability with Common-Cause Failures,*** IEEE Trans. Reliability, Vol.R-38, pp 406- 410, October 1989.

70. Parker K.P. and McCluskey E.J., ***Probabilistic Treatment of General Combinational Networks,*** IEEE Trans. Computers, Vol.C-24,pp 668-670, June 1975.

71. Pedar A. and Sarma V.V.S., ***Phased- Mission Analysis for Evaluating the Effectiveness of Aerospace Computing Systems,*** IEEE Trans. Reliability, Vol.30, December 1981.

72. Pedar A., ***Reliability Modelling and Architectural Optimization of Aerospace Computing Systems,*** Ph.D.Thesis, Indian Institute of Science, Bangalore, India, 1981.

73. Reibman A.L., ***Modeling the Effect of Reliability on Performance,*** IEEE Trans. Reliability, Vol.39, pp 314-320, August 1990.

74. Renu Bala and Aggarwal K.K., ***A Simple Method for Optimal Redundancy Allocation for Complex Networks,*** Microelectronics and Reliability, Vol.27, pp 835-837, 1987.

75. Rushdi A.M., ***Symbolic Reliability Analysis with the Aid of Variable Entered Karnaugh Maps,*** IEEE Trans. Reliability, Vol.R- 32, pp 134-139, June 1983.

76. Rushdi A.M., ***On Reliability Evaluation by Network Decomposition,*** IEEE Trans. Reliability, Vol.R-33, pp 379-384, December 1984.

77. Satyanarayana A. and Prabhakar A., ***New Topological Formula and Rapid Algorithm for Reliability Analysis of Complex Networks,*** IEEE Trans. Reliability, Vol.R-27, pp 82-100, June 1978.

78. Sharma J. and Venkateswaran K.V., ***A Direct Method for Maximizing System Reliability,*** IEEE Trans. Reliability, Vol.R-20, pp 256- 259, November 1971.

79. Shashwati Guha and Aggarwal K.K., ***Extension of Minimum Effort Method for Nonseries Parallel Systems,*** International Journal of Quality and Reliability Management, Vol.6, pp 19-26, 1989.

80. Shen K. and Xie M., ***On the Increase of System Reliability by Parallel Redundancy,*** IEEE Trans. Reliability, Vol.39, pp 607-611, December 1990.

81. Singh B. and Proctor C.L., ***Reliability Analysis of Multi-State Device Networks,*** Proc. Annual Reliability and Maintainability Symposium, pp 31-35, 1976.

82. Singh N. and Kumar S., ***Reliability Bounds for Decomposable Multicomponent Systems,*** IEEE Trans. Reliability, Vol.29. pp 22-23, April 1980.

83. Soi I.M.N. and Aggarwal K.K., ***A Review of Computer Communication***

Classification Schemes, IEEE Communication Magazine, vol.19, pp 24-32, 1981.

84. Soi I.M.N. and Aggarwal K.K., ***Reliability Indices for Topological Design of Reliable CCNs,*** IEEE Trans. Reliability, Vol.30, pp 438-443, 1981.

85. Soi I.M.N., ***Topological Optimization of Large Scale Reliable Computer Communication Networks,*** Ph.D Thesis, Kurukshetra University, Kurukshetra, India, 1982.

86. Suresh Rai, ***Some Aspects of Reliability of Computers and Communication Networks,*** Ph.D Thesis, Kurukshetra University, Kurukshetra, India, 1979.

87. Suresh Rai and Arun Kumar, ***Recursive Technique for Computing System Reliability,*** IEEE Trans. Reliability, Vol.R-36, pp 38-44, April 1987.

88. Suresh Rai and Aggarwal K.K., ***An Efficient Method For Reliability Evaluation of a General Network,*** IEEE Trans. Reliability, Vol.R- 27, pp 206-211, August 1978.

89. Tillman F.A., Hwang C.L., Fan L.T. and Lal K.C., ***Optimal Reliability of Complex System,*** IEEE Trans. Reliability, Vol.R-19, pp 95-100, August 1970.

90. Tillman F.A., Hwang C.L. and Kuo W., ***Optimization Techniques for System Reliability with Redundancy- A Review,*** IEEE Trans. Reliability, Vol.R-26, pp 148-155, August 1977.

91. Veeraraghavan M. and Trivedi K.S., ***An Improved Algorithm for Symbolic Reliability Analysis,*** IEEE Trans. Reliability, Vol.40, pp 347-358, August 1991.

92. Vinod Kumar and Aggarwal K.K., ***Determination of Path Identifiers for Reliability Analysis of a Broadcasting Network using Petrinets,*** International Journal of Systems Science, Vol.19, pp 2643-2653, 1988.

93. Wilkov R.S., ***Analysis and Design of Reliable Computer Communication Netwroks,*** IEEE Trans. Communication, Vol.COM-20, pp 660-678, June 1972.

94. Wilson J.M., ***An Improved Minimiging Algorithm for Sum of Disjoint Products,*** IEEE Trans. Reliability, Vol.R-39, pp 42-45, April 1990.

SUBJECT INDEX

D

E

F

G

H

I

J

K

L